BROOKFIELD ZOO

MW01408503

BROOKFIELD ZOO LIBRARY
3300 GOLF ROAD
BROOKFIELD, IL 60513
708-688-8957

THEORETICAL
POPULATION
GENETICS

TITLES OF RELATED INTEREST

A biologist's advanced mathematics
D. Causton

Basic growth analysis
R. Hunt

Chromosomes today, volume 9
A. Stahl, J. M. Luciani & A. M. Vagner-Capodano (eds)

Cell movement and cell behaviour
J. M. Lackie

Distribution free tests
H. Neave & P. Worthington

The eukaryote genome in development and evolution
B. John & G. Miklos

Estimating the size of animal populations
J. G. Blower, L. M. Cook & J. A. Bishop

Introduction to vegetation analysis
D. Causton

Rates of evolution
P. S. W. Campbell & M. F. Day

Yeast biotechnology
D. Berry, G. Stewart & I. Russell (eds)

THEORETICAL POPULATION GENETICS

J. S. GALE
Department of Genetics, University of Birmingham

London
UNWIN HYMAN
Boston   Sydney   Wellington

© J. S. Gale, 1990
This book is copyright under the Berne Convention. No reproduction without permission. All rights reserved.

Published by the Academic Division of
Unwin Hyman Ltd
15/17 Broadwick Street, London W1V 1FP, UK

Unwin Hyman Inc.,
8 Winchester Place, Winchester, Mass. 01890, USA

Allen & Unwin (Australia) Ltd,
8 Napier Street, North Sydney, NSW 2060, Australia

Allen & Unwin (New Zealand) Ltd in association with the
Port Nicholson Press Ltd,
Compusales Building, 75 Ghuznee Street, Wellington 1, New Zealand

First published in 1990

British Library Cataloguing in Publication Data
Gale, J. S.
Theoretical population genetics.
  1. Population genetics
  I. Title
  375.1'5

   ISBN 0-04-575026-2
   ISBN 0-04-575027-0

Library of Congress Cataloging in Publication Data
Gale, J. S.
   Theoretical population genetics/J. S. Gale.
      p.   cm.
Bibliography: p.
Includes index.
ISBN 0-04-575026-2. — ISBN 0-04-575027-0 (pbk.)
1. Population genetics. 2. Distribution (Probability theory)
I. Title.
   [DNLM: 1. Genetics, Population.   2. Models, Genetic.
3. Probability.   QH 455 G151t]
QH455.G35 1989
575. 1'5—dc20
DNLM/DLC
for Library of Congress
                                                            89-14653
                                                                 CIP

Typeset in Great Britain by APS Ltd., Salisbury, Wiltshire
Printed in Great Britain by Cambridge University Press

*In memory of E. B. Ford,
who inspired us to study the relation
between Mendelism and evolution*

Preface

The rise of the neutral theory of molecular evolution seems to have aroused a renewed interest in mathematical population genetics among biologists, who are primarily experimenters rather than theoreticians. This has encouraged me to set out the mathematics of the evolutionary process in a manner that, I hope, will be comprehensible to those with only a basic knowledge of calculus and matrix algebra. I must acknowledge from the start my great debt to my students. Equipped initially with rather limited mathematics, they have pursued the subject with much enthusiasm and success. This has enabled me to try a number of different approaches over the years. I was particularly grateful to Dr L. J. Eaves and Professor W. E. Nance for the opportunity to give a one-semester course at the Medical College of Virginia, and I would like to thank them, their colleagues and their students for the many kindnesses shown to me during my visit.

I have concentrated almost entirely on stochastic topics, since these cause the greatest problems for non-mathematicians. The latter are particularly concerned with the range of validity of formulae. A sense of confidence in applying these formulae is, almost certainly, best gained by following their derivation. I have set out proofs in fair detail, since, in my experience, minor points of algebraic manipulation occasionally cause problems. To avoid loss of continuity, I have sometimes put material in notes at the end of chapters. In such a tightly knit subject, the need to refer back repeatedly from later to earlier chapters can be a serious obstacle, as beginners have often told me. To minimize this difficulty, I have risked trying the reader's patience by repeating earlier material on occasion.

I long hesitated on whether to give detailed references in the main text; the traditions in biological and mathematical writing are rather different here. Finally, I decided to follow the usual biological practice and give extensive references, in the hope that this would encourage the reader to consult the original work, the reading of which has given me so much pleasure.

I am most grateful to Dr A. J. Birley, Dr I. J. Mackay, Dr C. S. Haley, Dr C. P. Werner, Dr R. C. Jones, Dr A. J. Girling and Professor P. D. S. Caligari for information, comments and advice. I am much indebted to Mrs E. A. Badger for typing the manuscript and Mrs P. Hill for preparing the diagrams. Finally, I should like to thank the publishers for unfailing helpfulness.

J.S.G.
Birmingham

Contents

	Preface	*page* ix
1	Introduction	1
	Notes and exercises	9
2	Wright–Fisher, Moran and other models	12
	On simple models	12
	Stochastic models: deterministic models	13
	Random genetic drift: general approach	14
	Distribution of allele frequency under drift	14
	Change of allele frequency in one generation	16
	The mean in any generation	17
	Change of allele frequency in one generation: the variance	20
	Is the binomial distribution of allele number relevant?	22
	Effective population size	31
	Variance of allele frequency in any generation	36
	Inbreeding effective size	39
	Populations under systematic 'pressure'	42
	A continuous approximation to the conditional distribution	43
	Diffusion methods	44
	Moran's model	46
	Notes and exercises	49
3	On the description of changes in allele frequency	56
	A bewildering abundance of descriptions	56
	Probability distribution of allele frequency	59
	A numerical example	67
	Probability of fixation	69
	Dominant latent root (maximum non-unit eigenvalue)	71
	Modified process	75
	Probability distribution of absorption time	78
	Mean absorption time	82
	Variance of absorption time	84
	Mean sojourn times	86
	Variance of sojourn time	89
	Probability of ever reaching a given frequency	89
	A matrix representation of mean sojourn time	90
	Notes and exercises	91

CONTENTS

4 Survival of new mutations: branching processes — 106

On special methods — 106
Probability of survival — 107
The need for accuracy — 107
Are diffusion methods appropriate? — 108
A fundamental simplification — 110
Probability generating functions — 111
Independent propagation: branching processes — 113
Utility of the branching process approach — 114
Fundamental equations for branching processes — 115
Calculating the probability of survival — 117
The case $N_e < N$ — 119
A problem in genetic conservation — 123
Probability of survival when t is fairly large — 124
Pattern of survival in natural populations — 126
A caution — 126
Genic selection — 127
Probability of ultimate fixation — 130
Calculating the probability of fixation — 131
Population subdivision — 137
Cyclical variation in population size — 138
A note on epistasis and linkage — 141
Spread of a fungal pathogen — 143
Notes and exercises — 144

5 Probability of fixation: the more general case — 152

Fundamental equation — 152
Uniqueness of the solution — 153
Probability of fixation for Moran's model — 155
Moran's method for the Wright-Fisher haploid model — 159
Population subdivision — 165
The case $N_e < N$ — 167
Effect of selection on fixation probability — 170
Kimura's method for finding the probability of fixation — 172
Population subdivision reconsidered — 180
Fluctuating viabilities — 182
Rate of evolution — 183
Notes and exercises — 189

6 Some notes on continuous approximations — 194

The problem stated — 194
Probability of reaching a given allele frequency — 195

CONTENTS

Events at the boundaries	200
The problem of representing the initial frequency	203
Ad hoc method	203
Objections to our *ad hoc* method	204
Dirac's delta function	205
Notes and exercises	208

7 Mean sojourn, absorption and fixation times — 213

Retrospect	213
Fundamental equations	215
Uniqueness of solutions	217
Mean times for Moran's model	219
A tentative approach for the Wright-Fisher model	222
A generating function for mean sojourn times	225
Calculating mean sojourn times for the Wright-Fisher model	227
Mean fixation times	235
Continuous approximations	237
Calculating mean sojourn and absorption times	243
Interpreting the results	249
Mean sojourn times in the modified process	251
Mean fixation time	252
Effect of selection	253
Mean times under genic selection in the modified process	257
Effect of selection: some results	258
Calculation of mean times under selection: a numerical example	261
Mean fixation time under selection	263
Variance of sojourn, absorption and fixation times	265
Concerted evolution of multigene families	267
Notes and exercises	269

8 Introduction to probability distributions: probability flux — 277

On the need to calculate probability distributions	277
Continuous approximation of the allele frequency	279
Approximation of zero and unit frequencies	281
Continuous time approximation	284
Probability flux	286
Components of flux	290
Calculating the probability flux	292
Probability flux for discontinuous frequency	296
Expressions for M and V	297
Values of $P(x, t)$ when x takes limiting values	300
Notes and exercises	301

CONTENTS

9 Stationary distributions: frequency spectra — 303

Long-term distributions — 303
Wright's formula for the stationary distribution — 306
Stationary distributions (case of no selection): mean and variance — 307
Variation at nucleotide sites: effective number of bases — 310
Finding the stationary distribution in the case of no selection — 313
Stationary distribution (continued): allele frequencies zero and unity — 316
Accuracy of the results — 320
Nucleotide sites (resumed): probability of monomorphism — 322
Multiple alleles: infinite alleles model — 324
Mean number of alleles present — 325
Mean number of alleles in a given frequency range — 329
The frequency spectrum — 331
Infinite sites models — 334
Frequency spectrum for transposable elements — 336
Stationary distributions under selection — 336
Harmful recessives — 340
Notes and exercises — 344

10 Diffusion methods — 347

Aims: notation — 347
Forward equation — 347
Backward equation — 348
Generality of the backward equation — 350
Kimura–Ohta equation — 353
Initial, terminal and boundary conditions — 355
An alternative approach — 357
Finding $\phi(p, x, t)$ in the neutral, no-mutation case — 359
Finding $v(p, t)$ and $u(p, t)$ in the neutral, no-mutation case — 370
Voronka–Keller formulae — 374
Moments of the distribution — 375
Fisher's view on the reliability of diffusion methods — 379
Two-way mutation: no selection — 383
Selection — 387
Notes and exercises — 387

11 General comments and conclusions — 397

A note on methods — 397
General summary and conclusions — 399

References — 403

Index — 411

1

Introduction

> May the Deluder of Intelligences never trouble the profundity of thine apprehension.
> J. B. S. Haldane, *My Friend Mr Leakey*

'Any theory of evolution must be based on the properties of Mendelian factors, and beyond this, must be concerned largely with the statistical situation in the species'; thus wrote Wright (1931) in the introduction to his famous paper 'Evolution in Mendelian populations'. Similar statements can be found in the writings of Fisher (e.g. Fisher 1921) and Haldane (e.g. Haldane 1924a). The attempt, by these three workers, to analyse the evolutionary process quantitatively proved a brilliant success and provided an intellectual standard by which all subsequent discussions of evolution are judged. Their findings, which were accepted remarkably rapidly by almost all biologists, had a profound effect on evolutionary thinking. Most important was the discovery that natural selection, even of very modest intensity, will have a critical effect on both the direction and speed of the evolutionary process, particularly in large populations. It remained to be demonstrated whether, in natural populations, the intensity of selection at most loci would in fact be sufficiently large for the effects of natural selection to overwhelm those of other factors under most circumstances. Although initially there was much controversy on this question, observations on natural populations by ecological geneticists, led by Dobzhansky in the USA and by Ford in England, appeared to settle the matter. These workers demonstrated the presence of remarkably large intensities of selection in a number of cases in which natural selection was by no means obvious *a priori*. It was natural to conclude, provisionally, that intensities of selection in general, while not necessarily as large as in the cases just mentioned, would usually be sufficiently large for natural selection alone to determine the genetic composition of natural populations.

Under these circumstances, many biologists were reluctant to undertake the labour required to understand the theory in detail. Since the general principles were agreed, a brief acquaintance with the theory would, it was often felt, suffice. Moreover, the theory was, at that time, exceptionally difficult to follow. Bartlett's (1955) description of the work of Fisher, Haldane and Wright, 'as technical a body of research as that in statistical mechanics, say, and requiring as detailed a study', was, of course, intended as a compliment, but must, one feels, have struck a chill in almost any biologist who came across it. Brilliant intuitions, daring approximations, arguments set out so briefly that one was not always sure precisely what was being argued, however much diluted by passages of limpid lucidity, posed a very formidable task for the reader.

Fortunately, this is no longer so. One of the most attractive features of the revival of theoretical population genetics, which began in the early 1950s with the work of Feller and of Kimura, has been a systematic examination of the writings of the founders of the subject (of course, much new work has also been done). As a result, the subject has become very much more accessible. The arguments have been set out in detail and discussed in a very much more rigorous manner than in the original; as is usually the case in mathematics, this rigorous treatment has made the whole subject decidedly easier. Finally, although *faute de mieux* informal arguments are still fairly prominent, the advent of computers has made the checking of results from such arguments a very straightforward matter. Wright's famous paper on self-sterility alleles (Wright 1939a) is a particularly interesting example. This paper has a history reminiscent of that of the Greek bronze horse in the Metropolitan Museum. Long admired as a masterpiece, it was declared non-genuine on the basis of perfectly reasonable arguments, only to be restored to favour in the end. More generally, computer checking has revealed that many formulae are valid over a much wider range of circumstances than might be supposed from their derivation, thus giving much comfort to the biologist who wishes to use such formulae in situations arising in his or her work.

We should note that both the pioneers and the more recent workers, while producing a very impressive range of results, have relied on quite a limited repertoire of mathematical techniques, many of which are standard in applied mathematics. In the author's experience, lack of familiarity with these standard methods is the main source of difficulty for less mathematical biologists wishing to understand our subject. They have difficulty in locating elementary accounts of these methods and find it confusing when, as very often happens, such accounts are illustrated solely by examples in physics or engineering. Were this not so, they might be more willing to accept the view of the mathematician who described Kimura's (1964) famous review article as 'very readable'. At any rate, these methods are not very difficult to follow; the present author is much indebted to the many writers (e.g. Sagan 1961, Sneddon 1961, Mackie 1965, Stephenson 1968, Spiegel 1971) who have shown how these methods can be set out in a simple manner, which he will endeavour to follow.

These simplifications are indeed fortunate, in view of recent developments in evolutionary theory. We recall the demonstration, critical to evolutionary thinking in the 1950s and 1960s, of relatively large intensities of selection in natural populations. Now, these studies of natural selection were carried out on characters visible to the naked eye, or visible under the microscope. In the absence of evidence to the contrary, it was natural to assume that loci controlling such morphological characters evolve in a manner typical of loci as a whole. The development of new techniques for detecting variation, first in proteins and much more recently in DNA itself, provided an opportunity for testing this view. If the DNA controlling morphological characters evolves in a manner representative of the whole DNA, we would expect the two to change roughly in parallel over evolutionary time. Thus a (relatively) rapid change of environment would produce, barring extinction, a (relatively) rapid advance towards a new adaptive pattern, reflected in (relatively) rapid changes at many levels of the genome. In a static environment, neither morphology nor anything else would change at any but a very slow rate. In fact, this simple picture is not correct (for detailed summaries see Kimura 1983, Li *et al.* 1985, Nei 1987). Consider, for example, changes in coding sequences that lead to changes in corresponding proteins. Any given protein changes in amino-acid composition at a characteristic rate per year, which does not differ very substantially from one line of descent to another, even when different lines are grossly discrepant in rate of evolution of morphology. Analogous results are found for some other changes in the genome. The changes studied in detail fall into three classes: (a) changes in pseudogenes (genes that apparently once coded for a functional protein but have lost this capacity owing, for example, to a frame-shift mutation); (b) changes in introns (non-coding sequences of DNA bases that lie within the coding sequences); and (c) synonymous changes (base-pair changes in coding sequences that do not lead to a change in protein). For each of these classes, evolutionary change occurs at a rate *characteristic of the class*, at least to a first approximation (but see Li *et al.* 1985, 1987, Kimura 1987 for possible qualification of this statement). In fact, no portion of the genome has yet been identified whose rate of change is coupled with that of a morphological character or group of these characters. Possibly morphological characters evolve mainly by changes at regulatory loci, the evolution of which is not understood. Whether or not this is so, the old assumption that one could, *without further discussion*, extrapolate results from field studies of morphological characters to DNA in general is clearly unjustified. Even if the differences in morphology under study arise solely from differences in protein structure, there is no guarantee that the results would apply to protein structural differences in general, still less to differences in the DNA as a whole. It is generally accepted that non-morphological differences must be investigated in their own right.

Of course, one may still argue that these 'purely molecular' changes, although not the same as the morphological changes in some important respects, are still of the same substance, in that natural selection is the main factor responsible for either type of change. In this view, virtually any change in DNA composition is either sufficiently disadvantageous or sufficiently advantageous for its future to be decided by selection. Thus any noticeable increase in frequency of the changed DNA would be attributed to 'positive' selection, that is to a selective advantage conferred by that change. Alternatively, one might propose a more limited role for natural selection, e.g. all-important for evolution of proteins but unimportant for evolution of pseudogenes or perhaps significant but not all-important for most molecular changes.

However, in the view of Kimura, first formulated in Kimura (1968a), 'only a minute fraction of DNA changes in evolution are adaptive in nature' (Kimura 1983). Disadvantageous mutations are normally eliminated by natural selection ('negative' selection); nearly all other mutations are supposed neutral or almost so. While adaptive changes proceed in the familiar manner, a noticeable increase in frequency of changed DNA would normally be due to random genetic drift, although of course positive selection would be critical in some cases. Thus, on this view, the theory of changes in frequency of neutral alleles under drift, often considered in the past as mere fun for the mathematically minded, is essential for an understanding of the evolution of a substantial portion of the genome. For example, such familiar questions as 'How many generations are required for a given change in allele frequency?' or 'How can we explain the high level of polymorphism in natural populations?' must be discussed in terms of the neutral theory.

For a rigorous test of the neutral theory, we should have to consider all classes of DNA evolution, including the evolution of regulatory sequences. Such general information is not yet available. Kimura has, therefore, had to argue his case on the basis of a restricted set of DNA changes, in the hope that the evolution of this set is typical of the evolution of the whole genome. Thus the most that could be demonstrated at the moment is that a biologically significant fraction of the genome evolves according to the neutralist scheme. There is strong evidence that this is so. Consider pseudogenes. While it is possible that these may have some function related to their length, it seems very unlikely that their base content has any functional significance. Thus changes in base composition of pseudogenes are unrestrained by negative selection. On the neutral theory, therefore, we expect pseudogenes to evolve relatively rapidly (on an evolutionary timescale!) and this has in fact happened; pseudogenes evolve more rapidly than any other class of DNA whose rate of evolution has been examined.

Now consider synonymous mutations. These, on balance, evolve rather more slowly than pseudogenes; this can, however, be explained by mild negative selection operating against synonymous mutations. When a synonymous mutation occurs, a different transfer RNA may be required for translation; since different transfer RNAs differ in abundance within the cell, synonymous mutations may sometimes be slightly disadvantageous.

A neutralist explanation is also proposed for evolution of introns. These also evolve rather more slowly than pseudogenes. Mutations, such as those found in some cases of β-thalassaemia, that upset excision of RNA corresponding to an intron from the primary RNA transcript, are disadvantageous, but other mutations in introns are supposed inconsequential (we exclude here introns in mitochondria and those nuclear introns which are known to assume a non-intronic role under some circumstances). Thus intron evolution is presumed to be restrained only by mild negative selection (on balance). The argument depends critically on the belief that introns have no function, or at least no function related to their base content; all that is required of them is a grateful exit when appropriate.

> He nothing common did or mean
> Upon that memorable scene:
> But with his keener eye
> The axe's edge did try.

While this may be so, all that can be said with certainty is that nuclear introns cannot normally code for polypeptides, since they generally have termination codons in all reading frames (Lewin 1987); other possible functions have not been excluded.

Perhaps most controversial of all is the case of DNA changes that lead to protein differences. Of course, most of these will disrupt or at least diminish protein activity, so that the intensity of negative selection against this class of mutations is very much greater than for other classes so far considered, with a correspondingly greatly reduced rate of evolution. In some specific cases, a rough idea of this selection intensity can be obtained from a study of the biochemical functioning of the protein concerned; the greater the fraction of mutants likely to be disadvantageous, the slower the rate of evolution. While observed rates conform to this prediction, this result is compatible with both the neutralist and selectionist views, although the roughly constant rate of evolution, for a given protein, accords best with the neutral theory. On the other hand, no-one doubts that *some* cases of protein evolution have occurred under positive selection; the controversy relates to the proportion of such cases. The relative importance of positive selection and drift for protein variation and evolution have been discussed many times (e.g. Gale 1980), with no generally agreed conclusion, and we shall not weary the reader with a repetition of familiar arguments. Our main point is made: there is strong evidence that a significant fraction of the genome evolves in neutralist fashion and the revival of interest in the neutral theory seems justified. More generally, random genetic drift may play a more conspicuous part in

INTRODUCTION

evolutionary theory than has been the case in the recent past. For example, the experimental work of Hartl & Dykhuizen (reviewed 1985) on enzymes mediating carbohydrate metabolism in *Escherichia coli* provides strong evidence that, under normal circumstances for this species, variants of these enzymes are neutral (or almost neutral) *inter se*; only under rather exceptional circumstances would natural selection discriminate between them. We should note that, in standard selection theory, the role of drift is greater than beginners sometimes suppose. To demonstrate the conditions under which the long-term composition of natural populations is decided by natural selection only is rather an intricate matter, which we shall discuss in due course. We shall merely note for the moment that situations may exist in nature in which the long-term effects of drift and of natural selection are comparable in magnitude; possibly these situations are common, as has long been proposed by Wright.

On the other hand, some geneticists still favour the strongly selectionist view. They do not, of course, contest the role of negative selection and indeed accept the view (which long pre-dates current controversies) that the more marked the effect of a mutation on the phenotype, the more likely is the effect to be harmful. The selectionist argument (Fisher 1930b, Clarke 1971) is that the less marked the effect of a mutation, the more likely is the effect to be *advantageous* (rather than neutral). It is supposed that all, or nearly all, the genome is functional. The fraction of mutations that are advantageous would be greatest for pseudogenes, smaller for introns and synonymous mutations, and smallest for non-synonymous mutations. Differences between classes of DNA in rate of evolution would then reflect these class-to-class differences in the supply of advantageous mutations. A possible difficulty for this view has been pointed out by Kimura (1983). The rate of evolution for a given class under selection would depend not only on the supply of advantageous mutations for that class but also on the magnitude of the selective advantage; the smaller the latter, the slower the rate, other things being equal. Kimura suggests that advantageous mutants with small effect would confer only a small advantage and that this could more than offset the effect of increased supply on the rate of evolution. He gives an interesting quantitative discussion of this problem, which we shall mention later.

However, the neutral theory also has its difficulties, although we can only indicate the most important of these at this early stage of the discussion. We shall show repeatedly that the selection theory is very much easier to accept if population sizes are typically very large, whereas the neutral theory requires rather moderate or small population sizes. However, the term 'population size' requires very careful specification. In an early controversy with Wright, Fisher (1929) maintained that the world population size of the species was usually the relevant quantity when discussing neutral theory, rather than the local population size, as proposed by Wright; Fisher's view was later justified by the work of Maruyama (e.g. 1970a, b, 1977). At first sight, this would suggest a very large population size. To overcome this difficulty, the neutralist advocates a substantial discount of the world population size, on the grounds that the bulk of the population at any time may contribute very little to the long-term gene pool of the species. Thus large differences between adults in fecundity and periodic marked reductions in the overall size of the adult population would entail a fairly large mark-down of population size (Wright 1938a, 1939b). A further possibility has been proposed recently by Maruyama & Kimura (1980). Suppose the species is made up of subpopulations and that subpopulation extinction occurs fairly often; when a subpopulation becomes extinct, its habitat is recolonized from another subpopulation of the species. Then, at any time, the number of breeding adults in the whole species may be very large, but this is deceptive, since many individuals are subject to long-term genetic death. It is possible in principle, but uncertain in practice, that these various factors are sufficient to reduce population sizes to the level required by the neutral theory; the data available are quite inadequate for a decision on this point. At the moment, we have to *assume* the neutral theory correct in order to estimate appropriate population sizes, so that the estimates given, while possibly correct, provide no independent support for the theory. Thus the role of random genetic drift in the evolution of natural populations remains uncertain.

On the other hand, there can be no doubt of the importance of drift in *applied* genetics. In animal and plant breeding, population sizes are often small and this has a crucial effect on ultimate response to selection (Robertson 1960, 1970, 1977, Hill 1970, Hill & Robertson 1966). Similarly in the conservation of genetic resources; for many crops, seed has a limited longevity under storage, so that at intervals (perhaps every five years) a seed bank must be 'regenerated' by raising and crossing plants to obtain new batches of seed. At every regeneration, alleles may be lost by drift. Obviously, the conservation programme should be designed to keep this loss as small as practically possible; as we shall show, this is easily done once a few standard points in population genetics theory have been established.

Finally, we may ask; 'How should the population geneticist approach recent discoveries on the nature of the genome?' Obviously, our theory first developed under very primitive notions of the nature of the hereditary material. It is occasionally suggested that, in the light of modern findings (e.g. introns, transposable elements, clustering of genes of identical or closely related function), the old theory is altogether outmoded and that a complete reconstruction of the theory is required. While there can be no certainty here, it seems very unlikely that we shall have to jettison the old theory in the comprehensive manner proposed. This theory was, of course, based on the standard Mendelian scheme, so well supported by the observations of the early geneticists. Hence there must be an area of reality that is well delineated by our older theory. In fact, this theory will work well enough for many purposes, provided we follow the usual modern practice of defining the gene as the unit of function. For example, the 'gene for alcohol dehydrogenase' is that region of the DNA that includes the sequences coding for alcohol dehydrogenase together with the corresponding leader sequence, trailer sequence and introns (note that the latter, even if functionless, must be included, since a mutation upsetting splicing will destroy gene function). Clearly, from our earlier discussion, we sometimes need to distinguish different classes of DNA within such a gene. Often, however, this is not necessary. For example, in describing the spread of a fungal pathogen, it would be sufficient to note a mutation to virulence, without specifying where in the gene the mutation occurred. In such cases, the classical notion of the gene is quite adequate. Generally, we use the simplest representation of the gene consistent with the problem under study.

While this approach leads to an easier theory, it may seem perverse ever to employ an out-of-date model of the hereditary material; to ignore, say, the fact that coding sequences are split or to discuss mutation without mentioning that many mutations arise from the insertion of a transposable element. Here a well known analogy from physics may help. For well over 200 years, physicists described the world in terms of Newtonian mechanics. Later, it emerged that some of the assumptions of the Newtonian scheme, such as the absoluteness of time, cannot be reconciled with observation. For example, if time were absolute, the half-life of a given unstable elementary particle would be independent of circumstances and muons formed in the upper atmosphere would have disintegrated before reaching the Earth's surface; to explain their behaviour, we need the relativistic mechanics of Einstein. But this does not mean that Newtonian mechanics should never be used and that those who teach this subject are practising an impudent fraud. For many practical purposes, the Newtonian assumptions are near-enough correct and it is quite unnecessary to complicate the problem by using a more 'truthful' but more difficult and no more relevant approach. Similarly, even the most contentious of cricketers would hardly contest a groundsman's statement that he had rolled the pitch 'flat' on the argument that the groundsman had ignored the curvature of the Earth, although strictly the cricketer would be right.

It is indeed fortunate that the simple Mendelian model is often appropriate, since we can be confident that this model may be applied whatever the species under study. On the other hand, as Lewin (1987) reminds us, 'the eucaryotic kingdom is extremely broad, and at present we have detailed information about the genetic organization of only a few types of species . . . we are dealing with features that may be represented to widely varying degrees in different individual genomes'. When considering, then, the population genetics of the many recently discovered features of the eucaryote genome, we must bear in mind that these features are not necessarily universal.

Notes and exercises

The following problem, adapted and extended from a problem in Kemeny et al. (1965), will introduce the reader to some leading features of the genetics of populations.

A man lives on a street M paces long. At one end of the street is his house, at the other end a lake. Somewhere on the street, n paces from the lake, is a bar. He leaves the bar in a regrettable state.

Let P be the probability that, on any occasion he moves, he goes one pace towards home, and let Q be the corresponding probability that he goes one pace towards the lake; $P + Q = 1$.

Let y_n be the probability that, starting n paces from the lake, he eventually reaches home, rather than falling into the lake. We want to find y_n for several cases of interest.

(a) For definiteness, take $y_0 = 0$ i.e. if he starts at the lake, he falls in.
(b) Also, $y_M = 1$, i.e. if he starts at home, he reaches home with probability 1.
(c) For all other n, we 'decompose' his movements into first step, remaining steps.

On his first step, he *either* moves one pace homewards, with probability P, after which he is $n + 1$ paces from the lake, so that his probability of eventually reaching home is y_{n+1}; *or* he moves one pace lakewards, with probability Q, after which his probability of reaching home is y_{n-1}. Thus

$$y_n = Py_{n+1} + Qy_{n-1}$$

It may be shown that this equation, taken in conjunction with our conditions $y_0 = 0$, $y_M = 1$, determines y_n uniquely.

The reader will easily verify that the standard formula

$$y_n = \frac{n}{m} \quad \text{in cases where } P = Q$$

$$= \frac{1 - (Q/P)^n}{1 - (Q/P)^M} \quad \text{in cases where } P \neq Q$$

satisfies our equation and conditions and therefore is the solution to our problem.

1 Suppose $n = \frac{1}{2}M$, i.e. our drinker starts halfway to home.
   (a) When $P = Q$, we have $y_n = \frac{1}{2}$. This is indeed obvious, but the next result is not.

INTRODUCTION

(b) Consider the case $P = 0.51$, $Q = 0.49$, i.e. there is a slight homeward bias on every step.

Show that, if $M = 100$, $y_n = 0.8808$ (compare this with the value 0.5 above).

By considering various values of M (e.g. 10, 50, 100, 200, 400) convince yourself that the longer the street, the greater the probability of reaching home, this probability approaching 1 as the length of street increases.

2 Now suppose $n = 1$, i.e. our friend starts very close to the lake.

Find y_n for different combinations of M and P (e.g. values of M as above, values of $P = 0.5, 0.51, 0.7, 0.9$) and convince yourself of the following:

(a) (y_n when $P > Q$)/(y_n when $P = Q$) increases as M increases,
(b) even when $P \gg Q$, y_n is not 1,
(c) when $P > Q$, $y_n \to 1 - Q/P$ (i.e. independent of M) as $M \to$ large.

3 Treat the above as a parable.

M is the number of alleles, at a locus, in a population (i.e. for a diploid, M is twice the population size). n is the initial number of allele A, $M - n$ that of allele a. n/M is the initial frequency of allele A. Reaching home $= A$ becoming fixed eventually. Falling into the lake $= A$ becoming lost eventually. $P:Q =$ relative fitnesses conveyed by alleles A and a (thus $P = Q$ implies neutral alleles, $P > Q$ implies a selective advantage of A over a).

4 Try to guess the answer to the following:

(a) What is the probability of fixation for a neutral allele having initial frequency p?
(b) Suppose A is at a selective advantage over a and that the initial frequency of A is intermediate, say $\frac{1}{2}$. How does the probability that A becomes fixed depend on the population size?
(c) Suppose now that there is just one A present initially, having just arisen by mutation, i.e. initial frequency of $A = 1/M$.
  (i) What is the probability of fixation if A is a neutral allele?
  (ii) For any given selective advantage of A over a, how does the probability of fixation of A depend on the population size?
  (iii) Is an advantageous mutant necessarily fixed eventually?
  (iv) How does

$$R = \frac{\text{probability of fixation for an advantageous allele}}{\text{probability of fixation for a neutral allele}}$$

depend on the population size?

NOTES AND EXERCISES

(v) Suppose we take, as a criterion for a small selective advantage to matter, that R shall be large. Will natural selection be more important in the long run if the population is large or if it is small?

Answers

(a) p.
(b) Probability increases to a limit of unity as the population size increases.
(c) (i) $1/M$. (ii) Decreases to a non-zero limit as population size increases. (iii) No. (iv) Increases, dramatically and without limit, as population size increases. (v) Large.

2
Wright–Fisher, Moran and other models

> I paint objects as I think them, not as I see them.
> Picasso

On simple models

Perhaps the most remarkable of the achievements of the early workers in theoretical population genetics was the realization that results of far-reaching importance, amounting to general laws, could be obtained from very simple representations of biological reality. It is by no means obvious *a priori* that the results derived from such models would be relevant to the highly complex situations found in natural populations. Indeed, even today it is occasionally suggested that any attempt to find general laws governing the genetic composition of natural populations must fail unless account is taken, simultaneously, of all the factors that might affect that composition. In a milder version of this argument, it is proposed that, while some features of the natural situation may be ignored, other features, felt by the critic on intuitive grounds to be essential, must necessarily be included. Such very reasonable but nevertheless mistaken arguments are common in the history of science. Thus the chemists of the early 19th century opposed Avogadro's hypothesis with very sensible arguments, which, of course, proved ultimately to be ill-founded. In the present context, we have only to think of the Hardy–Weinberg law, with its demonstration of the 'variation-preserving' character of the hereditary mechanism, to appreciate that very simple models can give useful results. As Ewens (1969, p. 131) remarks:

> irrespective of any possible unreality of the assumptions under which the Hardy–Weinberg law was derived, the general importance of this law is undeniable. The Mendelian system of heredity ensures the existence of a vast and continuing store of variability upon which the forces of natural

selection can act ... a population must be fairly small before such a conclusion can no longer be drawn.

In general, the following procedure has proved very successful. The investigator first derives biologically significant results from a simple model. Later, he or she (or others) introduces modifications into the model to see whether the earlier results still hold. Perhaps surprisingly often (but not, of course, always) they continue to hold, near enough, save under rather special circumstances. Moreover, modifications that have an important effect on the outcome in some circumstances may be of little consequence in others. We shall attempt to provide evidence for these statements in due course.

With these considerations in mind we shall, for any given problem under study, choose the simplest mathematical representation of biological reality that seems appropriate. Provided we can assure ourselves, at some stage of the investigation, that our model will not give misleading results for the problem in hand, we need feel no anxiety at the omission of phenomena that may be all-important in a different context.

Stochastic models: deterministic models

We shall, in general, adopt a 'stochastic' approach; that is, we take account of factors leading to random changes in allele frequencies. The most important of these factors are random genetic drift and random fluctuation in intensity of selection. Although it is possible to allow for the simultaneous action of both of these factors, such a treatment is rather complicated and probably best avoided in the first instance. Moreover, the effects of drift have been explored very much more thoroughly than those of random change in selection intensity. We proceed, therefore, as follows: except where stated otherwise, we shall take drift to be the sole source of random change in allele frequency, but will note on occasion how conclusions may have to be modified to allow for random change in intensity of selection.

A further simplification, common in elementary treatments, is to take a 'deterministic' approach, that is assume that random changes in allele frequency may be ignored. Once we have analysed the effects of drift, we shall be able to set down circumstances in which a deterministic treatment is justified. However, even in the stochastic approach, deterministic arguments are often relevant. Suppose, say, that the frequency of a given allele A in a given generation has known value p and that the frequency of this allele one generation later, calculated deterministically, is p^*. In the stochastic treatment, this p^* appears as the *probability* that a randomly chosen allele in the later generation is A. We shall, in the course of our enquiry, seek any quantities, arrived at by the stochastic treatment, that could have been calculated deterministically.

Random genetic drift: general approach

To allow for drift, we regard the population under study as if it were drawn, at random, from an indefinitely large set of potential populations. In any given generation, all populations in the set are identical to the population under study in respect of size, mating system, sex ratio, distribution of progeny number, etc., and also of factors such as selection intensity or mutation rate that affect the allele frequency as calculated deterministically. We shall discuss the case where initially (generation 0) the frequency of any given allele, say A, is the same in all populations of the set. Then, in any given later generation, any differences in genetic composition between populations of the set are attributable solely to the action of random genetic drift and its interaction with factors such as selection. The outcomes in the different populations represent 'what might happen'; to say that a particular outcome in the population under study is highly probable is the same as saying that this outcome will result in most of the populations in the set. Thus in any generation we may meaningfully speak of the probability that the allele frequency in the population under study takes some stated value, v say; we interpret this probability as the proportion of populations in the set in which the allele frequency is v. More generally, we may speak of the probability distribution of allele frequencies in any generation, this being the probability distribution of allele frequencies over populations in the set.

Distribution of allele frequency under drift

Let the frequency of A among the newly formed zygotes in generation 0 be p_0 (in all populations of the set) and the corresponding frequency one generation later, calculated deterministically, be p_0^*. Then, as we have said, p_0^* is the probability that an allele, randomly chosen from the newly formed zygotes in generation 1, is A. Let the population size be N. Then any population in generation 1 can be regarded as a random sample, size N, drawn from an infinite population in which (among other things) the frequency of A is p_0^*. We have one such sample for every parent population in the set. However, as all these populations have the same frequency in generation 0, we can regard all the samples as drawn from a single infinite population – a very familiar situation. Thus the probability distribution of the frequency of A in generation 1 (or at least some properties of this distribution) can be calculated fairly easily, as we shall see shortly.

DISTRIBUTION OF ALLELE FREQUENCY UNDER DRIFT

In later generations, where populations in the set differ in allele frequencies, the situation is a little more complicated. We could, however, in any generation, t say, confine ourselves, temporarily, to parent populations that happen to have the same frequency p_t of A in that generation and then proceed as we did when going from generation 0 to generation 1. A probability for generation $(t + 1)$ thus calculated is clearly a *conditional* probability, being calculated subject to the allele frequency taking a stated value in generation t, whereas we really want the *unconditional* probability (usually, for short, just called the probability) in which this arbitrary restriction for generation t is not made. The two sorts of probability are connected in the following way:

(unconditional) probability that the frequency is p_{t+1} in generation $(t + 1)$
 = $\Sigma\{$[(unconditional) probability that the frequency is p_t in generation t]
 × [probability that the frequency is p_{t+1} in generation $(t + 1)$ conditional on the frequency being p_t in generation t]$\}$

The sum is taken over all possible values of p_t. Thus, after all, the conditional probability distribution, despite its apparent artificiality, is seen to be relevant to the computation of the (unconditional) probability distribution in any generation, sometimes referred to as the 'classical problem' of theoretical population genetics. At least in principle, the equation just given should lead to a solution of this problem, provided the conditional distribution is known. In practice, obtaining this solution, even in biologically very simple situations, is by no means straightforward; the first complete solution was not obtained until the work of Kimura (1955a.) However, many interesting properties of the distribution, obtainable without knowing the distribution itself, were derived earlier, for several situations of biological importance, by Wright, Fisher and Haldane. The reader will probably accept at once that the conditional distribution is relevant to the determination of these properties. For example, it is easily shown (see note 3 at the end of this chapter) that to obtain the mean of the distribution in $(t + 1)$ we take the mean of the distribution calculated *conditional* on the allele frequency being p_t in t and average the answer over all values of p_t. Generally, the classical problem is so central to any description of the process of change of allele frequency that it seems almost obvious that any quantity depicting this process, for example the mean number of generations required for the allele frequency to change from stated value x to stated value y, is in principle calculable from the solution of the classical problem, even though in practice a different method of calculation may prove more convenient. This being so, we may conclude with some confidence that, to make any progress, we must first obtain some properties of the conditional distribution for several situations of biological interest. We begin with the mean.

Change of allele frequency in one generation

For simplicity, we write p for the frequency of A in some generation, t say, p^* for the deterministically calculated frequency one generation later and p' for the actual frequency in the later generation. We wish to calculate the mean value of p', conditional on the frequency being p in generation t; this mean will be written $Ep'|p$ (the vertical line being read 'conditional on' – where this line does not appear, the mean will be unconditional). We use a device which, although artificial at first sight, will be helpful on several occasions.

Imagine, for every population separately, that the $2N$ alleles at our locus in generation $(t + 1)$ present in that population are written down in random order and numbered $1, 2, 3, \ldots, 2N$. Concentrate on some particular place in the list, say the ith place; which particular allele appears in this place will vary with population. The probability that A appears in this ith place (= the proportion of populations in which A appears there) equals p^*. Let X_i ($i = 1, 2, 3, \ldots, 2N$) be a random variable such that, in any given population,

$$X_i = 1 \quad \text{if the allele in the } i\text{th place is } A$$
$$ = 0 \quad \text{otherwise}$$

Let $q^* = 1 - p^*$. Then the mean of X_i, written $EX_i|p$, is (by definition of mean)

$$1(p^*) + 0(q^*) = p^* \qquad \text{for any } i$$

Moreover, for any population,

$$r = X_1 + X_2 + X_3 + \cdots + X_{2N}$$

is the number of A alleles present in that population. Since (by a standard statistical result) the 'mean of a sum' equals the 'sum of the means', we have at once the mean of r:

$$Er|p = E\left(\sum_{i=1}^{2N} X_i\right)\bigg|p = \sum_{i=1}^{2N}(EX_i|p) = \sum_{i=1}^{2N} p^* = 2Np^*$$

Further, for any constant k and random variable x,

$$E(kx) = kEx$$

whence

$$Ep'|p = E\left(\frac{r}{2N}\right)\bigg|p = \frac{1}{2N} Er|p = p^*$$

that is, the mean allele frequency in $(t + 1)$ is p^*, so that the mean change in frequency in going from t to $(t + 1)$ is

$$p^* - p$$

The mean in any generation

It may surprise the reader that we have bothered to prove a result that is intuitively obvious. However, experience indicates that it is worth proving everything we can, since intuition can be very misleading on occasions. For example, the argument just given shows that, since in generation 0 the frequency of A is a fixed quantity p_0, the mean frequency in generation 1 is equal to the deterministically calculated frequency. Is it true, for later generations, that the *unconditional* mean frequency in any generation just equals the deterministically calculated frequency for that generation? One might suppose that, as in some sense random variations cancel out on average, the usual elementary deterministic calculations, in which drift is ignored throughout, should at least give the (unconditional) mean. It is a remarkable fact that, in general, this is not correct, although it is true in some special cases.

To see this, let us write $P_1, P_2, P_3, \ldots$ for the frequency of A in generations $1, 2, 3, \ldots$ as calculated deterministically; any P_t will be a function of the frequency P_{t-1} one generation earlier, say

$$P_t = f(P_{t-1})$$

For example, if allele frequencies are changing by mutation only, A mutating to other alleles at rate u, and if back-mutation is negligible, then

$$f(P_{t-1}) = P_{t-1}(1 - u)$$

Generally, we may then write, in the deterministic approach

$$P_1 = f(p_0) \qquad P_2 = f(P_1) \qquad P_3 = f(P_2) \qquad \text{and so on}$$

Now write $p_1, p_2, p_3, \ldots$ for the frequency of A in generations 1, 2, 3, ... when drift is taken into account. We have shown that, as p_0 is given,

$$Ep_1 = P_1$$

but we need care in discussing later generations. If the (unconditional) mean is to equal the deterministically calculated value, it must change in the same way as the latter, giving

$$Ep_1 = f(p_0) \qquad Ep_2 = f(Ep_1) \qquad Ep_3 = f(Ep_2) \qquad \text{and so on}$$

The first of these relationships is certainly correct. How about the others? We have shown that

$$Ep_2|p_1 = f(p_1)$$

whence, from the rule for calculating the unconditional mean, which was given in the previous section,

$$Ep_2 = Ef(p_1)$$

(the average on the right being taken over all values of p_1), which certainly appears different from the result $f(Ep_1)$ necessary for the mean to equal the deterministic value. Generally, for these to be equal, we must have, in any generation t,

$$\mathop{E}_{p_t} f(p_t) = f\!\left(\mathop{E}_{p_t} p_t\right)$$

Notice that, on the left-hand side of this equation, we take the frequency of A in every population, calculate the function f for every such frequency and then average the result over all populations. On the right-hand side, we take the frequency of A in every population, average these frequencies and then calculate the function f of the average.

The reader is probably aware of the fact that, if we have a random variable x, taking a function of x and then averaging does *not*, in general, give the same result as averaging first and then taking the function of the answer. For example, E log x is not the same as log(Ex). This being so, the mean allele frequency will not, in general, be the same as the deterministically calculated frequency except in generation 1.

THE MEAN IN ANY GENERATION

There is, however, an important class of exceptions to this general rule. If the function of x is a *linear* function

$$f(x) = Ax + B$$

(A and B being constants), then

$$Ef(x) = E(Ax + B) = AEx + B = f(Ex)$$

so that we obtain the same result whether we average first or take the function first. Thus if p^* is a linear function of p (in which case the population is said to be 'under linear pressure'), as for example in the mutation case given earlier

$$p^* = p(1 - u)$$

deterministic calculation gives the correct mean in any generation. We have at once in this mutation case

$$Ep_{t+1} = Ep_t(1 - u)$$

whence

$$Ep_t = p_0(1 - u)^t$$

Results of this kind are obtained for populations changing under mutation or migration. On the other hand, if natural selection is operating, 'pressures' are no longer linear. Sometimes, however, it is possible to use a linear function as an approximation to the true function. Consider a population of haploid organisms, with two alleles present, allele frequencies being p of A and q of a in some generation. A and a are supposed to have relative fitnesses $(1 + \alpha):1$, so that $p^*:q^* = (1 + \alpha)p:q$, giving

$$p^* = \frac{p(1 + \alpha)}{p(1 + \alpha) + q} = \frac{p(1 + \alpha)}{1 + \alpha p}$$

(since by definition $p + q = 1$). When allele A is very rare, the denominator $(1 + \alpha p)$ will be very close to unity, so that p^* is almost a linear function of p in this case. In fact

$$p^* = p(1 + \alpha)(1 + \alpha p)^{-1}$$
$$= p(1 + \alpha)(1 - \alpha p + \alpha^2 p^2 - \cdots)$$
$$= p(1 + \alpha) + (\text{terms in } p^2, p^3, \ldots)$$

Then if p is sufficiently small for the terms in $p^2, p^3, \ldots$ to be neglected, we have to a very close approximation the linear relationship

$$p^* = p(1 + \alpha)$$

This relationship was exploited extensively by Fisher (1930b) when dealing with the spread of a very rare advantageous allele (Ch. 4).

In cases where such a linear approximation is not possible, deterministic calculation *sometimes* yields a good approximation to the mean, but very careful analysis is required before it can be established whether such an approximation is justified. An interesting case has been stressed by Robertson (1962). Consider a recessive lethal a with mutation rate from A to a equal to μ. According to a well known deterministic argument, the frequency of a ultimately settles down at $\sqrt{\mu}$. However, unless the population is quite large, the mean ultimate frequency, as we shall see later, is substantially less than $\sqrt{\mu}$; only if the population size is larger than (about) $1/\mu$ does the mean approach $\sqrt{\mu}$.

We now sum up this section. If, in any generation, the allele frequency has known value p, then if we ignore drift and calculate the allele frequency one generation later we obtain the mean of the frequencies that would arise in that generation when drift was acting. However, the allele frequency has known value only at the start (generation 0). If we take this initial value, ignore drift and calculate allele frequency in successive generations, we do not necessarily obtain the mean allele frequency for the corresponding generation (except for generation 1).

Change of allele frequency in one generation: the variance

We turn now to the variance of the conditional distribution. This has been investigated most fully for the case of two alleles at a locus A, a in a population where allele frequencies are changing as a result of drift only. Then if we write p, q for the known frequencies of A, a in generation t, we have $p + q = 1, p^* = p, q^* = q$. We write p', q' for the corresponding frequencies in generation $(t + 1)$. As, in this section, all results will be conditional on the known frequencies p, q, we omit the conditional symbol $|$ in the interests of simplicity. As before, we write the number of A alleles in $(t + 1)$ as

$$r = X_1 + X_2 + X_3 + \cdots + X_{2N}$$

where $X_i = 1$ with probability p and $X_i = 0$ with probability q in the present case. Writing $V(r)$ for the variance of r, $V(X_i)$ for the variance of X_i and

Cov(X_i, X_j) for the covariance of X_i and X_j we have, by a standard statistical result,

$$V(r) = \sum_{i=1}^{2N} V(X_i) + \sum_{i \neq j} \text{Cov}(X_i, X_j)$$

(the second sum is taken over all possible pairs of values of i and j - both i and j going from 1 to $2N$ - but leaving out cases where $i = j$).

The variance terms are straightforward. We have, by the usual standard result,

$$V(X_i) = EX_i^2 - (EX_i)^2$$

Then $EX_i^2 = 1^2(p) + 0^2(q) = p$ (for any X_i) and from the previous section

$$EX_i = p$$

Therefore

$$V(X_i) = p - p^2 = pq$$

and

$$\sum_{i=1}^{2N} V(X_i) = 2Npq$$

The covariance terms are more difficult. According to Wright (1931), explicitly, and Fisher (1930b), implicitly, we obtain a good representation of biological reality by supposing that the alleles appearing in any population in $(t + 1)$ are 'chosen' *independently* of one another. If this is so, all covariances are zero and we have

$$V(r) = 2Npq$$

and since, for any constant k and random variable x, $V(kx) = k^2 V(x)$, the variance of the frequency of A in $(t + 1)$ is

$$V(p') = V\left(\frac{r}{2N}\right) = \left(\frac{1}{2N}\right)^2 2Npq = \frac{pq}{2N}$$

In fact, with this assumption of independence the number of A alleles will necessarily be bionomially distributed; we have a random sample of $2N$

alleles, every allele being drawn, independently of all the others, from an infinite population with frequencies of p of A and q of a, whence the probability of obtaining r A alleles is

$$\frac{(2N)!}{r!(2N-r)!} q^{2N-r} p^r$$

Indeed, the method we used to derive V(r) is one standard procedure by which the variance of the bionomial distribution may be obtained.

Is the binomial distribution of allele number relevant?

It is by no means obvious that this 'Wright-Fisher' model (i.e. the binomial distribution just given) is in any way appropriate. We shall discuss in some detail apparent objections to this model and the extent to which such objections can be surmounted (Wright 1931, 1938a, 1939b, 1969, Malécot 1948, Li 1955, Kimura & Crow 1963, Crow & Kimura 1970, Ewens 1979).

One difficulty that sometimes worries beginners is that, in our model, all the sampling appears to take place at one time, whereas in fact the sampling process is spread across the life cycle. However, this is not a problem. Provided alleles are sampled independently at every stage, we must end up after one complete life cycle with a sample of $2N$ independently chosen alleles, with probability p that any such allele is A and probability q that any such allele is a, and the binomial distribution stands.

The assumption of independent sampling of alleles presents much more of a problem. The following test of independence is helpful. If sampling is independent, then if we were told that by chance an A has been chosen, this information tells us nothing about the probability that another A will be chosen in the same population, this probability remaining at p, as it was before we had the information. At least as far as diploids are concerned, the assumption of independence must often be false. If, say, M zygotes are formed initially, to be reduced, by chance, to N zygotes in the course of the life cycle, then clearly alleles are being lost, by chance, a pair at a time. If alleles are associated into zygotes *at random*, this pairwise loss does not vitiate the independence assumption. Consider, however, a plant species reproducing partly by selfing, partly by random mating. It is easier to think of the case where the amount of selfing is substantial (although the same considerations will apply to a lesser extent with any departure from random mating). Then the population consists mainly of AA and aa; in terms of our test of independence, if we know that an A has been lost at a given moment, we can be fairly sure that another A was lost at the same time, so that the assumption of independence must be wrong. To obtain a quantitative formulation of this point, we must describe the infinite population from which our sample is taken with more care than we did when calculating the mean, when it was sufficient to state that in this infinite population the frequency of A is p^* ($= p$ in the present case). We imagine an infinite population of *zygotes* in which allele frequencies are pA and qa, and in which the quantity f is defined (in the usual way) by

$$f = 1 - \frac{\text{frequency of } Aa}{2pq}$$

(although this is standard, it may be worth stressing that this f is *not* the coefficient of inbreeding in either Wright's or Malécot's sense; we use f rather than F to emphasize this distinction).

Then the frequency of Aa is $2pq(1 - f)$ and as we must have

$$(\text{frequency of } AA) + \tfrac{1}{2}(\text{frequency of } Aa) = p$$

we have as usual:

	AA	Aa	aa
frequency	$I = p^2 + fpq$	$J = 2pq(1-f)$	$K = q^2 + fpq$

To form generation $(t + 1)$, N zygotes are drawn independently and at random from this infinite population. Then the probability of drawing i AA, j Aa, k aa individuals is

$$\frac{N!}{i!j!k!} I^i J^j K^k$$

that is, i, j and k follow the multinomial distribution. Now

$$p' = \frac{i + \tfrac{1}{2}j}{N}$$

so that p' is a linear function of multinomially distributed variables, say

$$p' = Ai + Bj$$

where $A = 1/N$, $B = 1/(2N)$. From the usual 'rule of means'

$$Ep' = AEi + BEj$$

and from standard properties of the multinomial distribution

$$V(p') = A^2Ei + B^2Ej - (1/N)(Ep')^2$$

Since

$$Ei = NI \qquad Ej = NJ$$

then

$$Ep' = I + \tfrac{1}{2}J = p$$

$$V(p') = \frac{I}{N} + \frac{J}{4N} - \frac{p^2}{N} = \frac{pq}{2N}(1+f)$$

equivalent to taking the binomial variance $pq/(2N)$ and replacing N by $N/(1+f)$. This result is due to Li (1955), who used rather an intuitive approach. Only in the case of random mating ($f = 0$) do we obtain the variance $pq/(2N)$ appropriate to the case where alleles are sampled independently. In fact, since

$$V(p') = \left(\frac{1}{2N}\right)^2 [2Npq + 2N(2N-1)\operatorname{Cov}(X_i, X_j)]$$

$$= \frac{pq}{2N}\left(1 + \frac{2N-1}{pq}\operatorname{Cov}(X_i, X_j)\right)$$

then

$$\operatorname{Cov}(X_i, X_j) = \frac{fpq}{2N-1}$$

In the foregoing analysis we have (for simplicity!) tacitly assumed that all parents have the same chance of contributing to the next generation. If this is not so, sampling of alleles will not be independent, even if mating is at random (the same is true for populations of haploids). To see this, consider again the Wright–Fisher model where

$$\text{Cov}(X_i, X_j) = 0$$

Now in some populations in $(t + 1)$ it will happen that the alleles in the ith and jth places come from the same parent in t; this being so, they will be more alike (on average) then the general run of pairs of alleles in those places, so that $\text{Cov}(X_i, X_j)$ is positive for populations of this kind. On the other hand, in those populations where the two alleles happen to come from different parents, they will be slightly less alike (on average) than the general run, giving rise to a small negative correlation. Overall, with many more cases of different parents than of the same parents, the positive and negative covariances cancel when all populations are taken into account, *for this model*. Such cancellation cannot obtain in general. The more uneven the parental fecundities, the greater the chance that two randomly chosen alleles come from the same parent and therefore the greater the contribution of positive covariances and the smaller the contribution of negative covariances to the overall covariance. Cancellation, then, must correspond to a rather special distribution of parental fecundities.

For simplicity, we consider random mating populations. Strictly speaking, our discussion will be correct only for haploids or for monoecious diploids in which the odd chance selfing may occur. However, if the population size is at all large, such selfing will be very infrequent and thus will contribute very little to the next generation; we may surmise then (correctly in fact – Wright 1931) that our results will hold to a very close approximation for self-incompatible monoecious species (unless the population size is very small). In a similar way, we reasonably expect very similar results for populations of dioecious species in which *the sex ratio is* 1:1. Again, this expectation is justified, although a rigorous treatment of the dioecious case is not straightforward (Norman 1975a, Ewens 1979).

We consider in detail the case where any differences between parents in potential fecundity are caused by differences in parental environment. We suppose it a matter of chance which parents have received an environment conducive to high fecundity and which have received a poor environment. More precisely, we imagine environments as varying continuously with regard to effect on fecundity and suppose, for any generation, environments to be allocated at random to parents.

Traditionally (e.g. Crow & Kimura 1970) the problem is discussed in terms of the variation in the number of alleles contributed by a *parent* to the next generation. However, it is very much easier to think in terms of the variation in the number of descendant alleles contributed by a *parental allele* to the next generation. Concentrate on any parental allele; since alleles are neutral, no harm is done by thinking of this allele as A and of all other $(2N - 1)$ parental alleles as a, so that $p = 1/(2N)$. Thus the probability that our parental allele gives r progeny alleles is just the probability that a parental population with frequencies

$$\frac{1}{2N} \text{ of } A \qquad \left(1 - \frac{1}{2N}\right) \text{ of } a$$

gives a progeny population having r A alleles. By definition of the Wright-Fisher model, this probability is

$$\frac{(2N)!}{r!(2N - r)!}\left(1 - \frac{1}{2N}\right)^{2N-r}\left(\frac{1}{2N}\right)^r$$

a binomial distribution with

$$\text{mean} \quad 2N \times \frac{1}{2N} = 1$$

$$\text{variance} \quad 2N \times \frac{1}{2N} \times \left(1 - \frac{1}{2N}\right) = 1 - \frac{1}{2N} = V(g)$$

say (from the usual properties of the binomial distribution). If N is large, this distribution will be very close to a Poisson distribution with mean 1. Notice that our distribution applies to any parental allele, so that our model requires 'equal potential fecundity' of all parental alleles; i.e. if an allele is picked at random from the progeny generation, it is equally likely to descend from any of the $2N$ parental alleles and thus has probability $1/(2N)$ of descending from any specific parental allele.

Note that our variance $V(g)$ is the variance of the number of descendants of a *single* parental allele, in *repeated* sampling from the same population. $V(g)$ must be distinguished from V_w, the variance of the number of descendants of *different* parental alleles on a *single* sampling. V_w will vary at random from one sampling to another. While, for any model, the *mean* of V_w over repeated samplings equals $V(g)$, the value of V_w on any single sampling could, for some models, depart fairly markedly from $V(g)$, unless N were fairly large. However, *given our binomial distribution*, the variance of V_w over repeated samplings proves to be

$$\frac{1}{2N}\left[11 - \frac{27}{2N} + \frac{22}{(2N)^2} - \frac{6}{(2N)^3} - (2N - 1)\left(\frac{9}{2N} - \frac{14}{(2N)^2} + \frac{6}{(2N)^3}\right)\right]$$

(or about $1/N$ when N exceeds 50 or so), which is quite small for values of N encountered in practice. Hence, when testing data for agreement with the binomial model, we shall not go far astray in supposing that a single occasion V_w is, near-enough, the same as $V(g)$.

We have shown that the Wright–Fisher model, when applied to haploids, implies a distribution of progeny individuals that (unless N is quite small) is essentially Poisson with mean 1. For diploids, the result just given is often reinterpreted in a slightly different way. If a progeny allele has probability $1/(2N)$ of deriving from any particular parental allele, it must have a probability

$$\frac{1}{2N} + \frac{1}{2N} = \frac{1}{N}$$

of coming from a particular *parent*. Then the probability that a parent gives rise to progeny alleles is

$$\frac{(2N)!}{s!(2N-s)!}\left(1-\frac{1}{N}\right)^{2N-s}\left(\frac{1}{N}\right)^s$$

a binomial distribution with mean 2, variance $2(1 - 1/N)$; essentially a Poisson distribution with mean 2. Crow & Kimura (1970, p. 346) give our result in this form. Alternatively, we can think in terms of the distribution of offspring of a particular parent. Again, given our assumptions, this is essentially Poisson with mean 2.

Now it is well known (e.g. Wright 1931, Crow & Morton 1955) that these Poisson distributions do not apply in practice; large (sometimes very large) differences in potential parental fecundities appear to be the norm. For example, in a natural population of long-headed poppies, Mackay (personal communication) found

mean seed output per plant = 245

variance of seed output per plant = 1488 630

However we attempt to adjust these figures (Crow & Morton 1955) to allow for the fact that most of the seed will never give rise to an adult plant, we shall never arrive at anything remotely resembling a Poisson distribution. Thus, in practice, if p, q are the known frequencies of alleles A, a in some generation, the variance of the frequency of A in the next generation is clearly *not* $pq/(2N)$. To find the true variance, we follow the neat treatment of Ewens (1979).

Write down the 2N parental alleles in some order and label them 1, 2, 3, ..., 2N. In repeated sampling, a given allele retains its label but environments are reallocated at random to alleles on every occasion of sampling (for diploids, this means that, on every occasion, alleles are paired anew at random and environments allocated anew at random to the resulting zygotes). In the sequel, all variances are variances over repeated sampling; similarly with covariances and correlations.

Let g_i be the number of progeny alleles descended from the ith parental allele. Write, as before, $V(g)$ for the variance of g_i; $V(g)$ will be the same for any parental allele. Note that, since

$$\sum_{i=1}^{2N} g_i = 2N$$

the g_i's will be negatively correlated. Given our sampling scheme, $\text{Cov}(g_i, g_j)$ will be the same for all i, j (provided $i \neq j$). Now, since N is a constant,

$$V\left(\sum_{i=1}^{2N} g_i\right) = V(2N) = 0$$

whence, from the standard formula for the variance of a sum,

$$\sum_{i=1}^{2N} V(g) + \sum_{i \neq j} \text{Cov}(g_i, g_j) = 0$$

where i, j each go from 1 to $2N$. Thus

$$2N V(g) + 2N(2N - 1)\text{Cov}(g_i, g_j) = 0$$

and so

$$\text{Cov}(g_i, g_j) = -\frac{V(g)}{2N - 1}$$

Suppose then that there are $l = 2Np$ A alleles in the parental generation. Label these 1 to l. The number of A alleles in the progeny generation will then be

$$\sum_{i=1}^{l} g_i$$

with variance

$$\sum_{i=1}^{l} V(g) + \sum_{i \ne j} \text{Cov}(g_i, g_j)$$

where now i, j each go from 1 to l, so that our last formula becomes

$$lV(g) + l(l-1)\text{Cov}(g_i, g_j) = lV(g) - l(l-1)\frac{V(g)}{2N-1}$$

$$= \frac{V(g)l(2N-l)}{2N-1}$$

$$= \frac{V(g)pq}{2N-1}(2N)^2$$

on recalling that $l = 2Np$, so that $2N - l = 2Nq$.

Thus the variance of the frequency of A in the progeny generation is

$$V(p') = \left(\frac{1}{2N}\right)^2 \frac{V(g)pq}{2N-1}(2N)^2 = \frac{pq}{2N-1}V(g)$$

for a haploid population size $2N$ or for a diploid population size N. In the case $V(g) = 1 - 1/(2N)$, our formula reduces to $pq/(2N)$; in the general case, we take $pq/(2N)$ and replaces N by $\frac{1}{2}(2N-1)/V(g)$, as noted, for haploids, by Kimura (1964).

For diploids, an equivalent but slightly different formulation is usual. Let k be the number of descendant alleles *per parent* (rather than per parental allele, as above). When a diploid parent produces a gamete, the probability that a particular one of its alleles is incorporated into that gamete equals $\frac{1}{2}$. Then if the parent contributes k alleles to the next generation, the probability that a parental allele contributes g alleles to the next generation is

$$\frac{k!}{g!(k-g)!}(\tfrac{1}{2})^{k-g}(\tfrac{1}{2})^g$$

a binomial distribution with mean $\frac{1}{2}k$, variance $\frac{1}{4}k$, these being *conditional on* k. Then the *unconditional* mean of g equals $\frac{1}{2}Ek$.

We now use the standard statistical result (with |, as usual, representing 'conditional on')

$$V(x) = \mathop{E}_{y}\{V(x|y)\} + \mathop{V}_{y}\{Ex|y\}$$

Hence the *unconditional* variance of g is

$$V(g) = \mathop{E}_{k} \text{ (conditional variance of } g) + \mathop{V}_{k} \text{ (conditional mean of } g)$$

$$= \mathop{E}_{k} (\tfrac{1}{4}k) + \mathop{V}_{k} (\tfrac{1}{2}k)$$

$$= \tfrac{1}{4}Ek + (\tfrac{1}{2})^2 V(k)$$

For a population of constant size $Ek = 2$, whence

$$V(g) = \tfrac{1}{2} + \tfrac{1}{4}V(k)$$

and so

$$V(p') = \frac{pq}{2(4N-2)}[2 + V(k)]$$

equivalent to taking the binomial variance $pq/(2N)$ and replacing N by $(4N-2)/[2 + V(k)]$, a formula due to Wright (1938a).

With equal potential fecundity of parents

$$V(k) = 2\left(1 - \frac{1}{N}\right) \qquad \text{and so} \qquad \frac{4N-2}{2 + V(k)} = N$$

Otherwise, in *natural* populations, $(4N-2)/[2 + V(k)]$ will be less than N. But, in *experiments*, we may arrange that every individual contributes *exactly* two alleles to the next generation; if so, $V(k) = 0$, whence $(4N-2)/[2 + V(k)] = 2N - 1$.

So far we have assumed that the population size is constant. However, our argument is readily adapted to the case where population size differs between parent and progeny generations, provided we take care in deciding which of these sizes is appropriate at each stage of the argument. Write N for the population size in generation t, N' for the size in generation $(t + 1)$. We leave it to the reader to show that

$$V(p') = \frac{pq}{2N-1} V(g) \frac{N^2}{(N')^2}$$

As before

$$Eg = \tfrac{1}{2}Ek \qquad V(g) = \tfrac{1}{4}[Ek + V(k)]$$

but now

$$Eg = \frac{N'}{N} \qquad Ek = 2\frac{N'}{N} \qquad \text{so} \qquad \frac{N}{N'} = \frac{2}{Ek}$$

whence

$$V(p') = \frac{pq}{(2N-1)Ek}\left(1 + \frac{V(k)}{Ek}\right)$$

that is, we replace N by

$$\frac{(2N-1)Ek}{2[1 + V(k)/Ek]}$$

(Kimura & Crow 1963, Crow & Kimura 1970).

In the (very special) case of equal potential fecundity of all parental alleles but unequal population sizes in the two generations, the probability that a parental allele gives rise to s progeny alleles equals

$$\frac{(2N')!}{s!(2N'-s)!}\left(1 - \frac{1}{2N}\right)^{2N'-s}\left(\frac{1}{2N}\right)^s$$

so that

$$V(g) = 2N' \times \frac{1}{2N} \times \left(1 - \frac{1}{2N}\right) = \frac{N'}{N}\frac{2N-1}{2N}$$

giving

$$V(p') = \frac{pq}{2N'}$$

the usual binomial variance with population size equal to that in the *progeny* generation.

Effective population size

'The poet nothing affirmeth, and therefore never lies' may, perhaps, be a reasonable view for the literary critic but obviously will not do for us. The very beautiful theory developed from the Wright–Fisher model will be of very little interest unless we can rescue this model from the attack developed in the previous section.

We begin as follows. There can be no doubt, in view of our many different formulae for $V(p')$, that the conditional distribution of p' will depend very much on the biological circumstances of the population concerned. Nevertheless, it is clear that these different distributions have some properties in common. For example, we showed earlier that the mean $Ep'|p$ of the conditional distribution takes the same value p in all cases. What other properties do these distributions share?

On inspection of our various formulae for $V(p')$, which we shall now write in its correct form $V(p'|p)$, it appears that $V(p'|p)$ is (exactly or near enough) inversely proportional to the population size and will therefore in general be quite small. An exception, *in principle*, is the case where $V(g)$ is so large as to be comparable in magnitude with the population size, as would happen, for example, if one parental allele gave rise by chance to the whole of the next generation (see note 4). The evidence (for all its well known inadequacies!) strongly indicates that, in natural or cage populations, values of $V(g)$, although decidedly larger than would be obtained with equal potential fecundities, are still very much smaller than the size of the population concerned. For example, a rough-and-ready calculation on Mackay's poppy data cited earlier indicates a value of $V(g)$ (when adjusted to allow for the fact that most seed will never give rise to a mature plant) of around 26, whereas the population size was in fact 2316 (see note 5). It follows, then, that in practice $V(p'|p)$ will be quite small. Then by common sense or, preferably, by standard statistical theory (Chebyshev's inequality) large deviations of p' from the mean $Ep'|p$ are improbable. Let μ_r, the rth *moment* of p' about its mean $Ep'|p$, be defined as

$$E(p' - Ep'|p)^r$$

Then $\mu_1 = 0$, $\mu_2 = V(p'|p)$ and since large deviations from the mean are improbable we may take it that all the 'higher' moments (i.e. $\mu_3, \mu_4, \mu_5, \ldots$) are negligible. In confirmation of this informal argument, it may be shown from standard theory that, for the Wright–Fisher model,

$$\mu_{r+1} = \frac{pq}{2N}\left(r\mu_{r-1} + \frac{d\mu_r}{dp}\right)$$

from which we see that all the higher moments are of order $(1/N^2)$ or less and hence may be neglected if N is at all large. More generally, for random mating populations we find

$$\mu_3 = \frac{pq(q-p)}{(2N-1)(2N-2)} \mu_{3g}$$

where $\mu_{3g} = \mathrm{E}(g - \mathrm{E}g)^3$. Thus μ_3 will be negligible in this case also, unless fecundities are extremely uneven. Similar considerations apply for the other higher moments. Now, under very general conditions, which will certainly apply to all cases arising in population genetics, a knowledge of the mean together with the moments about the mean of a probability distribution is equivalent to a knowledge of all the properties of that distribution. If two distributions have the same mean and moments about the mean, they will therefore be identical. We have shown that all our conditional distributions have the same mean and, to a reasonable approximation, the same higher moments, since in practice the latter will be close to zero in all cases unless the population is very small. Thus the differences between the various conditional distributions are summarized by the differences in their variances. How critical are these differences in variance? For random mating populations with unequal, environmentally determined, potential fecundity of parental alleles, we showed that the variance may be obtained by taking the Wright–Fisher variance $pq/(2N)$ and replacing N by a *constant*, which we shall write as N_e. Thus the variance and hence (to a close approximation) the conditional distribution in a population size N with unequal potential fecundities are the same as in a Wright–Fisher population of size N_e; hence N_e is known as the *variance effective population size*.

The concept of effective population size, due to Wright (1931), has in fact a very wide range of application; for many different biological situations, it is possible to calculate an N_e such that the variance of the conditional distribution is $pq/(2N_e)$. Consider, for example, the case where inequalities in potential fecundity are (wholly or partly) due to genetic variation (at loci other than the locus under study). As Nei & Murata (1966) point out, this differs from the case where inequalities in fecundity are determind by random allocation of environments to parents, independently in every generation, which we have discussed. For, when fecundity is inherited, the progeny of a highly fecund individual will tend also to produce a large number of progeny. Then a neutral allele that happens to be associated with, say, high-fecundity alleles in one generation will tend to be similarly associated for several later generations; thus the initial effect of fecundity differences is augmented. Nei & Murata (1966) show that in this case (to a close approximation)

$$N_e = \frac{N\mathrm{E}k}{1 + (1 + 3h^2)\mathrm{V}(k)/\mathrm{E}k} \qquad \text{in generation } (t+1)$$

where h^2 is the narrow heritability of fecundity (other symbols having the same meaning as before). For a population of constant size, $Ek = 2$, so that

$$N_e = \frac{4N}{2 + (1 + 3h^2)V(k)}$$

An analogous effect will obtain when parental and progeny environments are correlated.

Another important case considered by Wright (1931, 1939b) and very rigorously by Ewens (1979) concerns the effect of inequalities in the sex ratio. If, in a population of constant size N, there are, in any generation, N_1 males and N_2 females, it may be shown that

$$N_e = 4N_1N_2/N$$

to a close approximation. If $N_1 = N_2 = \frac{1}{2}N$, this formula reduces to $N_e = N$; otherwise N_e is always less than N and indeed will be substantially less if the sex ratio is very unequal.

So far we have discussed effective size in random mating populations. The case of mixed selfing and random mating is more difficult. We showed that (with constant population size and equal potential fecundities) we obtain the variance of the conditional distribution by taking $pq/(2N)$ and replacing N by $N/(1 + f)$, but this does not, at least at first sight, bring the variance into equivalence with the Wright–Fisher variance, since f is not a constant. In particular, it is not immediately obvious how much importance need be attached to *random* variation in the value of f. To obtain a clue to this, consider the change in f, over successive generations, in a population in which effects of drift are supposed negligible. Write (Aa), $(Aa)'$ for the frequency of Aa in parent and offspring generations, respectively, and similarly (f), $(f)'$ for the value of f in these successive generations. Let s be the proportion of seed arising from selfing. Then

$$(Aa)' = \tfrac{1}{2}s(Aa) + (1-s)2pq$$

Writing

$$(Aa) = 2pq[1 - (f)] \qquad (Aa)' = 2pq[1 - (f)']$$

we find

$$(f)' = \tfrac{1}{2}s + \tfrac{1}{2}s(f)$$

whence

$$\left((f)' - \frac{s}{2-s}\right) = \tfrac{1}{2}s\left((f) - \frac{s}{2-s}\right)$$

Thus the difference between f and $s/(2-s)$ declines by a factor $\tfrac{1}{2}s$ every generation and, since $\tfrac{1}{2}s < \tfrac{1}{2}$, f rapidly approaches its equilibrium value $s/(2-s)$, as shown by Haldane (1924b). This strong tendency for f to attain its equilibrium value, contrasting so strikingly with the weak tendency for change of any kind under drift, suggests that even when random changes are taken into account f will, in the course of a very few generations, attain a value close to $s/(2-s)$ and remain close to the latter thereafter. Wright (1969, pp. 194–6) gives a mathematical discussion confirming these intuitive ideas, which he shows to be appropriate provided s is much greater than $1/N$. It follows then that, after a few generations at most, we may write

$$N_e = \frac{N}{1 + s/(2-s)} = N\frac{2-s}{2}$$

and ignore any random change in f.

Suppose a population contains a newly arisen neutral mutant A. Then we have:

	AA	Aa	aa
number	0	1	$N-1$
frequency	0	$1/N$	$1 - 1/N$

with

$$p = \frac{1}{2N} \qquad q = 1 - \frac{1}{2N} \qquad f = 1 - \frac{1/N}{2pq} = -\frac{1}{2N-1}$$

so that the value of f is initially close to zero in this important case. For a few successive generations, we shall have to calculate f, which (unless N is very small) we can do using our deterministic formula

$$(f)' = \tfrac{1}{2}s + \tfrac{1}{2}s(f)$$

in order to find N_e, but thereafter we take N_e to be $N(2-s)/2$. This procedure, although not of course strictly correct, will give reliable results save in very small populations.

For simplicity, we have considered factors causing N_e to depart from N one at a time. However, the effects of different factors can be considered jointly. Suppose, for example, we have mixed selfing and outcrossing and also unequal potential fecundities and unequal population sizes. Crow & Kimura (1970) give a formula for the effective size in the progeny generation, which in our notation is

$$N_e = \frac{NEk}{(s^2/Ek)(1+f) + (1-f)}$$

where N and f are values in the *parental* generation and $s^2 = NV(k)/(N-1)$. They establish this formula on the supposition that parental contributions of male and female gametes to the next generation are determined independently, but their method of proof adapts readily to the case where male and female contributions are correlated. The resulting formula for N_e turns out to be the same as that just given, provided account is taken of the correlation when calculating $V(k)$. Write k^* for an individual's contribution of male gametes and k^{**} for that individual's contribution of female gametes; then

$$V(k) = V(k^*) + V(k^{**}) + 2\,\text{Cov}(k^*, k^{**})$$

The covariance term will often make a substantial contribution to $V(k)$. Clearly, for progeny produced by selfing, a parent's contribution of male and female gametes will be equal. Even for outcrossing, the effect of common environment and genotype on number of flowers per plant, flower size and relative growth of male and female parts within individual flowers must also lead to a positive correlation between male and female gametes contributed. In practice, it is not at all easy to estimate either $V(k^*)$ or the covariance term. For a species (such as the long-headed poppy) with a fairly high level of selfing and fairly strict control of relative growth, we may *perhaps* be justified in taking $V(k^*)$ to be roughly equal to $V(k^{**})$ and the overall correlation between male and female gametes contributed to be close to unity. We mention these details to illustrate the (well known) point that, save in very special cases, really reliable estimates of N_e have not yet been obtained. Nearly all estimates of N_e are open to objections, as indeed the authors of these estimates state very clearly. The advent of DNA fingerprinting (Jeffreys *et al.* 1985a, b, 1986), however, offers hope of a dramatic improvement here (see Burke & Bruford 1987, Wetton *et al.* 1987).

Variance of allele frequency in any generation

So far we have considered the variance of the frequency of allele A in the conditional distribution only. However, for a full discussion of effective

VARIANCE OF ALLELE FREQUENCY IN ANY GENERATION

population size, which takes account of variation in effective size from generation to generation, we must calculate the unconditional variance of the frequency of A for an arbitrary generation, t say. We continue to suppose two alleles at a locus, allele frequencies changing by drift only. Let P, Q be the frequencies of A, a in *any* generation. Irrespective of whether we are dealing with conditional or unconditional distributions, we have the identity

$$V(P) = EP^2 - (EP)^2 = EP - (EP)^2 - EP + EP^2$$
$$= EP - (EP)^2 - E(PQ)$$

(since $Q = 1 - P$). Now, as we showed earlier, EP, the mean allele frequency, equals the allele frequency one generation earlier if we are thinking of conditional distributions and equals the initial allele frequency p_0 if we are thinking of unconditional distributions. Thus knowing $E(PQ)$ we easily find $V(P)$ and vice versa. In practice, although our ultimate aim is to calculate the unconditional $V(P)$ for an arbitrary generation t, the intermediate stages of the calculation are most readily carried out in terms of $E(PQ)$. Write p_{t-1}, q_{t-1} for the frequencies of A, a in generation $(t-1)$ and p_t, q_t for the corresponding frequencies in t; also write $(N_e)_t$ for the variance effective population size in t. Then, for the *conditional* distribution in t

$$V(p_t|p_{t-1}) = \frac{p_{t-1}q_{t-1}}{2(N_e)_t}$$

We now use our identity to calculate the *conditional* mean of $(p_t q_t)$. Since we are dealing with a conditional distribution, the mean allele frequency in t is p_{t-1}, whence

$$E(p_t q_t | p_{t-1}) = p_{t-1} - p_{t-1}^2 - V(p_t|p_{t-1})$$
$$= p_{t-1}q_{t-1}\left(1 - \frac{1}{2(N_e)_t}\right)$$

(since $q_{t-1} = 1 - p_{t-1}$). Then, by the rule we have used several times earlier, the *unconditional* mean of $(p_t q_t)$ is given by

$$E(p_t q_t) = E(p_{t-1}q_{t-1})\left(1 - \frac{1}{2(N_e)_t}\right)$$

(where both means are unconditional). By a similar argument,

$$E(p_{t-1}q_{t-1}) = E(p_{t-2}q_{t-2})\left(1 - \frac{1}{2(N_e)_{t-1}}\right)$$

where $(N_e)_{t-1}$ is the variance effective size in generation $(t-1)$. Substituting this last result in the formula immediately preceding, we have

$$E(p_t q_t) = E(p_{t-2} q_{t-2})\left(1 - \frac{1}{2(N_e)_{t-1}}\right)\left(1 - \frac{1}{2(N_e)_t}\right)$$

which we conveniently write in the more compact form

$$E(p_t q_t) = E(p_{t-2} q_{t-2}) \prod_{j=t-1}^{t} \left(1 - \frac{1}{2(N_e)_j}\right)$$

the symbol Π ('product') being analogous to Σ ('sum'). By continued application of the argument, we find

$$E(p_t q_t) = p_0 q_0 \prod_{j=1}^{t} \left(1 - \frac{1}{2(N_e)_j}\right)$$

on noting that in generation 0 allele frequencies p_0, q_0 are constants, so that $E(p_0 q_0) = p_0 q_0$.

We now use our identity once again, recalling that $Ep_t = p_0$ for any value of t, to give us the unconditional variance of the frequency of A in t as

$$p_0 - p_0^2 - E(p_t q_t) = p_0 q_0 \left[1 - \prod_{j=1}^{t} \left(1 - \frac{1}{2(N_e)_j}\right)\right]$$

Now if $(N_e)_j$ were the same, N_e^* say, in all generations, the fomula just derived would simplify to

$$p_0 q_0 \left[1 - \left(1 - \frac{1}{2N_e^*}\right)^t\right]$$

Hence, if N_e varies with generation, the population behaves *as if* it had a constant effective size N_e^* given by

$$\left(1 - \frac{1}{2N_e^*}\right)^t = \prod_{j=1}^{t} \left(1 - \frac{1}{2(N_e)_j}\right)$$

so that

$$t \log\left(1 - \frac{1}{2N_e^*}\right) = \sum_{j=1}^{t} \log\left(1 - \frac{1}{2(N_e)_j}\right)$$

Now, for any x numerically less than unity

$$\log(1-x) = -x - \tfrac{1}{2}x^2 - \tfrac{1}{3}x^3 - \cdots$$

Then, provided the N_e's are not too small, we may use the approximations

$$\log\left(1 - \frac{1}{2N_e}\right) = -\frac{1}{2N_e}$$

to obtain the simplified formula

$$\frac{1}{2N_e^*} = \frac{1}{t}\sum_{j=1}^{t} \frac{1}{2(N_e)_j}$$

valid, to a close approximation, provided $t \ll 2N_e^*$. Thus N_e^* is, approximately, the *harmonic mean* of the $(N_e)_j$ (Wright 1939b). This formulation is particularly helpful if the population size changes in a cyclical manner, a shortish sequence of effective sizes $(N_e)_1, (N_e)_2, \ldots, (N_e)_t$ being endlessly repeated. Substitution of a few values for the $(N_e)_j$ will at once convince the reader that, if the population is sometimes large, sometimes small, it will behave as a small population, as pointed out by Wright.

In conclusion, then, N_e will be substantially less than the 'census' population size N when the variance in offspring number is very large, when the number of breeding individuals is very much smaller in one sex than in the other, and when the smallest size during a cycle is very small.

Inbreeding effective size

In the last section, we summarized the effects of random genetic drift by calculating the variance of the frequency of allele A for an arbitrary generation t. There is, however, an alternative (closely related) method of summarization; we calculate, for arbitrary generation t, the mean frequency of heterozygotes. It is easily shown (see note 6) that this mean is also the probability that a (randomly chosen) individual is heterozygous in generation t; for this reason, the reader may find the summary in terms of mean frequency of heterozygotes more attractive than the summary in terms of variance of allele frequency, although, as we shall see, it is easy to convert from one of these summary quantities to another.

We begin by noting, with Levene (1949), that, in a finite population, the expected frequency of heterozygotes, *conditional* on allele frequencies, p, q in that population, is not exactly $2pq$. Let N be the population size and let two alleles be chosen from the population at random. If the first allele chosen was A (this happening with probability p), there remain $(2N-1)$ alleles, of which $2Nq$ are a; if the first allele chosen was a (this happening with probability q), there remain $(2N-1)$ alleles, of which $2Np$ are A. Thus the probability that an individual is Aa, which is also the mean frequency of heterozygotes, equals

$$p\frac{2Nq}{2N-1} + q\frac{2Np}{2N-1} = 2pq\frac{2N}{2N-1}$$

rather than $2pq$.

This being so, let N_t be the population size in generation t. Then the mean frequency of heterozygotes in generation t, conditional on allele frequencies being p_t, q_t in that generation, equals

$$2p_tq_t\frac{2N_t}{2N_t-1}$$

Averaging over all values of p_t in the usual way, we find the (unconditional) mean frequency of heterozygotes in generation t

$$H_t = 2\mathrm{E}(p_tq_t)\frac{2N_t}{2N_t-1}$$

From the previous section

$$2\mathrm{E}(p_tq_t) = 2\mathrm{E}(p_{t-1}q_{t-1})\left(1 - \frac{1}{2(N_e)_t}\right)$$

Suppose now that the population has *constant* size N over generations $(t-1)$, t. Then

$$H_t = \frac{2N}{2N-1}2\mathrm{E}(p_tq_t) \qquad H_{t-1} = \frac{2N}{2N-1}2\mathrm{E}(p_{t-1}q_{t-1})$$

and if we assume that, when the census size remains a constant N, the effective size remains a constant N_e, we have the simple result

$$H_t = H_{t-1}\left(1 - \frac{1}{2N_e}\right)$$

We shall call this 'the standard formula'.

INBREEDING EFFECTIVE SIZE

In the slightly more general case, where the census size remains constant but the effective size varies,

$$H_t = H_{t-1}\left(1 - \frac{1}{2(N_e)_t}\right)$$

but this formula is *not* correct when the census size varies. For brevity, write N for N_{t-1}, N' for N_t. Then

$$H_t = \frac{2N'}{2N'-1} 2E(p_t q_t) \qquad H_{t-1} = \frac{2N}{2N-1} 2E(p_{t-1} q_{t-1})$$

whence

$$H_t = \frac{2N'}{2N'-1} \frac{2N-1}{2N} H_{t-1}\left(1 - \frac{1}{2(N_e)_t}\right)$$

Notice that the $(N_e)_t$ appearing in this formula is our usual variance effective size in generation t; to emphasize this, we shall, in this section, rewrite $(N_e)_t$ as $N_{e(v)}$, v standing for variance. We stress that this (admittedly inelegant) formula, although written in terms of $N_{e(v)}$, is correct.

However, as first noticed by Crow (1954), a technical difficulty arises if we attempt to generalize the standard formula to cases of variable census size, *while retaining the mathematical form of that formula*. Clearly, it will not do to replace N_e by $N_{e(v)}$. Rather, from the formula just derived for H_t, we must replace N_e in the standard formula by, say, $N_{e(i)}$, where

$$1 - \frac{1}{2N_{e(i)}} = \frac{2N'}{2N'-1} \frac{2N-1}{2N}\left(1 - \frac{1}{2N_{e(v)}}\right)$$

whence

$$\frac{1}{2N_{e(i)}} = 1 - \frac{2N'}{2N} \frac{2N-1}{2N'-1}\left(1 - \frac{1}{2N_{e(v)}}\right)$$

This formula may be used to relate $N_{e(i)}$ to $N_{e(v)}$ in any generation, provided we bear in mind that $N_{e(i)}$, $N_{e(v)}$ and N' all refer to that generation, whereas N refers to the generation preceding. Only if $N' = N$ will $N_{e(i)}$ equal $N_{e(v)}$.

Consider, for example, the case of unequal potential fecundities (in a random mating population). We found that

$$N_{e(v)} = \frac{(2N-1)Ek}{2[1 + V(k)/Ek]}$$

Substituting into the expression for $N_{e(i)}$ derived above, and writing $2N' = NEk$, we find

$$\frac{1}{2N_{e(i)}} = 1 - \frac{NEk}{2N}\frac{2N-1}{NEk-1}\frac{(2N-1)Ek - 1 - V(k)/Ek}{(2N-1)Ek}$$

$$= \frac{Ek - 1 + V(k)/Ek}{2(NEk - 1)}$$

and therefore

$$N_{e(i)} = \frac{NEk - 1}{Ek - 1 + V(k)/Ek}$$

(Kimura & Crow 1963, Crow & Kimura 1970), which is not the same as $N_{e(v)}$.

Finally, we note an awkward point of terminology. We have defined $N_{e(i)}$ as the replacement for N_e in the standard formula. This is just the definition of the *inbreeding effective size* adopted (in practice at least) by Wright (1969). This seems quite natural to the present author, since, from Mendel onwards, geneticists have regarded the decay of heterozygosity as the principal consequence of inbreeding. However, most workers (e.g. Crow & Kimura 1970, Ewens 1979) see the matter differently, and give definitions of the inbreeding effective size that are not always equivalent to Wright's (see Ewens 1982). Of course, this disagreement on terminology does not affect our earlier discussion. We shall not pursue this further, and in the rest of this book will write N_e as short for $N_{e(v)}$.

Populations under systematic 'pressure'

So far we have discussed effective population size for cases in which allele frequencies are changing as a result of drift only. For cases where allele frequencies are changing under systematic 'pressure' (mutation or selection) as well as drift, the theory is much less developed. However, as far as is known, the following intuitive considerations usually lead to reliable results (but see Ch. 6). Let p be the allele frequency in generation t, p^* the allele frequency in $(t + 1)$ calculated deterministically (conditional on p) and p' the actual allele frequency in $(t + 1)$. If we assume the population in $(t + 1)$ to be drawn as a random sample of $2N$ alleles, every allele being drawn independently of the others, from an infinite population with allele frequencies p^*, q^*, we have

$$V(p'|p) = p^*q^*/(2N)$$

However, if evolution is slow, p^* and q^* will be close to p and q, respectively, and we have, to a good approximation,

$$V(p'|p) = pq/(2N)$$

In that $V(p'|p)$ is, near enough, the same in the present case as in the case where frequencies were changing under drift only (given the binomial model), it seems reasonable to suppose that the variance effective size, as calculated earlier, can be substituted for N when systematic pressures as well as drift operate. This does indeed seem to be the case. While no general theory exists to justify this approach, it is possible to check results in some special cases; the outcome is encouraging, as we shall see later.

A continuous approximation to the conditional distribution

Too much must not be made of results on effective size. It is wrong to suppose that one can *always* take a formula derived from the Wright–Fisher model and 'put it right' by replacing N by N_e. For example, one cannot 'correct' the Wright–Fisher distribution of allele numbers

$$\frac{(2N)!}{r!(2N-r)!} q^{2N-r} p^r$$

by replacing N with N_e, for this would mean that r, the number of A alleles, could range only from 0 to $2N_e$, whereas of course r ranges from 0 to $2N$.

However, we can argue as follows. Since, for the models we have considered, events in a population of size N closely parallel events in a Wright–Fisher population of size N_e, the Wright–Fisher model should lead to (at least) qualitatively correct conclusions on all major points, for example the evolutionary significance of natural selection of low intensity. This is indeed so, and explains why so much discussion has been carried out in terms of this model. Of course, we bring in N_e whenever we can; the following procedure is often appropriate.

We note first that, while in reality an allele frequency varies in discrete steps of $1/(2N)$, such steps will be small if N is at all large, in which case we may treat allele frequencies as continuous variables, without risk of serious error. Then, with our usual notation, the conditional distribution of p', the frequency of A in $(t+1)$, can be approximated by a continuous distribution with mean p^*, variance $pq/(2N_e)$ (N_e being of course the *variance* effective size) and higher moments negligible. It will not usually be necessary to specify the continuous distribution in more detail than this.

By a process equivalent to the repeated application, generation after generation, of this continuous conditional distribution, we may obtain a continuous approximation to the frequency of A in any generation, given an initial frequency p_0. We discuss this in detail in a later chapter. For the moment, we merely note one feature of the process of change of allele frequency that is critical for the theory of this process. If we know the allele frequency in some generation, t say, and wish to predict the allele frequency in future generations $(t+1)$, $(t+2)$, ..., it will not help us in any way, in making the prediction, to know what happened to the allele frequency *before* generation t. A random variable whose change with time is memory-free in the way just described, in contradiction to Eliot's

> Time present and time past
> Are both perhaps present in time future

(*Burnt Norton*), is said to be *Markovian*. Strictly speaking, the allele frequency is Markovian only if, in making our prediction, allele frequencies are the only frequencies that matter and it is not necessary to consider, say, genotype frequencies. While it is easy to think of cases that are *not* Markovian in the strict sense (see note 7) we shall (except for a brief reference in the next chapter) ignore this delicate distinction, which, rather surprisingly, matters (in practice) for small populations only (Moran 1958b, Watterson 1962, Norman 1975a, Ewens 1979), and refer to the allele frequency as Markovian, even though in some cases 'quasi-Markovian' is more correct.

Diffusion methods

It proves convenient to use a continuous approximation not only for the allele frequency but also for the time. This is merely a matter of scaling (Feller 1951). Let N be the population size. We measure time, τ say, in such a way that one unit of time on the τ scale corresponds to $2N$ generations. Thus a single generation is only $1/(2N)$th of a unit of time, when measured on the τ scale, and when N is large we may, without serious error, treat time as a continuous variable; that is, we regard allele frequencies as though they were changing continuously in time rather than in discrete generations. Naturally, we shall have to adjust some of our formulae to allow for this change in timescale. However, if we are to regard allele frequencies as changing continuously in time, with the implication that for very small times we have only very small changes in frequency, we must confine ourselves to cases where evolution is fairly slow.

When we have a process in which a continuous random variable, such as an allele frequency in our approximate treatment, changes continuously in time in a Markovian manner, we are said to have a *diffusion process*. The name derives from processes in physics having the properties just described. In population genetics, the representation of change of allele frequency as a diffusion process is due to Fisher (1922). It is interesting to note that Fisher was concerned lest results from mathematical physics might unawares be carried over into population genetics, where they would not necessarily be appropriate; he favoured, therefore, a development of population genetic theory independent of work in physics (A. W. F. Edwards, personal communication), although he did occasionally note a physical parallel. For many years, work in this area of theoretical population genetics, due almost entirely to Fisher and Wright, proceeded independently of work in other fields. Referring to the work of these two authors, Feller (1951) remarked: 'They have attacked individual problems with great ingenuity and an admirable resourcefulness, and had in some instances to discover for themselves isolated facts of the general theory of stochastic processes'. The Russian statistician Kolmogorov was apparently the first to notice the relationship between Wright and Fisher's work and some of his own fundamental work in probability theory. Stimulated by the receipt of a reprint from Kolmogorov, Wright (1945) set out these relationships explicitly. Since that time, most writers on theoretical population genetics (e.g. Moran 1962, Kimura 1964, Ewens 1969, 1979, Crow & Kimura 1970, Kimura & Ohta 1971a, Maruyama 1977) have, whenever appropriate, set out their work in the context of general ideas relating to diffusion processes (of course, there are many problems specific to genetics and the application of these general ideas to genetics is not always straightforward). Usually, a rather informal approach is taken. The rigorous theory of diffusion processes is difficult and well beyond the scope both of the present book and of the present author's powers of exposition. An excellent account, with many genetical illustrations, is given in Karlin & Taylor (1981).

The reader will naturally ask: 'Given all the approximations we have mentioned, how reliable are the results?' There are several ways of approaching this problem. Watterson (1962) obtained sufficient conditions under which the diffusion approach would give accurate results. He showed that this approach is reliable when the population size is very large and the mean change in allele frequency over any generation is roughly comparable with, or less than, the variance of the change in frequency over that generation. However, these rather exacting conditions are not necessary. Thus in some *special* cases, it is possible to obtain exact or near-exact solutions to specific problems, algebraically or by computer simulations (or both), and to compare results with those obtained from the diffusion approach. In order to reassure the reader, we anticipate our account of this work by stating that results from the diffusion approach prove to be reliable under a very wide range of circumstances.

This being so, there are essentially two ways in which to proceed when discussing the effect of drift. The first is to go for the diffusion approach from the start and derive everything we can from this method. This is the approach of Maruyama (1977), who very early in his book develops some very general properties of diffusion processes in genetics and then applies his results to a series of apparently not very closely related problems, which, however, turn out to be special cases of the very general situation considered at the start. A second approach is to solve as many problems as possible by *ad hoc* methods and adopt the diffusion approach only when the need to do so becomes pressing; the earlier findings can then be recovered using the diffusion approach. This procedure is used by Ewens (1979). In the present author's opinion, there is much in favour of either method of presentation. To adopt the diffusion approach from the start gives a very compact theory. On the other hand, *ad hoc* methods often require fewer assumptions than are necessary for the diffusion approach, so that it is easier to see when results will be reliable. The two approaches, then, are complementary. For the non-mathematician, however, the *ad hoc* approach seems very much easier, presumably because very abstract preliminaries are not required. We shall, therefore, adopt this latter approach and will not use diffusion methods until a fairly late stage of our discussion.

Moran's model

All of the models we have discussed so far have much in common. In particular, in all of them, generations do not overlap. We may ask: 'Given a situation biologically very different from those considered so far, will the mathematics considered so far be appropriate?' In particular, can we calculate an effective population size for this new situation which will bring it into mathematical equivalence with the old? For the moment, we merely describe a model, due to Moran (1958a), apparently quite different from Wright-Fisher and, in a naive manner, calculate a variance effective size N_e.

Consider a haploid population, size $2N$. At times $t = 1, 2, 3, \ldots$ a 'birth-death event' occurs, as follows. A randomly chosen individual reproduces. Immediately after this, a randomly chosen individual (possibly the individual that has just reproduced but *not* the new offspring) dies. Nothing happens between events. For the moment, we suppose allele frequencies to change by drift only, although the model can easily be elaborated to allow for mutation and selection.

Suppose, before a given birth–death event, that allele frequencies are p A and q a. The probability that the individual chosen to reproduce is A equals p; similarly the probability that the individual chosen to die is A also equals p. Arguing in this way, we see that the event can take place in one of four genetically distinct ways, with probabilities as follows:

Reproducing	Dying	Probability	Effect on number of A alleles
A	A	p^2	none
a	a	q^2	none
A	a	pq	increased by one
a	A	pq	decreased by one

Then for p', the allele frequency after the event, we have:

p'	$p' - p$	Probability
$p + 1/(2N)$	$1/(2N)$	pq
$p - 1/(2N)$	$-1/(2N)$	pq
p	0	$1 - 2pq$

Then

$$E(p'|p) = \left(p + \frac{1}{2N}\right)pq + \left(p - \frac{1}{2N}\right)pq + p(1 - 2pq) = p$$

$$V(p'|p) = \left(\frac{1}{2N}\right)^2 pq + \left(-\frac{1}{2N}\right)^2 pq + 0(1 - 2pq) = pq\frac{2}{(2N)^2}$$

A great advantage to the geneticist of this model is that it yields exact solutions to many problems of genetical interest. Moreover, change of allele frequency under this model can be approximated by diffusion methods and the solution obtained to various problems compared to the exact solutions, thus providing a check on the validity of the diffusion approach. For the moment, we attempt to establish some equivalences between this overlapping generation model and the discrete generation model of Wright and Fisher.

WRIGHT-FISHER, MORAN AND OTHER MODELS

Since a fraction $1/(2N)$ of the population dies at an event in Moran's model, it seems obvious that the expectation of life of an individual is $2N$ intervals between events. To prove this, note that an individual lives for just s such intervals if he survives $(s-1)$ consecutive events following his birth and dies at the sth event (notice that all individuals survive for at least one interval). Thus the probability that an individual lives for just s intervals is

$$\left(1-\frac{1}{2N}\right)^{s-1}\frac{1}{2N}$$

and (by definition of mean) the mean life of an individual is

$$\sum_{s=1}^{\infty} s\left(1-\frac{1}{2N}\right)^{s-1}\frac{1}{2N}$$

intervals. Now, from the usual results for the sum of a geometric progression,

$$\sum_{s=0}^{\infty} x^s = \frac{1}{1-x}$$

provided x is numerically less than unity. Differentiating both sides, the left-hand side being differentiated term by term (a procedure that can be justified), we have

$$\sum_{s=0}^{\infty} sx^{s-1} \left(= \sum_{s=1}^{\infty} sx^{s-1}\right) = \frac{1}{(1-x)^2}$$

so that the mean life of an individual is

$$\frac{1}{[1/(2N)]^2}\frac{1}{2N} = 2N$$

intervals. This being so, $2N$ events in a Moran model correspond to a single generation in the Wright-Fisher model (although usually not all individuals present at the start of a 'generation' will have been replaced in this time). Let the allele frequencies at the start of a Moran 'generation' be p_0, q_0. After one event, the variance of the frequency of A will, as shown above, be

$$p_0 q_0 \frac{2}{(2N)^2}$$

The variance of the allele frequency at the start of the next generation can now be found by the argument used for the Wright-Fisher model to compute

48

the variance in generation t. Then after $2N$ events (one 'generation') the variance of the frequency of A is

$$p_0 q_0 \left[1 - \left(1 - \frac{2}{(2N)^2} \right)^{2N} \right]$$

$$= p_0 q_0 \left[1 - \left(1 - 2N \frac{2}{(2N)^2} \right) + \text{(terms negligible when } N \text{ is large)} \right]$$

$$= p_0 q_0 / N$$

to a close approximation when N is large. Writing

$$p_0 q_0 / (2N_e) = p_0 q_0 / N$$

we see that, formally at least,

$$N_e = \tfrac{1}{2} N$$

from which we may guess that events in a Moran population will parallel events in a Wright–Fisher population half as large. A similar 'result' follows if, Moran's model, we compute $V(g)$, the variance of the number of offspring produced by individual parents ($=$ parental alleles, since the population is haploid) and make a formal substitution in the Ewens formula

$$N_e = \frac{2N - 1}{2V(g)}$$

(Ewens (1979), following an equivalent statement in Moran (1962)). We shall see later whether there is anything in these (apparently) very wild speculations.

Notes and exercises

1 Deterministic arguments can sometimes give very misleading results. Suppose M is a dominant disadvantageous allele ($s > 0$):

	MM	Mm	mm
viability	$1 - s$	$1 - s$	1

m mutates to M at rate u, M to m at rate v.

49

Consider the case

$$s = 10^{-3} \qquad u = 10^{-5} \qquad v = 10^{-8}$$

Assume random mating. Show that, if we ignore the effects of drift, there are *two* stable equilibria, with frequency of M

$$q_1 = 0.0101 \qquad \text{and} \qquad q_2 = 0.9989$$

respectively (and an unstable equilibrium at frequency 0.9910). The following conclusion would appear to follow for large populations. Provided M has always been disadvantageous and hence always rare, its frequency will approach q_1. If, however, M has previously been advantageous, and therefore very common, but is now disadvantageous, its frequency will approach q_2; thus a disadvantageous allele remains very common, the genetic composition of the population being for ever marked by its past history. This conclusion seems very strange and is, in fact, wrong. Once we allow for the effects of drift, we can show that, in a large population, the frequency of M at equilibrium will almost certainly be close to q_1, irrespective of past history. Even if the frequency, at some stage, lies in the neighbourhood of q_2, it will, owing to drift, leave that neighbourhood and, almost certainly, will not return to it (see Ch. 9).

2 For another example of a misleading deterministic analysis, consider the case of an advantageous allele A ($s > 0$, $h \geq 0$):

	AA	Aa	aa
viability	1	$1 - hs$	$1 - s$

The case $h = 0$ represents complete dominance of A over a in viability.

Suppose, for definiteness, that A has initial frequency $\frac{1}{2}$. Assume random mating and s small. How many generations are required for the frequency of A to reach value p? Intuitively, one would expect the answer when h is very small to be close to the answer when $h = 0$.

Use standard deterministic arguments to show that the number of generations required is (approximately)

$$t = \frac{1}{s}\left[\log\left(\frac{p}{q}\right) + \frac{p}{q} - 1\right] \qquad \text{when } h = 0$$

$$t^* = \frac{1}{s}\left\{\frac{1}{1-h}\log\left(\frac{p}{q}\right) + \frac{1-2h}{h(1-h)}\log\left[1 + \left(\frac{p}{q} - 1\right)h\right]\right\} \qquad \text{when } h \neq 0$$

where $q = 1 - p$. Verify that, when $h = 0.0001$, $p = 0.9999$,

$$t = 10\,007/s \qquad t^* = 6939/s$$

which are clearly very different, despite the very small value of h.

Generally, if h is very small, we have the close approximations

$$1 - h = 1 \qquad 1 - 2h = 1 \qquad t^* = \frac{1}{s}\left\{\log\left(\frac{p}{q}\right) + \frac{1}{h}\log\left[1 + \left(\frac{p}{q} - 1\right)h\right]\right\}$$

Thus t^* will be close to t only when

$$\log\left[1 + \left(\frac{p}{q} - 1\right)h\right] \qquad \text{is close to} \qquad \left(\frac{p}{q} - 1\right)h$$

which, from standard properties of logarithms, is true only when $(p/q - 1)h$ is small, which condition cannot hold universally. However small the (non-zero) value of h, there will always be a range of p, close to unity, for which $(p/q - 1)$ is sufficiently large to invalidate our condition. In standard mathematical terminology, $t^* \to t$ as $h \to 0$ but this convergence of t^* to t is *non-uniform*, in that the value of h, which is sufficiently small to make t^* close to t, depends on p. Notice that, although the deterministic treatment of change of allele frequency under natural selection is often given in elementary courses on population genetics, the preceding analysis is not elementary!

In fact our argument, although mathematically correct, is quite inappropriate, since the effect of drift cannot be ignored when p is close to 1 (see Ch. 7). In practice, when h is very small the difference between t^* and t will (usually at least) be much smaller than the preceding argument indicates.

3 *Relationship between mean and conditional mean.* Let $f(p)$ be the probability that the allele frequency takes value p in generation t, $g(p')$ be the corresponding probability for value p' in generation $(t + 1)$ and $h(p'|p)$ be the corresponding probability for value p' in generation $(t + 1)$, conditional on value p in generation t. Then

$$g(p') = \sum_p f(p)h(p'|p)$$

As a matter of definition, the mean allele frequency in $(t + 1)$ is

$$Ep' = \sum_{p'} p'g(p')$$

and the corresponding conditional mean in $(t+1)$ is

$$E(p'|p) = \sum_{p'} p'h(p'|p)$$

On substituting for $g(p')$ and rearranging, we find

$$Ep' = \sum_{p'} p' \sum_{p} f(p)h(p'|p) = \sum_{p} \left(\sum_{p'} p'h(p'|p)\right) f(p)$$

$$= \sum_{p} [E(p'|p)] f(p)$$

by definition of $E(p'|p)$. Thus (by definition of mean) Ep' is just $E(p'|p)$ averaged over all p.

4 *A very extreme case.* If all $2N$ progeny alleles descend from a single randomly chosen parental allele, with mean descendant number per parental allele equal to 1, the number of descendants of any *specific* parental allele is distributed as follows:

Number	Deviation from mean	Probability
$2N$	$2N - 1$	$1/(2N)$
0	-1	$1 - 1/(2N)$

This distribution has variance

$$V(g) = (2N-1)^2 \frac{1}{2N} + (-1)^2 \left(1 - \frac{1}{2N}\right) = 2N - 1$$

Then if p is the frequency of A in t, the variance of the frequency in $(t+1)$ is

$$V(p') = \frac{pq}{2N-1} V(g) = pq$$

and thus not small in this extreme case.

Show, in the notation of the main text, that in this case

$$\mu_{3g} = (2N-1)(2N-2) \qquad \text{giving} \qquad \mu_3 = pq(p-p)$$

so that, as expected, μ_3 is not negligible here.

NOTES AND EXERCISES

5 *Effective population size in long-headed poppies.* We argue *very* tentatively (note that we are dealing with a hermaphrodite species). If we consider a sufficiently large area, the long-term number of poppy plants presumably remains roughly constant. In view of seed dormancy, any germination of a given batch of seed will tend to be spread over many years, tending to damp out effects of annual fluctuations in number of adult plants. Plausibly, then, we assume that, on averaging over all plants, just one seed per plant will eventually produce a mature adult. We adjust the data on seed output to a mean of unity, using the method of Crow & Morton (1955). Writing M for mean, V for variance and using a prime to indicate an adjusted value, we have

$$\left(\frac{V'}{M'} - 1\right) = \left(\frac{V}{M} - 1\right)\left(\frac{M'}{M}\right)$$

and therefore

$$\left(\frac{V'}{1} - 1\right) = \left(\frac{1\,488\,630}{245} - 1\right)\left(\frac{1}{245}\right)$$

giving an adjusted variance V' of 25.80.

Writing, as in the main text, k^* for the number of male and k^{**} for the number of female gametes per plant, which eventually appear in mature offspring, we have thus estimated $V(k^{**})$ as 25.80.

Plausibly, $V(k^*)$ and $\text{Cov}(k^*, k^{**})$ will each be fairly close to $V(k^{**})$ in view of (a) the large proportion of seed arising from selfing (estimated at 75% - Humphreys & Gale 1974) and (b) the highly correlated production of anthers and ovules on the same plant (the numbers of either depend very strongly on flower size and number of flowers per plant, which in turn depend strongly on plant size - Mackay found a correlation of 0.98 between anther number and seed number for plants growing under non-competitive conditions). We have then, very roughly,

$$V(k) = V(k^*) + V(k^{**}) + 2\,\text{Cov}(k^*, k^{**}) = 103.2$$

whence

$$V(g) = \tfrac{1}{2} + \tfrac{1}{4}V(k) = 26.3$$

Further, with frequency of selfing 0.75,

$$f = \frac{0.75}{2 - 0.75} = 0.6$$

and finally, with $Ek = 2$,

$$\frac{N_e}{N} = \frac{Ek}{V(k)(1+f)/Ek + (1-f)} = 0.024$$

Although subpopulation extinction is regular in this 'fugitive' species, the existence of a large seed bank ensures that subpopulation replacement, which occurs when the soil is turned over, arises almost entirely from local rather than migrant seed, so that subpopulation extinction should not affect the value of N_e/N significantly.

The calculations here are open to many objections, but may, perhaps, give some notion of the magnitude of N_e/N.

6 Let H be the probability that a (randomly chosen) individual is heterozygous. For any population, list the individual members in random order. Let

$$h = X_1 + X_2 + X_3 + \cdots + X_N$$

where

$X_i = 1$ if the ith individual is heterozygous

      $= 0$ otherwise

Hence prove that H is the mean frequency of heterozygotes.

7 *Non-Markovian processes – an example.* If the probability that an allele is transmitted from one generation to the next is affected by the zygote in which that allele appeared, the process is not, strictly speaking, Markovian as far as allele frequencies are concerned. Suppose, for example, that we have complete dominance in viability. We wish to predict, from our knowledge of the allele frequency in generation t, the allele frequency in generation $(t + 1)$. However, our predictions will not be identical for two populations with the same allele frequencies, but differing genotype frequencies, in generation t. Generally, for most patterns of natural selection, we should, *in principle*, work in terms of actual genotype frequencies, even if our sole interest lay in allele frequencies. This would be quite difficult and fortunately is necessary, *in practice*, only for very small populations. For all other populations, we *can* work in terms of allele frequencies, *conditional expected* but not *actual* genotype frequencies being used when making our prediction. Unfortunately, the rigorous proof of this remarkable result (which Kimura seems to have taken for granted in his earlier work!) is exceptionally difficult to follow (the

NOTES AND EXERCISES

difficulty is illustrated by the fact that Watterson's original, very subtle, proof contains a gap, filled later by Norman!).

8 In Moran's model, an individual has expectation of life equal to $2N$ events, but this does *not* mean that, after $2N$ events, the whole original population will have been replaced. To appreciate this distinction, consider the following.

In the Second World War, the probability that a member of the RAF would fail to return from a bombing mission over Germany was about 1/20. A writer in *The Times Literary Supplement* argued that in 'strict mathematics' nobody would survive 20 missions. Noting that many did so survive, he attributed this to increasing skill in evading the German defences.

Prove that, given no improvement in skill, the probability of surviving 20 missions is 0.3585, approximately e^{-1}.

3

On the description of changes in allele frequency

> I sing of arms and of the hero who first came from the shores of Troy, exiled by Fate, to Italy and its Lavinian shore.
> Virgil, *Aeneid*, I, 1

A bewildering abundance of descriptions

Following on our discussion of models in the last chapter, we should now be in a position to attempt a mathematical description of changes in allele frequency over successive generations. We are, however, faced with an immediate difficulty, to the resolution of which the present chapter is devoted: which description is most appropriate? The beginner who consults the literature will discover a very lengthy list of different quantities describing the process of change in allele frequency. Moreover, the descriptive quantities that were prominent in the early literature are not always preferred nowadays, for reasons that will emerge in the course of our discussion.

Consider, for example, the case where allele frequencies are changing under drift, in either the presence or the absence of natural selection. This case appears very early in the literature (Fisher 1922) and, although artificial in that mutation is supposed not to occur once the process of change has started, has proved very helpful in the elucidation of the effects of drift and selection. We note an essential feature: an allele must eventually be 'absorbed', that is either be fixed or be lost. This is easily seen for any model, except Moran's, considered in the previous chapter. For if, in any generation, the frequency of an allele is not 0 or 1, there is a definite chance (often very small but nevertheless exceeding zero) that the frequency will reach 0 or 1 one generation later; once an allele has reached frequency 0 or 1, it remains at that frequency thereafter. Since an occurrence whose probability exceeds zero must happen eventually, we conclude that ultimate absorption is certain; note that this is true even if selection is such as to lead to a balanced polymorphism in the absence of drift. For Moran's model, the argument is a little more complicated, but the same conclusion follows.

Suppose then that the initial frequency of an allele, A say, takes known value p_0. In many applications of the theory, A is a newly arisen mutant, so that

$$p_0 = 1/(2N)$$

where N is the population size in the generation in which A first appears. We wish to describe the process of change in frequency of A from this initial stage until absorption. There are many ways in which we might do this. Thus we could calculate, generation by generation, the probability distribution of p, the frequency of A. If this distribution turned out to be too complicated to be readily understood, we could summarize the distribution, perhaps in terms of its mean and variance in successive generations. Alternatively, we might concentrate on the absorption of A. Thus we could find the probability of ultimate fixation - or of ultimate loss - of A (these probabilities adding to unity since ultimate absorption is certain). We could calculate the 'mean absorption time', that is the mean number of generations until absorption occurs, and indeed find the complete probability distribution of the number of generations required for absorption. Further, for any given frequency (e.g. $1/(2N)$ or $2/(2N)$ or $3/(2N)$ or ...) we can consider the 'sojourn time', that is the number of generations spent by A at that frequency; we could calculate, for any given frequency, the mean sojourn time or even the complete distribution of the sojourn time.

Our list of descriptive quantities, all of which are clearly relevant, is already very long but is not exhaustive. Other descriptive quantities, whose biological significance is less readily apparent, have been prominent in the literature. Our immediate aim is to decide the relative merits of the different descriptive quantities, including those mentioned but not defined in the last paragraph. This is probably best done in the context of a specific example. We shall calculate the quantities for a special case, with only occasional references to other cases; the advantages and disadvantages of the various descriptions will emerge quite easily.

A few comments are in order about our choice of special case, which might otherwise strike the reader as faintly comic. It might seem that we should consider a large population, since most of the literature (and indeed most of this book) is devoted to deriving results for populations assumed to be large (although in practice the results usually apply well enough to populations of quite moderate size). However, large-population theory incorporates mathematical concepts that, while *not* particularly difficult to understand (in an informal way), may well be unfamiliar to the reader and will therefore have to be explained. If we deal with large-population theory at this stage, the problem of understanding the quantities we are calculating becomes intermingled with the problem of understanding the calculation, causing unnecessary difficulties for the reader. Moreover, results from large-population theory are usually obtained by use of approximate methods, so that before discussing these results we must face the intriguing problem of their accuracy. We shall therefore postpone discussion of large-population theory.

We can, however, take advantage of the fact that the allele frequency is a Markovian variable, changing in *discrete* time (generations), and use the standard statistical theory for such variables, the *theory of Markov chains* (if the allele frequency is not Markovian in the strict sense, we will need to modify our procedure slightly). This theory, in which no approximations need be used, can in principle be applied to populations of any size; however, practical considerations limit its application to small or smallish populations such as, for example, those considered in the theory of inbreeding (e.g. Fisher 1949, 1965) and sometimes in plant and animal breeding and in genetic conservation. In population genetics proper, the Markov chain approach has been particularly helpful in checking results from large-population theory. For populations that, while not *very* small, are still sufficiently small (25 say) for Markov chain methods to be practicable, results from large-population theory usually agree rather well with exact results using Markov chain theory; thus we can be confident that the numerous approximations used in large-population theory are justified and moreover that the population can be of quite moderate size and yet the results still hold. Thus a discussion of the Markov chain approach will be appropriate, provided we can set out the mathematics in a fairly straightforward way. Now, most accounts of Markov chain theory (e.g. Bartlett 1955, Kemeny & Snell 1960, Cox & Miller 1965) make extensive use of matrix algebra, requiring from the reader a fair amount of experience in 'reading' expressions involving matrices. For a general treatment, this is essential. If, however, we are content to deal with special cases, we can follow the simpler procedure set out by Fisher (1949, 1965), which requires only very limited matrix algebra (in fact, school matrix algebra will suffice for the following discussion). Moreover, once we have dealt with a special case, the reader will have little difficulty in accepting that the method given will (sometimes with minor modifications) apply very generally. This being so, we discuss the simplest case appropriate to our requirements. We shall suppose that allele frequencies are changing by drift only in a two-allele Wright–Fisher diploid population of size 2 (or haploid population of size 4). This will enable us to exhibit, calculate and (given just a little extra information from other cases) discuss all the descriptive quantities mentioned earlier.

Probability distribution of allele frequency

We begin with the classical problem: to find the probability distribution of the frequency of allele A in any generation. Let (A_i) denote a population having i A alleles, i ranging from 0 to 4. Consider first the conditional distribution. The probability that a diploid parent population, size 2, having i A alleles, gives rise to a progeny population having r A alleles is

$$\frac{4!}{r!(4-r)!}\left(1-\frac{i}{4}\right)^{4-r}\left(\frac{i}{4}\right)^r$$

We have then the probability distributions given in Table 3.1.

Table 3.1 Conditional distributions of progeny populations.

Parent population	Probability	Conditional probability of giving progeny population shown				
		(A_4)	(A_3)	(A_2)	(A_1)	(A_0)
(A_4)	u	1				
(A_3)	v	$\frac{81}{256}$	$\frac{108}{256}$	$\frac{54}{256}$	$\frac{12}{256}$	$\frac{1}{256}$
(A_2)	w	$\frac{1}{16}$	$\frac{4}{16}$	$\frac{6}{16}$	$\frac{4}{16}$	$\frac{1}{16}$
(A_1)	x	$\frac{1}{256}$	$\frac{12}{256}$	$\frac{54}{256}$	$\frac{108}{256}$	$\frac{81}{256}$
(A_0)	y					1

Now write u, v, w, x, y for the unconditional probability that the population is (A_4), (A_3), (A_2), (A_1), (A_0), respectively, in a given generation and u', v', w', x', y' for the corresponding unconditional probabilities one generation later. Then from Table 3.1,

$$u' = u + \tfrac{81}{256}v + \tfrac{1}{16}w + \tfrac{1}{256}x$$

$$v' = \tfrac{108}{256}v + \tfrac{4}{16}w + \tfrac{12}{256}x$$

$$w' = \tfrac{54}{256}v + \tfrac{6}{16}w + \tfrac{54}{256}x$$

$$x' = \tfrac{12}{256}v + \tfrac{4}{16}w + \tfrac{108}{256}x$$

$$y' = \tfrac{1}{256}v + \tfrac{1}{16}w + \tfrac{81}{256}x + y$$

Given a population of known genetic composition in generation 0, we can, by successive application of these equations, find the probability that the population has any stated composition in any given generation. In principle, this could be done by substituting, generation by generation, values for $u, v, \ldots, y$ and calculating the corresponding values one generation later. In practice, such a procedure is very cumbersome and we try a less direct approach.

We argue as follows. The probabilities $u, v, \ldots, y$ do not change in a simple way from one generation to the next; however, it is possible that a simple function of these probabilities will change in a simple way and can thus easily be calculated for any generation. We hope that it will be possible to express each of the probabilities $u, v, \ldots, y$ in terms of such simple functions, in which case $u, v, \ldots, y$ can all be readily calculated for any generation. Of course, we cannot tell at this stage whether such a procedure will work; we try it and see what happens.

We seek a linear function of the probabilities, say

$$L = au + bv + cw + dx + ey$$

(where $a, b, \ldots, e$ are constants), which changes in a simple way, according to the formula

$$L' = au' + bv' + cw' + dx' + ey'$$
$$= \lambda L$$

where λ is a constant, as yet unknown. Can we find values of $\lambda, a, b, \ldots, e$ such that the equation $L' = \lambda L$ holds? An obvious but helpful solution would be to put $a, b, \ldots, e$ all equal to zero, in which case $L = 0, L' = 0$, giving $L' = \lambda L$ for any value of λ. To search systematically for other possible solutions, we write L' in terms of $u, v, \ldots, y$. We have

$$\begin{aligned} L' &= au' + bv' + cw' + dx' + ey' \\ &= a(u + \tfrac{81}{256}v + \tfrac{1}{16}w + \tfrac{1}{256}x) + b(\tfrac{108}{256}v + \tfrac{4}{16}w + \tfrac{12}{256}x) \\ &\quad + c(\tfrac{54}{256}v + \tfrac{6}{16}w + \tfrac{54}{256}x) + d(\tfrac{12}{256}v + \tfrac{4}{16}w + \tfrac{108}{256}x) \\ &\quad + e(\tfrac{1}{256}v + \tfrac{1}{16}w + \tfrac{81}{256}x + y) \\ &= au + \frac{81a + 108b + 54c + 12d + e}{256}v + \frac{a + 4b + 6c + 4d + e}{16}w \\ &\quad + \frac{a + 12b + 54c + 108d + 81e}{256}x + ey \end{aligned}$$

PROBABILITY DISTRIBUTION OF ALLELE FREQUENCY

This will equal

$$\lambda L = \lambda au + \lambda bv + \lambda cw + \lambda dx + \lambda ey$$

for *all* possible values of u, v, w, x, y if (and only if) the coefficients of u, v, w, x, y in L' are the same as those in λL; that is

$$a = \lambda a$$

$$(81a + 108b + 54c + 12d + e)/256 = \lambda b$$

and so on. This gives us

$$(1 - \lambda)a = 0$$

$$81a + (108 - 256\lambda)b + 54c + 12d + e = 0$$

$$a + 4b + (6 - 16\lambda)c + 4d + e = 0$$

$$a + 12b + 54c + (108 - 256\lambda)d + 81e = 0$$

$$(1 - \lambda)e = 0$$

We can solve these simultaneous equations to give values for $a, b, \ldots, e$ once we have found λ. To do the latter, we put our simultaneous equations in matrix form. Recalling the rule for matrix multiplication, we write ($\cdot$ standing for zero)

$$\begin{pmatrix} 1-\lambda & \cdot & \cdot & \cdot & \cdot \\ 81 & 108-256\lambda & 54 & 12 & 1 \\ 1 & 4 & 6-16\lambda & 4 & 1 \\ 1 & 12 & 54 & 108-256\lambda & 81 \\ \cdot & \cdot & \cdot & \cdot & 1-\lambda \end{pmatrix} \begin{pmatrix} a \\ b \\ c \\ d \\ e \end{pmatrix} = \begin{pmatrix} 0 \\ 0 \\ 0 \\ 0 \\ 0 \end{pmatrix}$$

or

$$\mathbf{C} \qquad \mathbf{A} = \mathbf{0}$$

where $\mathbf{C}, \mathbf{A}, \mathbf{0}$ are the matrices indicated. Now, depending on the value of λ, $\mathbf{C}$ may or may not have an inverse $\mathbf{C}^{-1}$. This enables us to reject almost all the values λ might take as irrelevant. For if $\mathbf{C}$ has an inverse, it follows that a,

$b, \ldots, e$ must all be zero (the unhelpful solution we noted earlier). To see this, multiply through the matrix equation given above by $\mathbf{C}^{-1}$, giving

$$\mathbf{C}^{-1}\mathbf{C}\mathbf{A} = \mathbf{C}^{-1}\mathbf{0}$$

By definition of an inverse matrix, the product $\mathbf{C}^{-1}\mathbf{C}$ must be the unit matrix $\mathbf{I}$: further, multiplication of a matrix by $\mathbf{I}$ leaves the matrix unaffected. Thus the left-hand side of the equation just given reduces to $\mathbf{A}$. On the right-hand side $\mathbf{0}$ was defined above as a (single column) matrix all of whose elements are zero; then by the usual rule for matrix multiplication, the right-hand side reduces to a (single column) matrix all of whose elements are zero. Thus we obtain

$$\mathbf{A} = \mathbf{0}$$

that is all the elements of $\mathbf{A}$, namely a, b, c, d, e, equal zero.

Thus we are led to consider values of λ such that $\mathbf{C}$ has no inverse. For this to be so $|\mathbf{C}|$, the determinant of $\mathbf{C}$, must (from standard matrix theory) equal zero. Using standard algebraic methods for evaluating determinants, we find

$$|\mathbf{C}| = (\lambda - 1)^2(4\lambda - 3)(8\lambda - 3)(32\lambda - 3) = 0$$

This equation supplies the values of λ of interest to us. These are

$$1, \ 1, \ \tfrac{3}{4}, \ \tfrac{3}{8}, \ \tfrac{3}{32}$$

the solution $\lambda = 1$ being (by convention) written twice in view of the squared factor $(\lambda - 1)^2$ in our equation $|\mathbf{C}| = 0$. Values of λ obtained by equating $|\mathbf{C}|$, or its analogues in other cases, to zero are known, in modern terminology, as *eigenvalues* (normally these would be found using standard computer programs, but our aim here is to explain the algebra essential to our argument). Unfortunately the term eigenvalue is also used in a different mathematical context; in population genetics the two usages clash. To avoid this confusion, we shall, in the present context, use the more old-fashioned term *latent roots*. Feller (1951) has shown that for a two-allele Wright–Fisher diploid population of size N (or haploid population of size $2N$), in which allele frequencies are changing by drift only, the latent roots are given by

$$\lambda = \frac{(2N)!}{(2N-r)!}\left(\frac{1}{2N}\right)^r \qquad \text{for } r = 0, 1, 2, \ldots, 2N$$

(for a proof, which is rather lengthy, see note 1). On putting $2N = 4$ and $r = 0, 1, 2, 3, 4$ we obtain

$$1, \quad 1, \quad \tfrac{3}{4}, \quad \tfrac{3}{8}, \quad \tfrac{3}{32}$$

as before. It follows at once from Feller's formula that, while the latent root $\lambda = 1$ appears twice, each of the remaining $(2N - 2)$ latent roots appears once only, whatever the value of N; this will be helpful later.

Will these values of λ supply values of $a, b, \ldots, e$ other than all zeros? Readers familiar with the theory of equations will see at once that this can be done in the present and in analogous cases; we need not, however, discuss this point for our special case, where the answer to this question will emerge in the course of the calculation.

Consider first the case $\lambda = \tfrac{3}{4}$. Substituting this value in the simultaneous equations for $a, b, \ldots, e$ given above, we find (from the first and the last equation)

$$\tfrac{1}{4}a = 0 \qquad \tfrac{1}{4}e = 0$$

so that $a = 0$, $e = 0$ and the remaining three equations reduce to

$$-84b + 54c + 12d = 0$$

$$4b - 6c + 4d = 0$$

$$12b + 54c - 84d = 0$$

Notice that the second equation here is $(-18) \times$ the sum of the other two equations, so that we have only two independent equations determining the three quantities b, c, d. Thus these equations do not supply unique values for b, c, d; this is not a problem, since *any* set of values of b, c, d that satisfy these equations will do.

On subtracting the third of these equations from the first we find at once that $b = d$; substitution of this result in the second equation yields $4b = 3c$. Thus in

$$au + bv + cw + dx + ey$$

we have

$$a = 0 \qquad e = 0 \qquad \frac{b}{3} = \frac{c}{4} = \frac{d}{3}$$

CHANGES IN ALLELE FREQUENCY

whence
$$R = 3v + 4w + 3x$$

(or any multiple of R) declines by a factor $\frac{3}{4}$ every generation, that is
$$R' = \tfrac{3}{4}R$$

R is known as the *principal component of frequency* corresponding with the latent root $\frac{3}{4}$. Writing R_t for the value of R in generation t, we have at once
$$R_t = (\tfrac{3}{4})^t R_0$$

Substituting $\lambda = \frac{3}{8}$ in the simultaneous equations for $a, b, \ldots, e$ we find the corresponding principal component of frequency
$$S = v - x$$

giving
$$S_t = (\tfrac{3}{8})^t S_0$$

and taking $\lambda = \frac{3}{32}$ we find
$$T = 9v - 16w + 9x$$

with
$$T_t = (\tfrac{3}{32})^t T_0$$

Consider now the simultaneous equations

$$3v + 4w + 3x = R$$
$$v - \phantom{4w + {}} x = S$$
$$9v - 16w + 9x = T$$

By the usual elementary method for solving simultaneous equations, we find the (unique) solutions

$$v = \tfrac{2}{21}R + \tfrac{1}{2}S + \tfrac{1}{42}T$$
$$w = \tfrac{3}{28}R \phantom{ + \tfrac{1}{2}S} - \tfrac{1}{28}T$$
$$x = \tfrac{2}{21}R - \tfrac{1}{2}S + \tfrac{1}{42}T$$

PROBABILITY DISTRIBUTION OF ALLELE FREQUENCY

Thus (R, S, T being easily calculated for any generation) the probabilities v, w, x in any generation are readily found.

So far we have shown how to calculate the probability that the population belongs to any given 'segregating' type, that is (A_3), (A_2) or (A_1). We now consider the 'absorbing' types (A_4) and (A_0). To do this, we substitute the 'repeated' latent root $\lambda = 1$ in the simultaneous equations for $a, b, \ldots, e$. We find at once

$$0a = 0 \qquad 0e = 0$$

Thus *any* values of a and e will satisfy the simultaneous equations. Our choice of values for a and e is determined by the following considerations, which we state at first very informally. We wish to set up five simultaneous equations that will supply unique solutions for the five probabilities u, v, w, x, y (of course, we have already found three of these equations, i.e. those in R, S and T). From the theory of equations, such unique solutions will exist provided the five simultaneous equations are 'distinct', that is none of them is a rehash of some or all of the others. We know, of course, that the R, S, T equations are distinct, since otherwise they would not have supplied unique solutions for v, w, x. We need, therefore, to find two more equations distinct from the R, S, T equations and from one another. Now the R, S, T equations do not involve u or y. Hence any equations involving u or y will be distinct from the R, S, T equations. Finally, if one of these new equations involves u but not y and the other equation involves y but not u, the two new equations will be distinct from one another.

Thus we first take $\lambda = 1$, $e = 0$; a can take any non-zero value, say 1. Then

$$-148b + 54c + 12d = -81$$

$$4b - 10c + 4d = -1$$

$$12b + 54c - 148d = -1$$

giving $b = \tfrac{3}{4}$, $c = \tfrac{1}{2}$, $d = \tfrac{1}{4}$.

Since a is arbitrary, we could equally have taken a to be 4, thus avoiding fractional values for b, c, d. We have then

$$a = 4, \quad b = 3, \quad c = 2, \quad d = 1, \quad e = 0$$

giving the following principal component of frequency corresponding with the latent root $\lambda = 1$:

$$P = 4u + 3v + 2w + x$$

Similarly, taking $\lambda = 1$, $a = 0$, $e = 4$ we find another principal component, distinct from P, corresponding with $\lambda = 1$:

$$Q = v + 2w + 3x + 4y$$

We have $P_t = 1^t P_0 = P_0$, $Q_t = 1^t Q_0$ so that P_t and Q_t are constants.

To put the matter rigorously, our procedure has ensured that P, Q, R, S, T are *linearly independent*; that is, the equation

$$iP + jQ + kR + lS + mT = 0$$

cannot hold (for general P, Q, R, S, T) unless i, j, k, l, m are all zero. Then, from the theory of equations, unique solutions for u and y (as well as for v, w, x) are obtainable. By the usual elementary method, we find

$$u = \tfrac{1}{4}P - \tfrac{25}{168}R - \tfrac{1}{4}S - \tfrac{1}{168}T$$

$$y = \tfrac{1}{4}Q - \tfrac{25}{168}R + \tfrac{1}{4}S - \tfrac{1}{168}T$$

The probability h_t that the population is still segregating (i.e. has not yet reached absorption) in generation t is

$$1 - u_t - y_t$$

Since

$$\tfrac{1}{4}(P_t + Q_t) = u_t + v_t + w_t + x_t + y_t = 1$$

we have

$$h_t = \tfrac{1}{4}(P_t + Q_t) - u_t - y_t$$
$$= \tfrac{25}{84}R_t + \tfrac{1}{84}T_t$$

Before proceeding to numerical calculations, we make a few remarks about cases other than the one just considered. For Wright–Fisher populations changing by drift only, Feller's formula for the latent roots shows that the only repeated root is $\lambda = 1$. This, in fact, ensures that the principal components, corresponding to our R, S, T, that relate to probabilities for segregating populations are all linearly independent; the case $\lambda = 1$ can be dealt with in the way we have just described, so that the procedure we have given for our special case stands unmodified. In more biologically complicated situations, fairly minor modifications of our procedure are often necessary. In the theory of inbreeding, for example, latent roots (other than $\lambda = 1$) are often repeated (see e.g. Fisher 1965), sometimes appearing more than twice; this difficulty can be overcome, rather along the lines we used for $\lambda = 1$. In very special cases, however, the whole procedure fails, it being impossible to obtain the required number of lincarly independent principal components. An example is noted by Nagylaki (1977) and Renshaw (1982). As far as is known, such cases are very unusual.

A numerical example

Returning now to our special case, suppose that initially A is a newly arisen mutant, so that the population is necessarily (A_1) in generation 0; then

$$x_0 = 1 \qquad u_0 = v_0 = w_0 = y_0 = 0$$

whence

$$P_0 = 1, \quad Q_0 = 3, \quad R_0 = 3, \quad S_0 = -1, \quad T_0 = 9$$

Then

$$P_t = 1, \quad Q_t = 3, \quad R_t = 3(\tfrac{3}{4})^t, \quad S_t = -(\tfrac{3}{8})^t, \quad T_t = 9(\tfrac{3}{32})^t$$

so that

$$u_t = \tfrac{1}{4} - \tfrac{25}{56}(\tfrac{3}{4})^t + \tfrac{1}{4}(\tfrac{3}{8})^t - \tfrac{3}{56}(\tfrac{3}{32})^t$$

$$v_t = \phantom{\tfrac{1}{4} -{}} \tfrac{2}{7}(\tfrac{3}{4})^t - \tfrac{1}{2}(\tfrac{3}{8})^t + \tfrac{3}{14}(\tfrac{3}{32})^t$$

$$w_t = \phantom{\tfrac{1}{4} -{}} \tfrac{9}{28}(\tfrac{3}{4})^t \phantom{{} - \tfrac{1}{2}(\tfrac{3}{8})^t} - \tfrac{9}{28}(\tfrac{3}{32})^t$$

$$x_t = \phantom{\tfrac{1}{4} -{}} \tfrac{2}{7}(\tfrac{3}{4})^t + \tfrac{1}{2}(\tfrac{3}{8})^t + \tfrac{3}{14}(\tfrac{3}{32})^t$$

$$y_t = \tfrac{3}{4} - \tfrac{25}{56}(\tfrac{3}{4})^t - \tfrac{1}{4}(\tfrac{3}{8})^t - \tfrac{3}{56}(\tfrac{3}{32})^t$$

and

$$h_t = \phantom{\tfrac{3}{4} -{}} \tfrac{25}{28}(\tfrac{3}{4})^t \phantom{{} - \tfrac{1}{4}(\tfrac{3}{8})^t} + \tfrac{3}{28}(\tfrac{3}{32})^t$$

Numerical values are given in Table 3.2. The reader is strongly urged to repeat these calculations.

Table 3.2 Probabilities u_t, v_t, w_t, x_t, y_t that the population has 4, 3, 2, 1, 0 A alleles in generation t; h_t is the probability that the population is still segregating.

t	u_t	v_t	w_t	x_t	y_t	h_t
0	0.0000	0.0000	0.0000	1.0000	0.0000	1.0000
1	0.0039	0.0469	0.2109	0.4219	0.3164	0.6797
2	0.0336	0.0923	0.1780	0.2329	0.4632	0.5032
3	0.0748	0.0943	0.1353	0.1471	0.5484	0.3768
4	0.1137	0.0805	0.1017	0.1003	0.6038	0.2825
5	0.1459	0.0641	0.0763	0.0715	0.6422	0.2119
6	0.1712	0.0495	0.0572	0.0522	0.6698	0.1589
7	0.1907	0.0376	0.0429	0.0387	0.6901	0.1192
8	0.2054	0.0284	0.0322	0.0288	0.7052	0.0894
9	0.2165	0.0214	0.0241	0.0215	0.7164	0.0670
10	0.2249	0.0161	0.0181	0.0161	0.7248	0.0503
11	0.2312	0.0120	0.0136	0.0121	0.7311	0.0377
12	0.2359	0.0090	0.0102	0.0090	0.7358	0.0283
13	0.2394	0.0068	0.0076	0.0068	0.7394	0.0212
14	0.2420	0.0051	0.0057	0.0051	0.7420	0.0159
15	0.2440	0.0038	0.0043	0.0038	0.7440	0.0119
16	0.2455	0.0029	0.0032	0.0029	0.7455	0.0089
17	0.2466	0.0021	0.0024	0.0021	0.7466	0.0067
18	0.2475	0.0016	0.0018	0.0016	0.7475	0.0050
19	0.2481	0.0012	0.0014	0.0012	0.7481	0.0038
20	0.2486	0.0009	0.0010	0.0009	0.7486	0.0028
21	0.2489	0.0007	0.0008	0.0007	0.7489	0.0021
22	0.2492	0.0005	0.0006	0.0005	0.7492	0.0016
23	0.2494	0.0004	0.0004	0.0004	0.7494	0.0012
24	0.2496	0.0003	0.0003	0.0003	0.7496	0.0009
25	0.2497	0.0002	0.0002	0.0002	0.7497	0.0007
∞	0.2500	0.0000	0.0000	0.0000	0.7500	0.0000

The case $v_0 = 1$ presents no problems; on grounds of symmetry, we just swap column headings in Table 3.2, interchanging u_t with y_t, and v_t with x_t, and the results in the table stand. On the other hand, results for the case $w_0 = 1$ come out quite distinct from those for $x_0 = 1$. Thus even in the simple case $N = 2$, two quite complicated tables are required for a complete description of the process of change in allele frequency. Clearly, for populations in general, an account of this process in terms of the probability distribution of the frequency of A, given in numerical form, is very comprehensive but far too complicated to be readily understood. Of course, if we could find a simple mathematical formula giving the probability distribution of the frequency of A, all would be well. Unfortunately (as we might guess from the formulae we derived earlier), such a simple formula does not appear to exist, save for rather special cases; general formulae (Kimura 1955a, Voronka & Keller 1975, Piva & Holgate 1977) are difficult to assimilate unless we carry out extensive numerical calculations based upon them, thus leading us back to the difficulty of presentation just discussed.

We might attempt to overcome these difficulties in three ways. First (for populations in general), we might present the probability distribution graphically for a few select values of the initial frequency and of t. A convenient method (Kimura 1955a) is to construct a separate graph for each of the selected initial frequencies but enter distributions for the different selected values of t on the same graph. Undoubtedly, this approach is helpful in exhibiting some important properties of the process of change in allele frequency. However, its utility is confined mainly to the neutral case. Once we invoke natural selection, even of the simplest kind, we should need to consider the effect of different intensities of that selection and the whole situation becomes too complicated for graphical presentation to be of much help.

Secondly, we might seek biologically important special cases in which the probability distribution (or a close approximation to it) takes a simple mathematical form, so that its properties are easily understood. Such cases do exist and we shall refer to them on occasion; however, the utility of this approach is decidedly limited.

Finally, for populations in general, we can try to summarize the whole process of change in allele frequency in terms of easily presented, readily understood quantities. The most natural procedure would be to derive these quantities from the probability distribution of the frequency of A. However, in view of the complexity of that distribution, it is convenient to use a less direct procedure; we try to derive the quantities without troubling to calculate the distribution itself. We illustrate both procedures in the remainder of this chapter.

Probability of fixation

We begin by noting that, to the evolutionist, the long-term consequences of a given biological set-up are much the most important. In particular, we may ask: 'Given an allele A with initial frequency p_0, what is the probability that this allele will ultimately be fixed?' This question has, in fact, received much attention (e.g. Haldane 1927, Fisher 1930a, Wright 1931, Kimura 1957, Moran 1960b, 1961, Ewens 1967a, Maruyama 1977).

Consider first our special case. Take our formulae for $u_t, v_t, \ldots, y_t$ (for the case $x_0 = 1$, corresponding to $p_0 = \frac{1}{4}$) and let t tend to infinity. Then

$$(\tfrac{3}{4})^t \to 0 \qquad (\tfrac{3}{8})^t \to 0 \qquad (\tfrac{3}{32})^t \to 0$$

so that v_t, w_t, x_t all tend to zero, confirming our earlier statement that ultimate absorption is certain. Further,

$$u_t \to \tfrac{1}{4} \qquad v_t \to \tfrac{3}{4}$$

Thus the probability of ultimate fixation of A is $\tfrac{1}{4}$. Similarly, for the case $w_0 = 1$ (corresponding to $p_0 = \tfrac{1}{2}$), the probability of ultimate fixation is $\tfrac{1}{2}$; thus our probability proves to be p_0 in both cases.

Now consider an indirect approach. Let the initial frequency of A be p_0. Since we are dealing with neutral alleles, the frequency of A, calculated *deterministically*, is p_0 in any generation. Moreover, from the argument developed in Chapter 2, the mean allele frequency in any generation will, in the case of neutral alleles, equal the deterministically calculated frequency in that generation. Thus at any time, *including ultimate time*

$$\text{mean frequency of } A = p_0$$

Since absorption is certain, the ultimate frequency of A is either 1 (A fixed) or 0 (A lost). Let the probability of ultimate fixation be α and of ultimate loss β. Then, by definition of a mean, the ultimate mean frequency of A is

$$1\alpha + 0\beta = \alpha$$

But we showed above that the ultimate mean frequency of A equals p_0. Hence $\alpha = p_0$, that is

$$\text{probability of fixation} = \text{frequency at the start}$$

for any allele, provided all alleles in the population are neutral. Notice that at no stage in the argument did we refer to any particular population size or any particular model. Our result then must be true for any population (even if divided into subpopulations).

Clearly, the probability of ultimate fixation is a descriptive quantity of the type we seek. It describes a very important property of the process of change in allele frequency and, in the neutral case, is given by a simple formula of universal applicability. When natural selection operates (see later), this

Dominant latent root (maximum non-unit eigenvalue)

Having discussed the ultimate outcome, we consider how we might summarize other stages of the process of change of allele frequency. The following approach appears promising.

Consider our special case. The non-unit latent roots proved to be $\frac{3}{4}$, $\frac{3}{8}$ and $\frac{3}{32}$. In our formulae for $u_t, v_t, \ldots, y_t$ these latent roots appeared in the form $(\frac{3}{4})^t$, $(\frac{3}{8})^t$ and $(\frac{3}{32})^t$. Now, when t is at all large, the first of these is very much larger than the others. For example, when $t = 4$,

$$(\tfrac{3}{4})^t = 0.3164 \qquad (\tfrac{3}{8})^t = 0.0198 \qquad (\tfrac{3}{32})^t = 0.0001$$

Thus, for sufficiently large t, we can, without serious loss of accuracy, ignore terms in $(\frac{3}{8})^t$ and $(\frac{3}{32})^t$ altogether, giving us (for the case $x_0 = 1$) the greatly simplified formulae

$$u_t = \tfrac{1}{4} - \tfrac{25}{56}(\tfrac{3}{4})^t$$

$$v_t = \tfrac{2}{7}(\tfrac{3}{4})^t$$

$$w_t = \tfrac{9}{28}(\tfrac{3}{4})^t$$

$$x_t = \tfrac{2}{7}(\tfrac{3}{4})^t$$

$$y_t = \tfrac{3}{4} - \tfrac{25}{56}(\tfrac{3}{4})^t$$

In fact, for $t = 4$, these formulae give

$$0.1087, \quad 0.0904, \quad 0.1017, \quad 0.0904, \quad 0.6087$$

which compare quite well with the exact values given in Table 3.2; for $t = 6$ onwards, approximate and exact results agree to the second decimal place (at least). Thus, from a fairly early stage onwards, the probabilities for all segregating types decline by a factor $\frac{3}{4}$ every generation; note that this factor is the value of the numerically largest non-unit latent root, known as the *dominant latent root* (or maximum non-unit eigenvalue). The argument just given may be applied to populations in general, provided we take t sufficiently large. We see that the value of the dominant latent root is critical for the approach to absorption; the larger the value of this root, the greater the probability that the population is still segregating at a sufficiently late stage of the process. It would appear, then, that the value of the dominant latent root gives an excellent summary of an important aspect of the process of change in allele frequency. In addition, it would appear particularly helpful if, for general populations, we could obtain formulae giving probabilities for different types of segregating population at this sufficiently late stage, analogous to the simplified formulae v_t, w_t, x_t just derived.

This approach was taken early by Wright and Fisher (for details, see notes and exercises). Wright (1931) showed that, for a Wright–Fisher diploid population of size N, in which allele frequencies are changing by drift only, the dominant latent root is

$$1 - 1/(2N)$$

Now consider, for such a population, the probabilities for segregating types, at a stage sufficiently late for terms involving the remaining non-unit latent roots to be neglected. For our special case, these probabilities are in the ratio

$$\tfrac{2}{7} : \tfrac{9}{28} : \tfrac{2}{7}$$

roughly 1:1:1. By an extraordinary intuition, apparently based solely on results from populations of size 2 and 3, Wright (1931) surmised that this is true in general; at a sufficiently late stage, *all* segregating types are (to a close approximation) equally probable (particularly when N is large). Having made this surmise, he was able to confirm it by a simple mathematical argument. To sum up, at a sufficiently late stage in the process, the probability that the allele frequency takes value p is (near enough) the same for all values of p over the range $1/(2N)$ to $1 - 1/(2N)$, being

$$C\left(1 - \frac{1}{2N}\right)^t$$

where C is the same for all p. In fact, it is easily shown (see note 3) that C must be close to

$$6p_0q_0/(2N)$$

The probability that the population is still segregating at this stage is obtained by summing

$$\frac{6p_0q_0}{2N}\left(1 - \frac{1}{2N}\right)^t$$

over all $(2N-1)$ segregating types to give

$$(2N-1)\frac{6p_0q_0}{2N}\left(1-\frac{1}{2N}\right)^t$$

or approximately

$$6p_0q_0 e^{-t/(2N)}$$

as given by Kimura (1955a) in the course of a very comprehensive analysis.

These results are very simple (although the simplicity disappears once we bring in natural selection). It is understandable, therefore, that the early workers were much preoccupied by the probability distribution at this sufficiently late stage. Wright (1931) demonstrated that his postulated uniform distribution, while true to a close approximation, was not exactly correct; the near-exact form of the distribution when N is large was found, in masterly fashion, by Fisher (1930a, b).

Unfortunately, there is a serious difficulty for the argument just given, first pointed out by Kimura (1955a). Write λ_D for the dominant latent root and λ_E for the next largest root. If λ_D is substantially greater in magnitude than λ_E (as happened in our special case) then

$$\lambda_D^t \gg \lambda_E^t$$

when t is at all large, and the omission of terms in λ_E and still smaller latent roots is justified unless t is quite small. If, however, λ_D exceeded λ_E by only a small margin, the strong inequality just given would hold only when t is very large; thus the omission of terms in λ_E and smaller roots would not be justified until a very late stage. Now, for the general case, the early workers did not know values for λ_E or smaller latent roots. We, however, can write these down at once from Feller's (1951) formula quoted earlier. We have

$$\lambda_D = \left(1-\frac{1}{2N}\right) \qquad \lambda_E = \left(1-\frac{1}{2N}\right)\left(1-\frac{2}{2N}\right)$$

If N is at all large, these will be very close to one another. If, say, $N=50$, we find

$$\lambda_D = 0.99 \qquad \lambda_E = 0.9702$$

If, say, $t = 50$ we have

$$(0.99)^{50} = 0.6050 \qquad (0.9702)^{50} = 0.2203$$

so that the omission of terms in λ_E would not be justified; only if t were in the region of 100 or more could we omit terms in λ_E and smaller latent roots. It follows then that Wright's simple distribution

$$C\left(1 - \frac{1}{2N}\right)^t$$

will in practice be valid only when t is very large. In fact, using the approximations

$$\left(1 - \frac{1}{2N}\right)^t = e^{-t/(2N)} \qquad \left[\left(1 - \frac{1}{2N}\right)\left(1 - \frac{2}{2N}\right)\right]^t = \left(1 - \frac{3}{2N}\right)^t = e^{-3t/(2N)}$$

valid when N is large, and evaluating these for a few values of $t/(2N)$, we can conclude with some confidence that the simple distribution will not be valid if $t < 2N$. In fact Kimura gives a case where it takes about $4N$ generations for this distribution to be attained, by which time, as he shows, the probability that the population is still segregating is less than 10 per cent.

We conclude then that the dominant latent root is, in general, of major importance as a descriptive quantity only if we wish to concentrate on a late stage of the process of change of allele frequency, by which time the probability that absorption has already taken place is high. This is the case in the theory of inbreeding, where our main interest is to find the number of generations required to achieve a very high probability of absorption (absence of segregation); here the dominant latent root is fundamental. In population genetics proper, the dominant latent root has proved helpful in a rather unexpected context. Suppose we have a population that does not accord with the Wright–Fisher model. Suppose further that in this population allele frequencies are changing by drift only, and consider a late stage of the process, at which stage the dominant latent root becomes of major importance; write this dominant root as

$$1 - 1/(2N_e)$$

Here N_e is yet another effective population size, the *eigenvalue* effective size, which, like the other effective sizes, can be calculated for various situations of biological interest. Consider the case where the population is divided into subpopulations between which very limited migration occurs. The eigenvalue effective size can be calculated and compared with the actual overall population size for different rates of migration. This provides *one* way of assessing the effect of low rates of migration; if, at a given rate, the effective and actual sizes agree well, we conclude that subdivision of the population is of little consequence, at least when t is large (Moran 1962, Maruyama 1970a, Kimura & Maruyama 1971).

We conclude that, while it is helpful for some specific investigations to determine the dominant latent root, this quantity does not provide a summary of the process of change of allele frequency, its utility usually being confined to a late stage of that process.

It might be thought, however, that the *complete set* of latent roots would summarize the whole process well. In fact, this is not so, for a very subtle reason given by Ewens (1973), whose argument we now discuss.

Modified process

We note first that the evolutionist is very much more interested in those mutants which survive and reach fixation than in those which are lost; he will wish to describe how these successful mutants spread through the population. This can be done by solving the classical problem for *successful mutants only*; we derive the distribution of the allele frequency, confining ourselves to those cases where the allele A in whose fate we are interested is ultimately fixed. To put the matter more formally, we derive the probability distribution of the frequency of an allele A (with initial frequency p_0) *conditional* on that allele being ultimately fixed. The term 'conditional process' has therefore been used to describe the process of change in frequency of successful alleles; to avoid proliferation of the term 'conditional', we shall use 'modified process' instead. Processes closely similar to our modified process were first studied by Fisher (1930b); more recently, our modified process has received considerable attention (Kimura & Ohta 1969a, Kimura 1970a, Maruyama 1972a, 1977, Ewens 1973, 1979, Pollak & Arnold 1975, Tavaré 1979).

To find the probability distribution in the modified process we proceed in much the same way as for the unmodified process. We ask: 'Given a parent population with i A alleles, what is the probability, for the modified process, that its progeny population has r A alleles?'

To answer this question, we ask another: 'What is the probability that the progeny population has r A alleles *and* that A is ultimately fixed?' We can formulate our answer in terms of either the unmodified or the modified process. Thinking first in terms of the unmodified process and taking the points 'r A alleles' and 'A ultimately fixed' in their natural order, we see that the answer must be

$\quad\quad\quad$ Prob$\{r$ A alleles in the progeny$\}$

$\quad\quad\quad\quad \times$ Prob$\{$fixation, starting with r A alleles$\}$

CHANGES IN ALLELE FREQUENCY

which we shall write as

$$p_{ir} \times \pi_r$$

Here p_{ir} is the usual probability of obtaining r A alleles in the unmodified process and π_r the usual probability of fixation for initial frequency $r/(2N)$.

We can, however, look at the matter differently. We take the points in reverse order. Without referring at first to the progeny generation, we first find the probability of fixation, based solely on our knowledge that there are i A alleles in the parent population. We write this probability as π_i, the usual probability of fixation for initial frequency $i/(2N)$. This settles the matter of fixation. Now write p_{ir}^* for the probability of r A alleles in the progeny, *given that fixation ultimately occurs*. Then the answer to our question can also be written

$$\pi_i \times p_{ir}^*$$

Here p_{ir}^* is the probability of obtaining r A alleles in the *modified* process. Thus, since the two answers must be the same, we have

$$p_{ir}^* = p_{ir}\pi_r/\pi_i$$

In the neutral case

$$\pi_r = r/(2N) \qquad \pi_i = i/(2N)$$

and therefore

$$p_{ir}^* = p_{ir}(r/i)$$

We apply this formula to our special case. Taking the p_{ir} from Table 3.1, we readily calculate the p_{ir}^* given in Table 3.3. Following the procedure described earlier, we find that the latent roots are

$$1, \tfrac{3}{4}, \tfrac{3}{8}, \tfrac{3}{32}$$

Notice that these are the same as in the unmodified process, except that the unit root now appears only once. Thus we have two processes with the same set of latent roots. If it turns out, as would hardly be surprising, that these two processes have rather little in common, we must conclude that the complete set of latent roots does *not* provide an adequate description of either process.

In order to compare the two processes in detail, we shall calculate probabilities u_t, v_t, w_t, x_t for the modified process and compare them with

MODIFIED PROCESS

Table 3.3 Conditional distributions of progeny populations in a modified process.

Parent population	Probability	Conditional probability of progeny population shown			
		(A_4)	(A_3)	(A_2)	(A_1)
(A_4)	u	1			
(A_3)	v	$\frac{27}{64}$	$\frac{27}{64}$	$\frac{9}{64}$	$\frac{1}{64}$
(A_2)	w	$\frac{1}{8}$	$\frac{3}{8}$	$\frac{3}{8}$	$\frac{1}{8}$
(A_1)	x	$\frac{1}{64}$	$\frac{9}{64}$	$\frac{27}{64}$	$\frac{27}{64}$

corresponding values for the unmodified process. The algebra is essentially the same as in the unmodified case, although slightly easier since there are no repeated roots. Corresponding to the latent roots just given, the principal components prove to be

$$Q = u + v + w + x$$

$$R = v + 2w + 3x$$

$$S = v - 3x$$

$$T = 3v - 8w + 9x$$

respectively, yielding solutions

$$u = Q - \tfrac{25}{42}R - \tfrac{1}{3}S - \tfrac{1}{42}T$$

$$v = \tfrac{2}{7}R + \tfrac{1}{2}S + \tfrac{1}{14}T$$

$$w = \tfrac{3}{14}R - \tfrac{1}{14}T$$

$$x = \tfrac{2}{21}R - \tfrac{1}{6}S + \tfrac{1}{42}T$$

Suppose, as before, that $x_0 = 1$. Then

$$u_0 = v_0 = w_0 = 0$$

whence

$$Q_0 = 1, \quad R_0 = 3, \quad S_0 = -3, \quad T_0 = 9$$

giving

$$u_t = 1 - \tfrac{25}{14}(\tfrac{3}{4})^t + (\tfrac{3}{8})^t - \tfrac{3}{14}(\tfrac{3}{32})^t$$

$$v_t = \tfrac{6}{7}(\tfrac{3}{4})^t - \tfrac{3}{2}(\tfrac{3}{8})^t + \tfrac{9}{14}(\tfrac{3}{32})^t$$

$$w_t = \tfrac{9}{14}(\tfrac{3}{4})^t \quad\quad - \tfrac{9}{14}(\tfrac{3}{32})^t$$

$$x_t = \tfrac{2}{7}(\tfrac{3}{4})^t + \tfrac{1}{2}(\tfrac{3}{8})^t + \tfrac{3}{14}(\tfrac{3}{32})^t$$

and

$$h_t = \tfrac{25}{14}(\tfrac{3}{4})^t - (\tfrac{3}{8})^t + \tfrac{3}{14}(\tfrac{3}{32})^t$$

since $h_t = 1 - u_t$. Numerical values are given in Table 3.4.

On comparing these results with those for the unmodified process given in Table 3.2, it appears that the two processes are in some ways very different. For example, the probability that the population is still segregating falls off very much more rapidly in the unmodified case. Further, in the modified case $v_t : w_t : x_t$ does not approach approximate equality but

$$\tfrac{6}{7} : \tfrac{9}{14} : \tfrac{2}{7}$$

Thus, for our special case at least, the unmodified and modified processes have the same latent roots but differ in some important properties. Ewens (1973) has shown that this conclusion holds in general, the discrepancy between the two processes being particularly marked when the population is large. Thus the complete set of latent roots does not provide an adequate summary of either process.

More generally, a useful set of descriptive quantities must include at least some that are not the same for an unmodified process and the corresponding modified process. We shall, therefore, when discussing descriptive quantities not yet described, calculate these for both processes and shall be much encouraged when corresponding answers come out differently.

Probability distribution of absorption time

We ask: 'How many generations does it take to go from the beginning to the end of a given process of change of allele frequency?' Since this number of generations will be subject to random variation, our aim is to find the probability distribution of this number.

Table 3.4 Probabilities u_t, v_t, w_t, x_t that the population has 4, 3, 2, 1, A alleles in generation t of a modified process; h_t is the probability that the population is still segregating.

t	u_t	v_t	w_t	x_t	h_t
0	0.0000	0.0000	0.0000	1.0000	1.0000
1	0.0156	0.1406	0.4219	0.4219	0.9844
2	0.1343	0.2768	0.3560	0.2329	0.8657
3	0.2992	0.2830	0.2707	0.1471	0.7008
4	0.4547	0.2416	0.2034	0.1003	0.5452
5	0.5836	0.1923	0.1525	0.0715	0.4163
6	0.6850	0.1484	0.1144	0.0522	0.3150
7	0.7627	0.1128	0.0858	0.0387	0.2373
8	0.8216	0.0852	0.0644	0.0288	0.1784
9	0.8661	0.0641	0.0483	0.0215	0.1339
10	0.8995	0.0482	0.0362	0.0161	0.1005
11	0.9246	0.0362	0.0272	0.0121	0.0754
12	0.9434	0.0271	0.0204	0.0090	0.0566
13	0.9576	0.0204	0.0153	0.0068	0.0424
14	0.9682	0.0153	0.0114	0.0051	0.0318
15	0.9761	0.0114	0.0086	0.0038	0.0239
16	0.9821	0.0086	0.0064	0.0029	0.0179
17	0.9866	0.0064	0.0048	0.0021	0.0134
18	0.9899	0.0048	0.0036	0.0016	0.0101
19	0.9924	0.0036	0.0027	0.0012	0.0076
20	0.9943	0.0027	0.0020	0.0009	0.0057
21	0.9958	0.0020	0.0015	0.0007	0.0042
22	0.9968	0.0015	0.0011	0.0005	0.0032
23	0.9976	0.0011	0.0009	0.0004	0.0024
24	0.9982	0.0009	0.0006	0.0003	0.0018
25	0.9986	0.0006	0.0005	0.0002	0.0013
26	0.9990	0.0005	0.0004	0.0002	0.0010
27	0.9992	0.0004	0.0003	0.0001	0.0008
28	0.9994	0.0003	0.0002	0.0001	0.0006
29	0.9996	0.0002	0.0002	0.0001	0.0004
30	0.9997	0.0002	0.0001	0.0000	0.0003
∞	1.0000	0.0000	0.0000	0.0000	0.0000

Write (as before) h_t for the probability that the population is still segregating in generation t and h_{t-1} for the corresponding probability one generation earlier. Let g_t be the probability that absorption occurs in generation t, *but not earlier*; this is the probability we seek. Then

probability that the population is *not* segregating in t

= probability that the population is *not* segregating in $(t-1)$

+ probability that the population *is* segregating in $(t-1)$ but absorption occurs in t

that is,

$$(1 - h_t) = (1 - h_{t-1}) + g_t$$

whence

$$g_t = h_{t-1} - h_t$$

Of course, this will be true whether we are dealing with the unmodified or modified process. For convenience, however, we reserve the term 'absorption time' for the unmodified process and use 'fixation time' for the modified process. The set of probabilities $\{g_t\}$ will be called the probability distribution of absorption time or probability distribution of fixation time, whichever is appropriate. Numerical values for our special case (with $x_0 = 1$) are given in Table 3.5. We discuss some details of these distributions before making a general point.

The distributions are noticeably skewed to the right, a small proportion of populations failing to reach absorption (or fixation) until a relatively late stage. It is further apparent that fixation is the slower of the two processes; an elementary calculation shows that the median absorption time is about 2.0 generations whereas the median fixation time is about 4.4 generations. The source of this discrepancy is readily located, on recalling that the initial frequency of A was $\frac{1}{4}$. Obviously, it will be easier for the frequency to go to 0 than to 1; since absorption includes the case where the frequency goes to 0 whereas in fixation the frequency must go to 1, absorption is the faster process. We expect, then, that the discrepancy between the two distributions will be very marked if we start with a newly arisen mutant in a very large population; this is indeed so, as we shall see in a later chapter. Further, in the early generations the frequency of such a mutant will necessarily be very low, so that loss of the mutant will be very likely. This suggests that the pattern of

Table 3.5 Probability of absorption, for the unmodified process, and fixation, for the modified process, in generation t.

t	Probability of absorption	Probability of fixation
1	0.3203	0.0156
2	0.1765	0.1187
3	0.1264	0.1649
4	0.0942	0.1555
5	0.0706	0.1289
6	0.0530	0.1013
7	0.0397	0.0777
8	0.0298	0.0589
9	0.0223	0.0444
10	0.0168	0.0334
11	0.0126	0.0251
12	0.0094	0.0188
13	0.0071	0.0141
14	0.0053	0.0106
15	0.0040	0.0080
16	0.0030	0.0060
17	0.0022	0.0045
18	0.0017	0.0034
19	0.0012	0.0025
20	0.0009	0.0019
21	0.0007	0.0014
22	0.0005	0.0011
23	0.0004	0.0008
24	0.0003	0.0006
25	0.0002	0.0004

absorption in the unmodified process shown in Table 3.5, with high probability of absorption in the early generations, will hold generally if we start with a newly arisen mutant; we discuss this is the next chapter.

The evolutionist, however, will not require such intriguing details to see the importance of our probability distributions; given an evolutionary process, what could be more natural than to ask: 'How long will this process take?' However, presentation of the results is rather a problem, once we bring in natural selection. An obvious way to overcome this problem is to calculate the mean and variance of our probability distributions, preferably without first calculating the distributions themselves, since the latter is a very formidable problem when natural selection operates. We turn then to methods for calculating the mean absorption time, the mean fixation time and the corresponding variances. Since these methods are very much the same for the modified as for the unmodified process, we take the two together; although we shall usually write in terms of absorption time, the changes required to deal with fixation time will be minimal and easily stated.

Mean absorption time

We begin with a few historical remarks. Mean absorption times were first calculated by Fisher (1930a, b) and Wright (1931); so subtle was their treatment, however, that this was not appreciated (the present author being among those afflicted with this regrettable blindness).

> 'O dark, dark, dark, amid the blaze of noon'
> (Milton, *Samson Agonistes*)

Moran (1962) was apparently the first to realize that Fisher, while solving what appeared to be quite a different problem, had in fact found mean absorption times, and the whole matter was put beyond doubt by the work of Ewens (1963a, 1979). Since the early 1960s, many authors (e.g. Watterson 1962) have discussed the mean absorption time and related topics. Most of this work relates to genetic changes in relatively large populations; we consider this work in detail later. In the present chapter, we shall find the mean and variance of the absorption time by Markov chain methods; a particularly helpful reference is Kemeny & Snell (1960).

We first derive the mean directly from the probability distribution. Of course, this can be done by elementary methods, but the following procedure, given by Fisher (1949, 1965), is computationally very easy and will also be helpful later. As before, let h_t be the probability that the population is still segregating in generation t. We shall show that the mean absorption time is

$$\sum_{t=0}^{\infty} h_t$$

Let g_t, as before, be the probability that absorption occurs in generation t but not earlier; we have shown that

$$g_t = h_{t-1} - h_t$$

MEAN ABSORPTION TIME

By definition of a mean, the mean absorption time is

$$\sum_{t=1}^{\infty} tg_t = \sum_{t=1}^{\infty} t(h_{t-1} - h_t) = 1(h_0 - h_1) + 2(h_1 - h_2) + 3(h_2 - h_3) + \cdots$$

If we take, say, the first three terms of this and remove the parentheses we obtain

$$h_0 - h_1 + 2h_1 - 2h_2 + 3h_2 - 3h_3 = h_0 + h_1 + h_2 - 3h_3$$

and generally, for any number n, the sum of the first n terms is

$$\sum_{t=0}^{n} h_t - nh_n$$

Now let $n \to \infty$. If we recall that (for $x_0 = 1$)

$$h_t = \tfrac{25}{28}(\tfrac{3}{4})^t + \tfrac{3}{28}(\tfrac{3}{32})^t$$

and note the standard mathematical result

$$\lim_{n \to \infty}(n\lambda^n) = 0 \qquad \text{for } -1 < \lambda < 1$$

we see that, as $n \to \infty$, $nh_n \to 0$, at least for the case we are considering; in fact, this holds generally, since h_t always takes the form $\Sigma A_i(\lambda_i)^t$, as is fairly obvious from our earlier discussion of special cases. We are left then with

$$\sum_{t=0}^{\infty} h_t$$

This is easily evaluated. Consider the case $x_0 = 1$. From the standard rules for the sum of a geometric progression,

$$\sum_{t=0}^{\infty}(\tfrac{3}{4})^t = \frac{1}{1 - \tfrac{3}{4}} = 4 \qquad \sum_{t=0}^{\infty}(\tfrac{3}{32})^t = \frac{1}{1 - \tfrac{3}{32}} = \tfrac{32}{29}$$

so that the mean absorption time equals

$$\tfrac{25}{28} \times 4 + \tfrac{3}{28} \times \tfrac{32}{29} = 3.6897$$

generations. The same result holds for $v_0 = 1$, whereas for $w_0 = 1$ we obtain 4.5517 generations. Taking now the value of h_t for the modified process, we find the mean fixation times, for the cases $v_0 = 1$, $w_0 = 1$ and $x_0 = 1$, to be

CHANGES IN ALLELE FREQUENCY

Tables 3.8 & 3.9 below), the standard deviation is also relatively large, although sometimes rather less than the mean in magnitude.

Mean sojourn times

We can supplement the information given by mean absorption (or fixation) times as follows: for every frequency between 0 and 1, we calculate the 'mean sojourn time', that is, the mean number of generations spent by our allele at that frequency.

Suppose (in an unmodified process) that initially there are i A alleles. We write $p_{ij}(t)$ for the probability that there are j A alleles t generations later; consistent with this notation, $p_{ii}(0) = 1$ and $p_{ij}(0) = 0$ for $j \neq i$, while $p_{ij}(1)$ is the same as our p_{ij} used earlier. We calculate the mean sojourn time of A at chosen frequency $j/(2N)$.

Let X_t ($t = 0, 1, 2, 3, \ldots$) be a random variable such that, in generation t,

$$X_t = 1 \quad \text{if the allele frequency is } j/(2N)$$
$$= 0 \quad \text{otherwise}$$

Therefore

$$EX_t = 1[p_{ij}(t)] + 0[1 - p_{ij}(t)] = p_{ij}(t)$$

Now write

$$s_{ij} = X_0 + X_1 + X_2 + \cdots$$

Then s_{ij} is the time spent at frequency $j/(2N)$. Since the 'mean of a sum' is the 'sum of the means', we see that – given an initial frequency $i/(2N)$ – the mean sojourn time at $j/(2N)$ is

$$\sum_{t=0}^{\infty}(EX_t) = \sum_{t=0}^{\infty} p_{ij}(t)$$

which we shall write as $\bar{s}_{ij}$.

Suppose, for example, that the initial frequency of A was $\frac{1}{4}$ and we wish to calculate the mean sojourn time at frequency $\frac{3}{4}$ for our special case. Then $p_{ij}(t)$ is, in our earlier notation, v_t for the case $x_0 = 1$, which we found to be

$$\tfrac{2}{7}(\tfrac{3}{4})^t - \tfrac{1}{2}(\tfrac{3}{8})^t + \tfrac{3}{14}(\tfrac{3}{32})^t$$

Summing this from $t = 0$ to $t = \infty$ we obtain 0.5793 generations.

Note that h_t, the probability that the population is still segregating in generation t, is obtained by summing probabilities for all segregating types in generation t; that is, given initial frequency $i/(2N)$,

$$h_t = \sum_{j=1}^{2N-1} p_{ij}(t)$$

Further, from our 'rule of means', the mean absorption time is just the sum of the mean sojourn times. Write $\bar{s}_i$ for the mean absorption time, given initial frequency of A equal to $i/(2N)$. We have then

$$\bar{s}_i = \sum_{j=1}^{2N-1} \bar{s}_{ij} = \sum_{j=1}^{2N-1} \sum_{t=0}^{\infty} p_{ij}(t) = \sum_{t=0}^{\infty} \sum_{j=1}^{2N-1} p_{ij}(t) = \sum_{t=0}^{\infty} h_t$$

confirming the result we obtained earlier by a more direct method.

The procedure for the modified process is the same as that just given, provided we replace the probabilities $p_{ij}(t)$ by the corresponding probabilities, $p_{ij}^*(t)$ say, for the modified process. However, it is not necessary to carry out both sets of calculations, for a given biological set-up, since it is easy to convert results from the unmodified process to results from the modified process (Ewens 1973). We showed earlier that (with a trivial change in notation)

$$p_{ij}^*(t) = p_{ij}(t)\pi_j/\pi_i$$

(π_i and π_j being the probability of fixation given initial frequencies $i/(2N)$ and $j/(2N)$, respectively), for the case $t = 1$. A minor change in wording in our proof is sufficient to show that this relationship is true for *any* t. Then

$$\sum_{t=0}^{\infty} p_{ij}^*(t) = \left(\sum_{t=0}^{\infty} p_{ij}(t)\right)\pi_j/\pi_i$$

or, writing $\bar{s}_{ij}^*$ for the mean sojourn time in the modified process corresponding to $\bar{s}_{ij}$ in the unmodified process,

$$\bar{s}_{ij}^* = \bar{s}_{ij}\pi_j/\pi_i$$

reducing to

$$\bar{s}_{ij}^* = \bar{s}_{ij}j/i$$

in the neutral case.

CHANGES IN ALLELE FREQUENCY

Table 3.9 Variance of sojourn time and fixation time for modified process.

Initial state	Initial frequency	Variance of sojourn time at frequency shown			Variance of fixation time
		$\frac{3}{4}$	$\frac{1}{2}$	$\frac{1}{4}$	
$v_0 = 1$	$\frac{3}{4}$	2.5701	1.9049	0.6113	8.9765
$w_0 = 1$	$\frac{1}{2}$	2.8195	3.1534	1.5486	12.3615
$x_0 = 1$	$\frac{1}{4}$	2.8166	3.4031	2.5701	13.3649

Consider an unmodified process starting at $i/(2N)$. The mean time spent at $j/(2N)$, *given* that the frequency goes to that value, is $\bar{s}_{jj}$ (this is a little confusing at first; the difficulty disappears on appreciating that, by definition, $\bar{s}_{jj}$ takes account of an obligatory first visit to $j/(2N)$, since an allele starting at $j(2N)$ necessarily visits this frequency). Thus $\bar{s}_{ij}$, the mean time spent at $j/(2N)$, is made up of two parts; in a fraction of cases f_{ij} the mean time at $j/(2N)$ is $\bar{s}_{jj}$ whereas in a fraction of cases $(1 - f_{ij})$ the mean time at $j/(2N)$ is zero. Hence, taking all cases into account

$$\bar{s}_{ij} = f_{ij}\bar{s}_{jj} + (1 - f_{ij})(0)$$

whence

$$f_{ij} = \bar{s}_{ij}/\bar{s}_{jj}$$

A matrix representation of mean sojourn time

Our method for finding mean times, although illuminating, is very lengthy, as we had first to find the probability distribution of the frequency of A in successive generations. However, our algebra can be adapted to give a much briefer method (very suitable for practical computation), given by Kemeny & Snell (1960); we quote their results, which we shall prove in a later chapter.

Take the table of conditional probabilities (Table 3.1 or 3.3 in our case) and ignore any probabilities relating to absorbing types; write the remainder of the table as a matrix Q (say). For example, from Table 3.3 we find

$$Q = \begin{pmatrix} \frac{27}{64} & \frac{9}{64} & \frac{1}{64} \\ \frac{3}{8} & \frac{3}{8} & \frac{1}{8} \\ \frac{9}{64} & \frac{27}{64} & \frac{27}{64} \end{pmatrix}$$

Let **I** be the unit matrix of the same size as **Q**. Then the elements of

$$(\mathbf{I} - \mathbf{Q})^{-1}$$

are the mean sojourn times. For example, for the case above

$$\mathbf{I} = \begin{pmatrix} 1 & 0 & 0 \\ 0 & 1 & 0 \\ 0 & 0 & 1 \end{pmatrix} \quad (\mathbf{I} - \mathbf{Q}) = \begin{pmatrix} \frac{37}{64} & -\frac{9}{64} & -\frac{1}{64} \\ -\frac{3}{8} & \frac{5}{8} & -\frac{1}{8} \\ -\frac{9}{64} & -\frac{27}{64} & \frac{37}{64} \end{pmatrix}$$

Inverting **I** − **Q** (by computer or otherwise) we obtain

$$\begin{pmatrix} 2.1793 & 0.6207 & 0.1931 \\ 1.6552 & 2.3448 & 0.5517 \\ 1.7379 & 1.8621 & 2.1793 \end{pmatrix}$$

as we found earlier (Table 3.7). Note that every $\bar{s}_{ij}$ appears in $(\mathbf{I} - \mathbf{Q})^{-1}$ in its 'natural' place, that is the same place as the corresponding p_{ij} appears in **Q**.

Mean absorption or fixation times can, of course, be obtained by summing the elements in the appropriate row of $(\mathbf{I} - \mathbf{Q})^{-1}$. In fact, even if we were not interested in sojourn times at all, but merely wished to find mean absorption or fixation times, it is still easiest to find mean sojourn times first and then sum. As we shall see, a similar oddity of technique appears in large-population theory.

Notes and exercises

1 *Feller's formula for the latent roots.* We follow Ewens (1969). In the sequel, all matrices will have $(2N + 1)$ rows, $(2N + 1)$ columns; it will be convenient to number rows or columns $0, 1, 2, 3, \ldots, 2N$ (rather than the conventional $1, 2, 3, \ldots, 2N + 1$). We follow the usual notation, in which the element in the ith row, jth column of our matrices **P**, **A**, **Z** (defined below) is denoted p_{ij}, a_{ij}, z_{ij}, respectively.

Let p_{ij} be the probability of going from allele frequency $i/(2N)$ in the parental generation to frequency $j/(2N)$ in the progeny generation; i, j take any value from 0 to $2N$ inclusive. We wish to find the latent roots of **P**, defined as the matrix with our p_{ij} in ith row, jth column.

Let **Z** be a matrix that is non-singular, that is the determinant $|\mathbf{Z}| \neq 0$. Let

$$\mathbf{A} = \mathbf{Z}^{-1}\mathbf{P}\mathbf{Z}$$

Using standard properties of the binomial distribution and arguing as above, we find that all elements below the leading diagonal of **A** are zero and that, for $j > 1$, the diagonal element a_{jj} equals

$$\left(1 - \frac{1}{2N}\right)\left(1 - \frac{2}{2N}\right)\cdots\left(1 - \frac{j-1}{2N}\right)$$

so that ($*$ standing for terms that we need not bother to calculate) **A** has the form

$$\begin{pmatrix}
1 & * & * & * & * & \cdots \\
0 & 1 & * & * & * & \cdots \\
0 & 0 & 1 - \dfrac{1}{2N} & * & * & \cdots \\
0 & 0 & 0 & \left[1 - \dfrac{1}{2N}\right]\left[1 - \dfrac{2}{2N}\right] & * & \cdots \\
0 & 0 & 0 & 0 & \left[1 - \dfrac{1}{2N}\right]\left[1 - \dfrac{2}{2N}\right]\left[1 - \dfrac{3}{2N}\right] & \cdots \\
0 & 0 & 0 & 0 & 0 & \cdots \\
\cdot & \cdot & \cdot & \cdot & \cdot & \cdots \\
\cdot & \cdot & \cdot & \cdot & \cdot & \cdots \\
\cdot & \cdot & \cdot & \cdot & \cdot & \cdots
\end{pmatrix}$$

We subtract λ from every term on the leading diagonal of **A**, giving us a new matrix $(\mathbf{A} - \lambda\mathbf{I})$, where **I** is the unit matrix. By definition, the latent roots of **A** are the values of λ satisfying

$$|(\mathbf{A} - \lambda\mathbf{I})| = 0$$

This equation is very easily written down in detail, owing to the zeros in **A**. For, by the usual rules for evaluating a determinant, a matrix of form

$$\begin{pmatrix}
a_{00} - \lambda & * & * & * & * & \cdots \\
0 & a_{11} - \lambda & * & * & * & \cdots \\
0 & 0 & a_{22} - \lambda & * & * & \cdots \\
0 & 0 & 0 & a_{33} - \lambda & * & \cdots \\
0 & 0 & 0 & 0 & a_{44} - \lambda & \cdots \\
0 & 0 & 0 & 0 & 0 & \cdots \\
\cdot & \cdot & \cdot & \cdot & \cdot & \cdots \\
\cdot & \cdot & \cdot & \cdot & \cdot & \cdots
\end{pmatrix}$$

($*$ = anything), with all zeros below the leading diagonal, must have determinant

$$(a_{00} - \lambda)(a_{11} - \lambda)(a_{22} - \lambda)(a_{33} - \lambda)(a_{44} - \lambda)\cdots$$

and equating this to zero, we have the latent roots of $\mathbf{A}$

$$\lambda_0 = a_{00} = 1$$

$$\lambda_1 = a_{11} = 1$$

$$\lambda_2 = a_{22} = \left(1 - \frac{1}{2N}\right)$$

$$\lambda_3 = a_{33} = \left(1 - \frac{1}{2N}\right)\left(1 - \frac{2}{2N}\right)$$

$$\lambda_4 = a_{44} = \left(1 - \frac{1}{2N}\right)\left(1 - \frac{2}{2N}\right)\left(1 - \frac{3}{2N}\right)$$

and generally, for $r > 1$,

$$\lambda_r = a_{rr} = \left(1 - \frac{1}{2N}\right)\left(1 - \frac{2}{2N}\right)\left(1 - \frac{3}{2N}\right)\cdots\left(1 - \frac{r-1}{2N}\right)$$

which, taken in conjunction with $\lambda_0 = 1$, $\lambda_1 = 1$, is the same as

$$\lambda_r = \frac{(2N)!}{(2N - r)!}\left(\frac{1}{2N}\right)^r \qquad (\text{for } r = 0, 1, 2, \ldots, 2N)$$

which is Feller's formula.

2 *A simple method for finding the dominant latent root.* The argument in note 1 was very long; it is interesting, therefore, that, provided we assume that there is a dominant latent root, this root is easily found by elementary methods. Wright (1931) gave an intuitive argument in favour of this assumption; for a rigorous discussion, see Karlin & Taylor (1975, Appendix 2).

Let $f(p, t)$ be the probability that the allele frequency is p in generation t. Consider $Ep(1 - p)$. By definition of mean, this is

$$\sum_p p(1 - p)f(p, t)$$

CHANGES IN ALLELE FREQUENCY

Hence, approximately,

$$\sum_{s=0}^{2N-1} p_{sr} = \frac{(2N)!}{r!(2N-r)!} 2N \frac{(2N-r)!r!}{(2N+1)!}$$

$$= \frac{2N}{2N+1} = \frac{1}{1 + 1/(2N)} = \left(1 + \frac{1}{2N}\right)^{-1}$$

$$= \left(1 - \frac{1}{2N}\right) + \text{(terms negligible when } N \text{ is at all large)}$$

as we hoped to show. Thus, in large populations, when t is very large, the probability $f(p, t)$ that the allele frequency takes value p in generation t is well approximated by

$$C\left(1 - \frac{1}{2N}\right)^t$$

for all p (apart from $p = 0$ or 1). C is now chosen so that our formula will give (approximate) *absolute* probabilities. We have, by definition of mean,

$$Ep(1-p) = \sum_{p=1/(2N)}^{1-1/(2N)} p(1-p)C\left(1 - \frac{1}{2N}\right)^t$$

$$= \left(1 - \frac{1}{2N}\right)^t C \frac{1}{(2N)^2} \sum_{i=1}^{2N-1} i(2N - i)$$

where $i = 2Np$. From the standard formulae

$$\sum_{i=1}^{n} i = \tfrac{1}{2}n(n+1) \qquad \sum_{i=1}^{n} i^2 = \tfrac{1}{6}n(n+1)(2n+1)$$

we have

$$Ep(1-p) = \left(1 - \frac{1}{2N}\right)^t C\tfrac{1}{6}\left(2N - \frac{1}{2N}\right)$$

$$\simeq \left(1 - \frac{1}{2N}\right)^t \frac{2NC}{6} \qquad \text{when } N \text{ is large}$$

98

But we know that

$$\mathrm{E}p(1-p) = \left(1 - \frac{1}{2N}\right)^t p_0(1-p_0)$$

Therefore

$$C = \frac{6p_0(1-p_0)}{2N}$$

whence

$$f(p, t) = \frac{6p_0(1-p_0)}{2N}\left(1 - \frac{1}{2N}\right)^t$$

approximately, in large populations, when t is large.

In verifying Wright's uniform distribution, we approximated a sum by an integral. If (using the standard *Euler-Maclaurin formula*) we take account of the errors in this approximation, it turns out that the uniform distribution, while very accurate (for large populations) for intermediate allele frequencies, is rather less accurate for extreme frequencies, as Wright noted. We consider these in note 4.

4 *Probabilities for extreme allele frequencies* (Fisher 1930a, b). Note first that the correct probability distribution, while not strictly uniform, must still be symmetrical; in the notation of note 3

$$f_r = f_{2N-r} \qquad (r = 1, 2, \ldots, 2N-1)$$

For, given sufficiently large t, the relative values of the $f(p, t)$ are independent of the initial frequency. Hence we can take any initial frequency we like to help us determine these relative values. If, then, we take initial frequency 0.5, the symmetry of the distribution is obvious, since we are dealing with neutral alleles.

Let

$$\Phi(x) = f_1 x + f_2 x^2 + f_3 x^3 + \cdots + f_{2N-1} x^{2N-1}$$

Here x is a 'dummy' variable (any other letter not used in the discussion would do instead of x), which may take any value from 0 to 1 inclusive. The reader will readily verify that

$$f_r = \frac{1}{r!}\left[\frac{d^r \Phi(x)}{dx^r}\right]_{x=0}$$

CHANGES IN ALLELE FREQUENCY

is just $\Phi(x)$ with x replaced by e^{x-1} and may thus be written

$$\Phi(e^{x-1})$$

Hence, provided we confine ourselves to r small, N very large, we have the very accurate equation

$$\Phi(e^{x-1}) = \Phi(x) + \tfrac{1}{2}$$

This equation is due to Fisher, who solved it, very elegantly, as follows. Let u_n be defined by

$$u_0 = 0 \qquad u_{n+1} = e^{u_n - 1}$$

On putting $x = u_n$, we have

$$\Phi(u_{n+1}) = \Phi(u_n) + \tfrac{1}{2}$$

Thus Φ increases by $\tfrac{1}{2}$ for unit increase in n; also, from our definition of $\Phi(x)$, $\Phi(u_0) = \Phi(0) = 0$. Hence

$$\Phi(u_n) = \tfrac{1}{2}n$$

Fisher showed that, to a close approximation,

$$\tfrac{1}{2}n = \frac{1}{1 - u_n} + \tfrac{1}{6}\log(1 - u_n) + \text{(other terms)}$$

these other terms being such as to disappear after a few successive differentiations (see Ch. 7 for details). Replacing u_n by x and ignoring the other terms for the moment, we have

$$\Phi(x) = \frac{1}{1 - x} + \tfrac{1}{6}\log(1 - x)$$

So

$$f_r = \frac{1}{r!}\left[\frac{d^r\Phi(x)}{dx^r}\right]_{x=0} = \frac{1}{r!}\left[\frac{r!}{(1-x)^{r+1}} - \frac{1}{6}\frac{(r-1)!}{(1-x)^r}\right]_{x=0}$$

$$= 1 - \frac{1}{6r}$$

NOTES AND EXERCISES

For very small r, it is best to take account of the other terms and also to make allowance for the fact that, even when these other terms are included, our expression for $\frac{1}{2}n$ is very slightly inaccurate when n is very small. Fisher gave the result:

r	f_r
1	0.818 203
2	0.916 762
3	0.944 923
4	0.958 266
5	0.966 634
6	0.972 225

Fisher noted that these f_r are very close to his

$$1 - 1/(6r)$$

which expression will certainly be adequate for $r > 6$ but still fairly small. However, if one aims at *complete* accuracy, it is not strictly correct to extend this expression to intermediate r, implying, on our scaling of the f_r, values of f_r *minutely less* than unity when r is intermediate. For, with all f_r *less* than unity, the f_r cannot add to $2N$. Ewens (1965) has shown that, for intermediate r, the f_r are *minutely greater* than unity, f_N being the largest. An exact but very complicated formula for the f_r is given by Khazanie & McKean (1966).

In summary, the departure from the uniform distribution is very small over most of the range of allele frequencies; only for very extreme frequencies is the departure noticeable, and even then it is fairly small. Now, if we attempt to derive our f_r by diffusion methods, we also arrive at the uniform distribution (see Ch. 10). Our lengthy discussion has given us an early opportunity to assess the reliability of results obtained by diffusion methods.

5. To become familiar with the methods described in this chapter, consider selfing in autotetraploids (Haldane 1930, Bartlett & Haldane 1934, Fisher 1949, 1965).

An inbred line is represented by one individual every generation. Write u, v, w, x, y for the probability that the individual has genotype $A_4, A_3a, A_2a_2, Aa_3, a_4$ respectively in generation t and u', v', w', x', y' for corresponding probabilities in generation $(t + 1)$.

4

Survival of new mutations: branching processes

> Full many a flower is born to blush unseen,
> And waste its sweetness on the desert air.
> Thomas Gray, *Elegy in a Country Churchyard*

On special methods

In the previous chapter, we gave methods yielding exact solutions to some critical problems. While the calculations may have seemed laborious, they are easily adapted for the computer, provided the population size is fairly small. For large populations, these calculations are beyond the computer's capacity and different methods are required. Usually, such 'large-population' methods yield approximate answers only. However, other things being equal, the inaccuracy should be smallest when the population size is very large. If, then, large-population methods give reasonably accurate results for *fairly small* populations, for which exact results are available, we can be confident *a fortiori* that these methods are satisfactory for large populations; sometimes this can be confirmed in other ways. In practice, the accuracy varies with the problem under consideration. Hence a systematic checking, problem by problem, of results is essential.

It is sometimes worthwhile to develop special methods for a restricted (but biologically relevant!) set of circumstances, in the interests of greater simplicity or accuracy (or both). For example, the genetic conservationist is principally concerned with preserving alleles that are *rare* in his source material, such as alleles conveying resistance to a strain of pathogen that has not yet arisen; he or she wishes to design a conservation programme that minimizes the loss of such initially rare alleles during the programme. The evolutionist is often particularly interested in the fate of alleles that have just arisen by mutation and are thus necessarily rare at the start (generation 0). The development of a special theory to deal with the future of an allele, A say, which is rare initially, would, therefore, appear justified, provided of course such a theory conveys some advantages not provided by a more general theory. No doubt readers will form their own views on this in due course.

Probability of survival

The argument is most easily set out for the case of neutral alleles, so we shall discuss these first. We consider the fate of a neutral allele A, initially rare in a large (or fairly large) random mating population. We ask: 'What is the probability that A is still present in the population in a given later generation?'

The need for accuracy

Obviously, we should try to obtain as accurate an answer as possible for this probability of survival. The need for accuracy is particularly acute when designing programmes for genetic conservation. Suppose we wish to conserve an allele, A say, which is rare in our source material. In the case of cereals, seed obtained from source can be stored for a very long time, but for many crops seed has a limited life, even under the best-known storage conditions. Thus at intervals (say every 5-10 years) it is necessary to raise plants to supply new seed. Let us designate the generation of plants, raised directly from seed obtained from source, generation 0. If u is the frequency of A in the source seed and N plants are raised per generation, then for an outbreeding species the probability that R A alleles appear in generation 0 is

$$\frac{(2N)!}{R!(2N-R)!}(1-u)^{2N-R}u^R$$

provided plants are raised in such a way that differential survival of genotypes is negligible. For every value of R, we wish to calculate the probability, $P(R, t)$ say, that A is present in generation t. The overall probability that A is present in that generation, taking account of loss of A both in producing generation 0 and in producing later generations, is, then

$$\sum_{R=0}^{2N} \frac{(2N)!}{R!(2N-R)!}(1-u)^{2N-R}u^R P(R, t)$$

The conservationist requires an accurate value of this, and hence of $P(R, t)$, particularly for the early generations. We shall, therefore, give particular attention to the accuracy of results.

Now, this lag of one generation is inconsequential if, as is usually the case, the probability that the frequency of $A = 0$ (or 1) changes very slowly from one generation to the next, so that the probability for generation $(t + 1)$ is almost the same as for generation t. A study of special cases shows that, unless the population is rather small, rapid changes will occur in the early generations when the initial frequency of A is very small, but not otherwise. Hence if A is rare initially, the diffusion result will be essentially correct for the later but not for the earlier generations (for very small populations, this conclusion holds whatever the initial frequency – see Ewens 1963b).

We now sum up this rather sketchy yet involved discussion. If $R (= 2Np_0)$ A alleles are present in generation 0, R being very small and N large, the probability that A is still present in generation t is about

$$1 - e^{-2R/t}$$

provided t, while large, does not exceed N. In the case $N_e \neq N$, the corresponding formula is, in fact

$$1 - e^{-(2R/t)N_e/N}$$

when t is sufficiently large but does not exceed N_e. For smaller t, a different approach is required.

A fundamental simplification

It proves convenient to work in terms of the number of A alleles, rather than the proportion. Then

> probability that r A alleles are present in generation $(t + 1)$
> $= \Sigma\{$(probability that s A alleles are present in generation t)
> $\times$ [probability that r A alleles are present in generation $(t + 1)$ *conditional* on s A alleles being present in generation t]$\}$

the sum being taken over all possible values of s. The repeated use of this relationship, generation by generation, will supply the probability that r A alleles are present in any given generation; the probability that A is present in that generation is, then, $1 -$ (the probability that $r = 0$). In the previous chapter, we gave a procedure equivalent to this, but as we have noted, that procedure is not practicable when the population size is large. However, in latter case we can introduce a very helpful simplification. We consider first

A FUNDAMENTAL SIMPLIFICATION

the case $N_e = N$. Our *conditional* probability is then given, strictly speaking, by the binomial distribution. But the population size, and hence the total number ($2N$) of alleles at the locus under study, is large. Provided then that s is small we may, in the familiar manner, approximate this binomial distribution by the corresponding Poisson distribution. Then if s A alleles are present in generation t, A being neutral, the mean number of A alleles in $(t+1)$ will also be s and the probability that r A alleles are present in generation $(t+1)$ may be approximated by

$$e^{-s}s^r/r!$$

a decidedly simpler expression than the corresponding binomial probability.
It will prove helpful to verify this standard result, as follows.

Probability generating functions

We consider a simple standard method for representing any (discontinuous) probability distribution. Let π_r be the probability that the random variable concerned takes value r. Write

$$f(x) = \sum_r \pi_r x^r = \pi_0 + \pi_1 x + \pi_2 x^2 + \cdots$$

Here x is a 'dummy' variable; any letter could be used instead, apart of course from π or r. For convenience, we take $-1 \leqslant x \leqslant 1$. We see that π_r is the coefficient of x^r; $f(x)$ is known as the 'probability generating function' (pgf) for the distribution. Note the important property

$$\pi_0 = f(0)$$

the value of $f(x)$ when $x = 0$.

For many distributions, the pgf has a pleasing simplicity. Thus for the binomial distribution (in our usual notation)

$$\pi_r = \frac{(2N)!}{r!(2N-r)!} q^{2N-r} p^r$$

so that

$$f(x) = \sum_{r=0}^{2N} \frac{(2N)!}{r!(2N-r)!} q^{2N-r} p^r x^r = \sum_{r=0}^{2N} \frac{(2N)!}{r!(2N-r)!} q^{2N-r} (px)^r$$

problems in physics, without knowledge that the same mathematics had been used in biology. According to Fisher (personal communication) this happened because 'they did not think they had anything to learn from an ignorant country bumpkin'. We shall not make the same mistake but will pay due attention to Fisher and Haldane's rustic native wit.

Now, as the total number of alleles at a locus is fixed at $2N$, there can never be independent propagation of alleles in the strict sense, since the more descendants left by one of the $2N$, the fewer descendants will be left by the others. However, as long as A is rare, such mutual interference between individual A alleles will be very slight and we may take the individual A alleles as propagating independently.

It is apparent, then, that our Poisson distribution is just one of a whole class of distributions for different cases of independent propagation. As we have shown, the Poisson is correct for independent propagation when $N_e = N$; we shall seek the correct distribution for some cases where $N_e \neq N$ a little later.

Utility of the branching process approach

In practice, the branching process approach is most useful when the initial number of A alleles, R say, is small and the population size is large. From elementary theory of changes under drift, we know that in large populations neutral alleles change in frequency very slowly. Thus the number of A will remain small, compared with $2N$, for a fairly large number of generations; hence the assumption of independent propagation will continue to hold (near enough) for a fairly large number of generations from the start. This is particularly true if we consider only small frequencies of A. Since large changes in frequency over a generation have negligibly small probability, the probability that the frequency of A is small in any given generation will depend, almost entirely, on the probability that the frequency was small one generation earlier. Thus the assumption of (near enough) independent propagation will continue to hold for small frequencies even when it would not be justified for the whole range of frequencies under consideration. Thus the branching process approach should give an accurate value for the probability that the frequency of A equals 0 and hence of the probability that A is present, at least if N is large and t, the number of generations from the start, is not too large. The larger the value of N, the greater will be the value of t for which accurate values will be given for the probability of survival. On the other hand, whenever we use an approximation, however close, for the conditional distribution, there is the possibility that the errors of approximation will gradually cumulate over successive generations, so that values may perhaps be inaccurate when t is sufficiently large. Particular care must be taken when using the Poisson approximation (or its analogues in cases where N_e differs from N) to calculate the probability that A survives *ultimately*. This procedure is justified in cases where the fate of A is essentially decided for good when A is still rare. This is not the case when A is neutral; thus in the neutral case, our procedure gives the probability of *ultimate* survival only for a population in which propagation of A alleles is always independent, that is an infinite population.

Fundamental equations for branching processes

Suppose we have just *one* A allele in generation 0. We write $f_t(x)$ for the pgf for the number of A alleles present in generation t. Then the probability, l_t say, that no A alleles are present in generation t equals $f_t(0)$. For the moment we consider our case $N_e = N$. Then, given one A in generation 0, we have for generation 1

$$f_1(x) = f(x) = e^{x-1}$$

and so

$$l_1 = f_1(0) = e^{-1} = 0.3679$$

Now consider later generations. Let $p_r(t)$ be the probability that r A alleles are present in generation t. Then by definition

$$f_t(x) = \sum_{r=0}^{\infty} p_r(t) x^r \qquad f_{t+1}(x) = \sum_{r=0}^{\infty} p_r(t+1) x^r$$

Now

$$p_r(t+1) = \sum_{s=0}^{\infty} p_s(t) \times e^{-s} s^r/r!$$

probability of $\quad$ probability $\quad$ probability of r in
r in $(t+1)$ $\quad$ of s in t $\quad$ $(t+1)$ conditional
$\qquad\qquad\qquad\qquad\qquad\qquad$ on s in t

Therefore

$$f_{t+1}(x) = \sum_{r=0}^{\infty} \sum_{s=0}^{\infty} p_s(t) e^{-s} \frac{s^r}{r!} x^r$$

The terms involving r are

$$(sx)^r/r!$$

to be summed over all r from 0 to ∞, giving e^{sx}. Therefore

$$f_{t+1}(x) = \sum_{s=0}^{\infty} p_s(t)e^{-s}e^{sx} = \sum_{s=0}^{\infty} p_s(t)e^{s(x-1)}$$

But $e^{x-1} = f(x)$ so that $e^{s(x-1)} = [f(x)]^s$. Hence

$$f_{t+1}(x) = \sum_{s=0}^{\infty} p_s(t)[f(x)]^s$$

which is the same as

$$\sum_{r=0}^{\infty} p_r(t)[f(x)]^r$$

since we have just changed notation. We have the very simple conclusion: to get $f_{t+1}(x)$ take $f_t(x)$ and replace x by $f(x)$.

Although we have proved this only for the Poisson, a simple modification of our proof (note 2) shows that our conclusion holds for all cases of independent propagation.

An alternative formulation, very convenient for numerical calculation, is obtained as follows. We have

$$f_1(x) = f(x)$$

that is, f_1 and f are the same function. From our conclusion above, then,

$$f_2(x) = f_1(f(x)) = f(f_1(x))$$

and further, using this last result, we find $f_3(x)$ by substituting $f(x)$ for x in

$$f_2(x) = f(f_1(x))$$

to give

$$f(f_1(f(x)))$$

But we have shown that

$$f_1(f(x)) = f_2(x)$$

Hence

$$f_3(x) = f(f_2(x))$$

Similarly

$$f_4(x) = f(f_3(x))$$

suggesting the general rule

$$f_t(x) = f(f_{t-1}(x))$$

that is, to get $f_t(x)$ take $f(x)$ and replace x by $f_{t-1}(x)$; this rule is easily verified by induction (note 3).

Calculating the probability of survival

We recall that l_t, the probability of 0 A alleles in generation t, equals $f_t(0)$; hence putting $x = 0$ in our last equation we obtain

$$l_t = f(l_{t-1})$$

and thus (remembering that $f(x) = e^{x-1}$) we have

$$l_1 = f(0) = e^{-1} = 0.3679$$

$$l_2 = f(l_1) = e^{0.3679-1} = 0.5315$$

$$l_3 = f(l_2) = e^{0.5315-1} = 0.6259$$

and so on (Fisher 1930b). In Table 4.1 we give values for the probability that A is still present, namely $1 - l_t$, for values of t from 1 to 10. For comparison, we give exact values for a diploid population size 25 (or haploid population size 50); in principle, these exact values cover the possibility that A is fixed, but in practice, for the values of t chosen, the probability of fixation is too small to affect the results given.

The accuracy of the branching process method is quite astonishing. Even with a population as small as 25, the relative error, although of course rising

Table 4.1 Probability that a neutral allele, appearing once in generation 0, is still present in a given later generation in a diploid population of size 25 ($N_e = N$)

Generation, t	Exact probability	Probability calculated by branching process method
1	0.6358	0.6321
2	0.4734	0.4685
3	0.3795	0.3741
4	0.3178	0.3121
5	0.2740	0.2681
6	0.2412	0.2352
7	0.2157	0.2095
8	0.1953	0.1890
9	0.1786	0.1723
10	0.1646	0.1582

as t increases, is remarkably small, reaching only 3.9% when $t = 10$. We can, therefore, be confident that for natural populations, for which N will normally be large, the branching process approach will yield accurate values for $(1 - l_t)$ for smallish t; for larger t the diffusion approach will be appropriate.

Particularly striking is the rapid loss of a newly arisen neutral mutant. The probability that our mutant is still present in generation 10, calculated by the branching process method, is only 0.1582; the reader will readily confirm that for generations 15 and 31 the corresponding probabilities are 0.1127 and 0.0589, respectively.

So far we have assumed that just one A is present initially. If R A alleles are present in generation 0, then, given independent propagation of the A's, the probability that A is still present in generation t equals

$$1 - (l_t)^R$$

This formula works surprisingly well, as long of course as t and R are not too large in comparison with N. Thus for $R = 5, t = 5$ our formula gives (taking l_t to six places for greater accuracy)

$$1 - (0.731\,923)^5 = 0.7899$$

whereas for $N = 25$ the exact result proves to be 0.8083; in fact, for the range $R = 1$ to 10, $t = 1$ to 10, the error never exceeds 4.5% of the true value when

Table 4.2 Probability that a neutral allele, appearing R times in generation 0, is still present in generation t, when $N_e = N$

	R				
t	1	2	3	4	5
1	0.6321	0.8647	0.9502	0.9817	0.9933
2	0.4685	0.7175	0.8499	0.9202	0.9576
3	0.3741	0.6082	0.7548	0.8465	0.9039
4	0.3121	0.5268	0.6745	0.7760	0.8459
5	0.2681	0.4643	0.6079	0.7130	0.7899

$N = 25$. Some numerical results are given in Table 4.2. Note again that the probability of survival falls rapidly.

So far, we have considered only the pgf for the Poisson distribution, corresponding to the case $N_e = N$. We have done this since the Poisson is familiar; however, our assumption $N_e = N$ is unrealistic, even for plants raised on an experimental field and very much more so for natural conditions. Our problem is to find the pgf for cases $N_e \neq N$: once this has been done, the rest of the argument we have developed will continue to hold.

The case $N_e < N$

It would be difficult to obtain a pgf covering all factors causing N_e to differ from N. We consider only the case where reproductive output depends on parental environment; probably this environmental variation is the main contributor to the discrepancy between N_e and N in natural populations.

The basic approach is to suppose that, *for any given mean*, the distribution of the number of descendants of a single A allele remains Poisson, but that the mean number of descendants depends on the environment.

Consider first a haploid population. We suppose that micro-environments are allocated to parents at random. Then the *mean* number of descendants of a single A allele, λ say, may be treated as a continuous random variable drawn from some distribution. Ideally, the latter should be found empirically, but the data available are much too scanty to permit this. Let us suppose that the distribution of λ has probability density function

$$c^c \lambda^{c-1} e^{-\lambda c} / \Gamma(c)$$

NEW MUTATIONS: BRANCHING PROCESSES

where c is fixed at some positive value and (as usual)

$$\Gamma(c) = \int_0^\infty y^{c-1} e^{-y} \, dy$$

y being a dummy variable. While, as required, our distribution has in fact a mean of 1, the choice of this particular distribution may seem very arbitrary. However, it will be apparent from Figure 4.1 that a very wide range of probability density functions can be obtained by appropriate choice of parameter c.

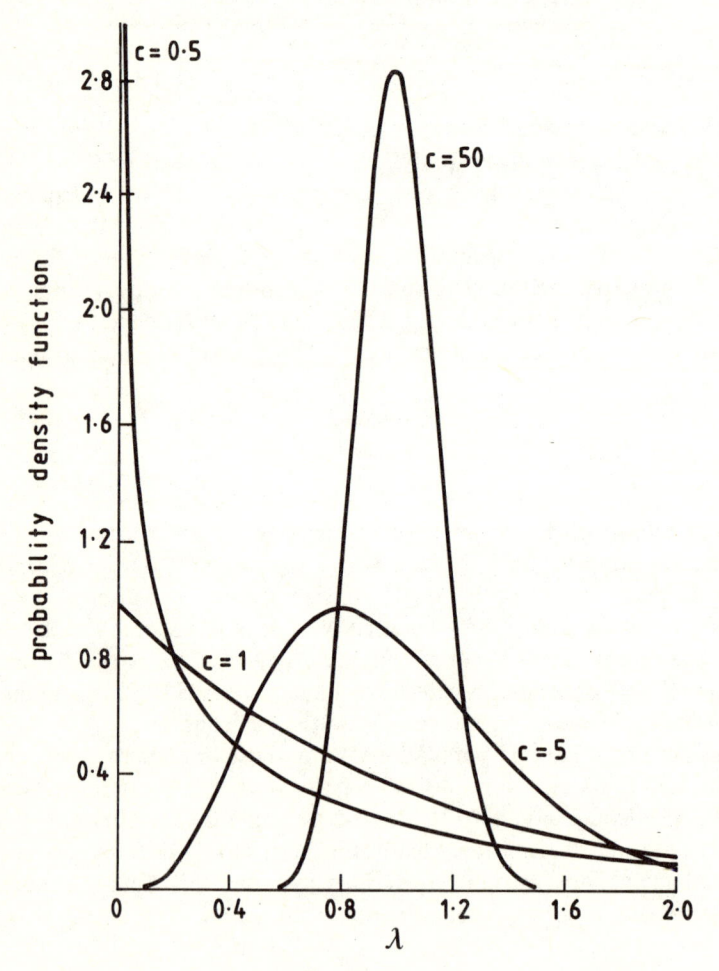

Figure 4.1 Some examples of the function $c^c \lambda^{c-1} e^{-\lambda c} / \Gamma(c)$.

THE CASE $N_e < N$

Combining our distribution of λ with the Poisson distribution for descendants conditional on λ, we obtain the probability that a single A gives r descendants

$$\int_0^\infty \frac{c^c \lambda^{c-1} e^{-\lambda c}}{\Gamma(c)} e^{-\lambda} \frac{\lambda^r}{r!} \, d\lambda$$

the integral appearing, rather than the sum, since λ is a continuous variable.

To obtain the corresponding pgf, $f(x)$ say, we multiply by x^r and sum over all r. The terms involving r in this sum will be

$$\sum_{r=0}^\infty (\lambda x)^r / r! = e^{\lambda x}$$

giving

$$f(x) = \int_0^\infty \frac{c^c \lambda^{c-1} e^{-\lambda c} e^{-\lambda} e^{\lambda x}}{\Gamma(c)} \, d\lambda = \frac{c^c}{\Gamma(c)} \int_0^\infty \lambda^{c-1} e^{-\lambda(c+1-x)} \, d\lambda$$

Making the substitution

$$y = \lambda(c + 1 - x)$$

and noting

$$d\lambda/dy = 1/(c + 1 - x)$$

we obtain, by the usual rules for integration by substitution,

$$f(x) = \frac{c^c}{(c+1-x)^c \Gamma(c)} \int_0^\infty y^{c-1} e^{-y} \, dy = \left(\frac{c}{c+1-x}\right)^c$$

since, as noted above, the integral equals $\Gamma(c)$ by definition.

Finally, let $a = c/(1 + c)$ and $b = 1 - a = 1/(1 + c)$, so that $c = a/b$. Our pgf can then be written

$$f(x) = \left(\frac{a}{a+b-bx}\right)^{a/b} = \left(\frac{a}{1-bx}\right)^{a/b}$$

since $a + b = 1$. This is a special case of the pgf

$$\left(\frac{a}{1-bx}\right)^k$$

121

where $a + b = 1$ but k is arbitrary. The corresponding distribution is known as the *negative binomial* distribution; its use in the present context is due to Moran (1961). Given our special case $k = a/b$, it may be shown (note 4) that the number of descendants of a single A has mean 1 (as required, since A is neutral) and variance $1/a$. Writing this variance as V(g), we obtain a simple interpretation of the 'mysterious' parameter a. For, from Chapter 2,

$$N_e = \tfrac{1}{2}(2N - 1)/V(g)$$

giving at once

$$a = 2N_e/(2N - 1)$$

which is very close to N_e/N if N is at all large.

Now consider diploids. In an individual of constitution AA, the two A's, being present in the same individual parent subjected to a particular microenvironment, will have the same value of λ; hence our assumption that λ's are allocated to individual A's independently at random no longer holds. However, as long as A is rare, this difficulty will not arise, since AA individuals will be very rare, and the theory just given continues to hold – for a more rigorous discussion of the diploid case, see Kojima & Kelleher (1962).

We now sum up. If the departure of N_e from N is due to variation in parental environment, the pgf for the number of descendants of a rare neutral allele is

$$\left(\frac{a}{1 - bx}\right)^{a/b}$$

where $a = N_e/N$ (unless N is very small) and $b = 1 - a$. In Table 4.3, we give numerical values for the case $a = 0.4$. Comparing the results with those in

Table 4.3 Probability that a neutral allele, appearing R times in generation 0, is still present in generation t, when $N_e = 0.4N$.

			R		
t	1	2	3	4	5
1	0.4571	0.7053	0.8400	0.9131	0.9528
2	0.2940	0.5016	0.6481	0.7516	0.8246
3	0.2162	0.3856	0.5184	0.6225	0.7041
4	0.1707	0.3123	0.4297	0.5271	0.6078
5	0.1410	0.2622	0.3662	0.4556	0.5324

Table 4.2, we see that a reduction of N_e/N from unity to 0.4 gives a marked increase in rate of loss of A. The reader is urged to repeat our calculations for some other values of a (answers for the case $a = 0.8$ are given in note 5). It will become apparent that the smaller the value of N_e/N, the more rapid is the loss of A. The rate of loss is very marked when N_e/N is small; thus when $a = 0.1$, the probability, given $R = 1$, that A is still present in generation 5 is only 0.0446.

A problem in genetic conservation

With our preceding calculations in mind, we take up again the problem of conserving a rare allele A in an *outbreeding* species. We note first that, when raising plants in any conservation programme, it is best to avoid natural selection whenever possible, since there is a serious risk that natural selection will lead to a marked decline in general variability. Further, it will rarely be the case that characteristics of the plant desirable to the breeder will coincide with characteristics favoured by natural selection; hence natural selection will normally lead to a deterioration in the economic qualities of the crop. The breeder, therefore, will take every opportunity to minimize natural selection; for example, plants are raised under optimum conditions to minimize the possibility of differential survival of genotypes.

How should adult plants be mated? If plants are allowed to open-pollinate, selection will occur owing to genetic differences in fecundity. Further, differences in fecundity, whether genetic or environmental, will affect the value of N_e, to an unknown extent in practice, since reliable estimates of N_e/N are very difficult to obtain. It is clear, from our earlier work, that we should try to make N_e as large as possible. From Chapter 2, we know that this is achieved when all parents contribute the *same* number of gametes to the next generation; in the notation of that chapter, we would then have $V(k) = 0$, giving

$$N_e = 2N - 1$$

Thus, both to avoid natural selection and to minimize chance loss of A, it is much the best (although admittedly time-consuming) to equalize the reproductive output of all parents. Every generation, plants are paired at random into $\frac{1}{2}N$ pairs; within every pair, plants are crossed reciprocally and one offspring raised from every parent.

Let us suppose then that, as a result of the breeder's efforts, natural selection is absent so that A is neutral (a small amount of selection at the A/a locus will not, in fact, affect our results materially – see below). If a parent has genotype AA, these two A's will not propagate independently; on our mating scheme, every parent contributes exactly two alleles to the next generation, so that the more of one of these A's contributed, the less of the other. However, as long as A is rare, AA homozygotes will be very rare, and we may regard all A's as propagating independently. An Aa parent contributes just two alleles to the next generation; of these, two are A with probability $\frac{1}{4}$, one A with probability $\frac{1}{2}$ and none A with probability $\frac{1}{4}$. Writing, as before, π_r for the probability that a single A has r descendant alleles one generation later, we have

$$\pi_0 = \tfrac{1}{4}, \quad \pi_1 = \tfrac{1}{2}, \quad \pi_2 = \tfrac{1}{4} \qquad \text{and } \pi_r = 0 \text{ for } r > 2$$

with corresponding pgf

$$f(x) = \tfrac{1}{4} + \tfrac{1}{2}x + \tfrac{1}{4}x^2$$

as long as A is rare. From this, we can show (Gale & Lawrence 1984) that, given $N = 25$, an allele with frequency 0.01 in the source seed will be present in generation $1, 2, \ldots, 5$ with probability 0.31, 0.26, 0.23, 0.20, 0.18, respectively, a rather discouraging result. For $N = 50$, the corresponding probabilities are 0.53, 0.46, 0.40, 0.36, 0.33. The details of the calculation are left to the reader as an exercise (note 6).

Probability of survival when t is fairly large

We return now to natural populations, for which N will normally be fairly large, so that the branching process approach will give reliable results for values of t considerably larger than those considered so far. Suppose just one A is present in generation 0; as before, let l_t be the probability that no A alleles are present in generation t. When $N_e = N$, we can of course continue to calculate l_t by repeated use of the formula

$$l_{t+1} = e^{l_t - 1}$$

but the following approach is more convenient.

Let $v_t = 1/(1 - l_t)$ so that $l_t = 1 - 1/v_t$. Since $l_0 = 0$, $v_0 = 1$; as t increases, v_t increases rather rapidly - for example $l_{10} = 0.8418$, giving $v_{10} = 6.3211$.

Substituting for l_{t+1} and l_t, we have

$$1 - \frac{1}{v_{t+1}} = e^{-1/v_t} = 1 - \frac{1}{v_t} + \frac{1}{2v_t^2} - \frac{1}{6v_t^3} + \frac{1}{24v_t^4} - \cdots$$

from the usual exponential series. Then

$$\frac{1}{v_{t+1}} = \frac{1}{v_t}\left[1 - \left(\frac{1}{2v_t} - \frac{1}{6v_t^2} + \frac{1}{24v_t^3} - \cdots\right)\right]$$

Taking reciprocals of both sides, we have

$$v_{t+1} = v_t\left[1 - \left(\frac{1}{2v_t} - \frac{1}{6v_t^2} + \frac{1}{24v_t^3} - \cdots\right)\right]^{-1}$$

Putting the expression in parentheses equal to y and using the standard mathematical result

$$(1 - y)^{-1} = 1 + y + y^2 + y^3 + \cdots$$

(valid when the absolute value of y is less than unity) we find, after some simple algebra,

$$v_{t+1} = v_t\left(1 + \frac{1}{2v_t} + \frac{1}{12v_t^2} + \cdots\right)$$

In fact, the terms in $1/v_t^3$ cancel and all other terms omitted are very small even when v_t is as small as 1. Hence the excellent approximation

$$v_{t+1} = v_t + \tfrac{1}{2} + 1/(12v_t)$$

Now the third term on the right-hand side is never very large, even for small t. Roughly speaking, v_t increases by $\tfrac{1}{2}$ every generation. Since $v_0 = 1$, we have then the approximation

$$v_t = 1 + \tfrac{1}{2}t$$

An improved approximation, very accurate when t is large, due to Fisher, is

$$v_t = \tfrac{1}{2}t + \tfrac{1}{6}\log t + 0.899\ 144$$

For $t = 15$, this gives $v_t = 8.8505$, $1 - l_t = 0.1130$ whereas the exact values are $v_t = 8.8770$, $1 - l_t = 0.1127$. However, the logarithmic term increases very slowly with t; for large t, either approximation is dominated by the term $t/2$ and the probability that A is still present, namely

$$1 - l_t = 1/v_t$$

will be well approximated by $2/t$ (Fisher 1930b). The more general formula

$$2N_e/tN$$

follows directly from the work of Kolmogorov in 1938 (quoted in Harris 1963); see note 7.

If R A alleles are present initially, R being very small, the probability that A is present will, therefore, be

$$1 - \left(1 - \frac{2}{t}\frac{N_e}{N}\right)^R$$

or about

$$1 - e^{-(2R/t)N_e/N}$$

(since, t being large, $2/t$ is small). The reader will recall that this is essentially the formula given by diffusion methods, for $t \leqslant N_e$. Hence there is no difficulty in deciding which method to use when calculating the probability of survival of an allele, which initially is very rare, in a natural population. We use branching process methods for the early generations, until t is large enough for results to agree with those obtained from diffusion methods, and use the latter for all larger t.

Pattern of survival in natural populations

The outcome is as follows. When the initial number of A alleles is very small, the probability of survival falls rapidly in the early generations; later this probability declines slowly but steadily. Of course, the probability of ultimate survival is p_0, as we showed in Chapter 3. We stress that a large majority of newly arisen mutants will be lost in the early generations, particularly when N_e/N is small.

A caution

So far, we have concentrated on the probability that the number of A alleles, s say, is zero. We could, however, attempt to calculate the complete probability distribution of s by branching process methods. Since, however, we are now considering *all* values of s, this approach will be valid for a more restricted range of t than was appropriate for the special case $s = 0$. It may be shown, from a study of the moments, that, for reliable results, t must be substantially less than $2N_e$.

Fisher (1930b) considered the case $R = 1$, $N_e = N$, t large, and showed that, under these circumstances, the distribution of s could be written in very simple form. However, this is valid only when t, while large, is still very much less than $2N$. Since t is large, Fisher's result should be obtainable by diffusion methods; this is indeed the case. We shall not, therefore, pursue the matter further at this stage.

Genic selection

We now discuss the case where A conveys a selective advantage. For simplicity, we suppose that the advantage lies in superior viability. We confine ourselves to the case where just two kinds of allele, A and a, are present at a given locus.

Consider first a haploid species. For consistency with our notation for diploids, we continue to write $2N$ for the total number of alleles at a locus, so that $2N$ rather than N is the population size in haploids. We have then:

	A	a	Total
number in generation t	s	$2N - s$	$2N$
viability	$1 + \alpha$	1	
mean relative number in generation $(t + 1)$	$s(1 + \alpha)$	$2N - s$	$2N + \alpha s$

Strictly, since these mean relative numbers add to $(2N + \alpha s)$, we should multiply them by a factor $2N/(2N + \alpha s)$ to obtain numbers adding to $2N$ in $(t + 1)$, but if N is large and s small our factor will be very close to unity and the mean number of A in $(t + 1)$ can be taken as

$$s(1 + \alpha)$$

If then $s \ll 2N$ and $N_e = N$, the probability that r A alleles appear in $(t + 1)$ will be well approximated by

$$e^{-s(1+\alpha)}[s(1 + \alpha)]^r/r!$$

In the case $s = 1$ this reduces to

$$e^{-(1+\alpha)}(1 + \alpha)^r/r!$$

with corresponding pgf

$$f(x) = \sum_{r=0}^{\infty} e^{-(1+\alpha)} \frac{(1+\alpha)^r}{r!} x^r = e^{-(1+\alpha)} e^{(1+\alpha)x} = e^{(1+\alpha)(x-1)}$$

The argument now follows familiar lines. Suppose just one A is present in generation 0. The probability that no A alleles appear in generation 1 will be

$$l_1 = f(0) = e^{-(1+\alpha)}$$

and l_t, the probability of 0 A alleles in generation t, will be given, just as in the neutral case, by

$$l_t = f(l_{t-1})$$

provided, of course, that the A's are propagating independently.

Suppose, for example, that $\alpha = 0.1$. Then

$$l_1 = e^{-1.1} = 0.3329$$

$$l_2 = e^{1.1(0.3329-1)} = e^{-0.7338} = 0.4801$$

$$l_3 = e^{1.1(0.4801-1)} = e^{-0.5719} = 0.5644$$

If R A alleles are present initially, then, given independent propagation of the A's, the probability that A is still present in generation t equals

$$1 - (l_t)^R$$

In the case $N_e < N$, we write, following Moran (1961),

$$f(x) = \left(\frac{a}{1-bx}\right)^{a(1+\alpha)/b}$$

where $a + b = 1$. The reader who has studied note 4 will easily verify that the corresponding (negative binomial) distribution has mean $1 + \alpha$, variance $(1 + \alpha)/a$ (reducing to the Poisson variance, $1 + \alpha$, as $a \to 1$). Hence, if a single A is present in a given generation t, the number of A in $(t + 1)$ will have variance $(1 + \alpha)/a$. With s independently propagating A's in t, the corresponding variance will be $s(1 + \alpha)/a$ from the usual rule for the variance of the sum of independent variables; in the Poisson case, the variance will be $s(1 + \alpha)$. Since the ratio of these last two variances equals $1/a$, we may write $a = N_e/N$ (but see note 8).

GENIC SELECTION

Now consider diploids but think of the viability of alleles rather than of genotypes. Suppose s A alleles and $(2N - s)a$ alleles are present in generation t. By analogy with the haploid case, imagine A to have viability $1 + \alpha$, a viability 1. Then in $(t + 1)$ we have mean relative numbers:

A	a	Total
$s(1 + \alpha)$:	$2N - s$	$2N + \alpha s$

Given random mating and hence random association of alleles, the allele numbers just given will correspond to relative zygotic numbers:

	AA	Aa	aa
generation t	s^2 :	$2s(2N - s)$:	$(2N - s)^2$
generation $(t + 1)$	$s^2(1 + \alpha)^2$:	$2s(2N - s)(1 + \alpha)$:	$(2N - s)^2$

Hence, given viabilities:

AA	Aa	aa
$(1 + \alpha)^2$	$1 + \alpha$	1

we can, conveniently, suppose selection to be acting on alleles ('genic selection'); when this is so the treatment of haploids and diploids will be identical. The terms 'genic selection' and 'the haploid model' are used interchangeably.

Notice that, if α is small, we may ignore α^2 and write the viability of AA as $(1 + 2\alpha)$, so that genic selection is virtually the same as no dominance in fitness in this case. Otherwise, the viability $(1 + \alpha)^2$ for AA may seem very artificial. Suppose, however, that N is large. When A is rare, AA will be very infrequent, so that the viability of AA will be immaterial and can be written as $(1 + \alpha)^2$ without causing serious error. Hence we may suppose genic selection, irrespective of the viability of AA, provided A is rare.

In Table 4.4 we give survival probabilities, for generations 1 to 10, when one A is present initially. As expected, the larger the selective advantage α, the greater the change of survival. Nevertheless, even with α as large as 0.1, the probability of survival falls rapidly in the early generations; when α is small (0.01 or less), the fall is almost as rapid as in the neutral case. Particularly dramatic is the effect of a low value of N_e/N in reducing the probability of survival. We conclude that unless $\alpha N_e/N$ is quite large, the survival of a newly

Table 4.4 Probability that an advantageous allele, appearing once in generation 0, is still present in generation t; α = selective advantage.

	$\alpha = 0.1$		$\alpha = 0.05$		$\alpha = 0.01$	
t	$N_e/N = 1.0$	0.1	1.0	0.1	1.0	0.1
1	0.6671	0.2453	0.6501	0.2356	0.6358	0.2277
2	0.5199	0.1328	0.4947	0.1243	0.4738	0.1176
3	0.4356	0.0916	0.4051	0.0839	0.3803	0.0778
4	0.3807	0.0709	0.3465	0.0635	0.3190	0.0578
5	0.3421	0.0585	0.3050	0.0514	0.2754	0.0459
6	0.3136	0.0504	0.2740	0.0434	0.2428	0.0381
7	0.2918	0.0447	0.2500	0.0377	0.2175	0.0325
8	0.2745	0.0405	0.2309	0.0335	0.1972	0.0284
9	0.2607	0.0372	0.2153	0.0303	0.1806	0.0252
10	0.2493	0.0347	0.2023	0.0277	0.1667	0.0227

arisen advantageous mutant in these early generations is *almost entirely a matter of chance.* Clearly, this phenomenon will have a major effect on the rate of evolution under natural selection.

Probability of ultimate fixation

If we continue to apply the formula

$$l_t = f(l_{t-1})$$

generation after generation, to calculate l_t, we find that eventually l_t reaches a limiting value, l say, and remains unchanged thereafter. Let $y = 1 - l$. Then y is the probability of ultimate survival of our newly arisen advantageous mutant A, under genic selection, in an infinite population.

Consider now a finite population, with A, as before, a new advantageous mutant. Suppose that

viability of $AA \geqslant$ viability of $Aa >$ viability of aa

Under these circumstances, A and a cannot coexist indefinitely (see Ch. 3); ultimately A is either lost or fixed. Thus the probability of ultimate survival of A is also the probability that A is ultimately fixed. It is remarkable that, under a wide range of circumstances, this probability is given (to a close approximation) by our quantity y. This will be so, provided the fate of A is decided for good while A is still rare.

Suppose that, at some fairly early stage, A, having escaped extinction so far at least, has left D descendants in a given generation. While it would be difficult to give a rigorous proof at this stage of our discussion, the reader who has studied our notes to Chapter 1 will readily accept that, provided α is not too small, the probability of fixation will be close to unity when D is still fairly small, substantially less then $2N$ when N is large. The larger the value of α, the smaller will be this critical value of D. In practice, then, the branching process approach for calculating the probability of fixation will work best when the selective advantage is *large*. This is very convenient, since diffusion methods work best when evolution is slow (see Ch. 2) and are thus appropriate when selective advantages are *small*. Thus the two approaches are complementary – if α is small and N sufficiently large, the two approaches give the same result, as we show in Chapter 5.

Further, given that the future of A is settled while A is still rare, the viability of AA will have little effect on the outcome and we may suppose genic selection almost irrespective of the viability of that genotype. An important exception is the case where the advantageous allele is completely recessive in viability; obviously, we do not have independent propagation in that case. Intuitively, an abrupt break between this and the case where the advantageous allele is almost recessive seems unlikely and this latter case is best excluded at the moment.

To sum up: branching process methods give a reliable value for the probability of fixation in a large population provided the selective advantage is not too small (we exclude cases where the advantageous allele is recessive, or almost recessive, in viability). Results will be particularly accurate when the selective advantage is large.

In practice l_t approaches its limiting value fairly rapidly, especially when α is large. Thus random extinction of A will usually occur fairly soon after the mutation happens. If the mutant survives for a number of generations, it will normally increase in numbers and thus be comparatively safe from random extinction (Ewens 1969).

Calculating the probability of fixation

Since l_{t-1} and l_t have common limit l, the latter must satisfy the equation

$$l = f(l)$$

so we consider solutions of this equation. Since

$$f(1) = \pi_0 + \pi_1 + \pi_2 + \cdots = 1$$

NEW MUTATIONS: BRANCHING PROCESSES

$l = 1$ will be a solution. If, for any positive α, we plot the curves

$$Z = l \qquad Z = f(l) = e^{(1+\alpha)(l-1)}$$

as in Figure 4.2, we find that these curves intersect, of course, at $l = 1$ but also at a value of l between 0 and 1; thus the equation $l = f(l)$ will have a solution between 0 and 1, in addition to the solution $l = 1$, at least for the cases we have plotted. In fact, it follows at once from a standard result in the theory of branching processes that (given $\alpha > 0$) there will always be two solutions of

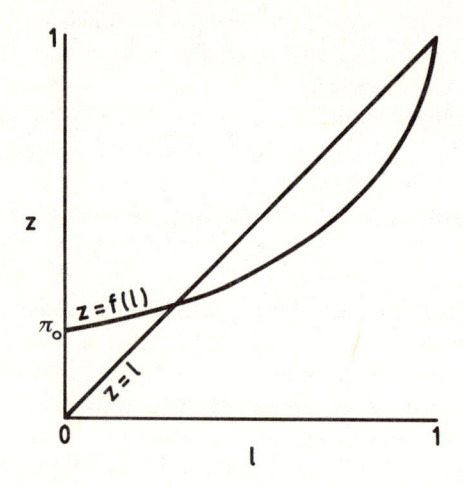

Figure 4.2 The curves $Z = l$, $Z = f(l) = e^{(1+\alpha)(l-1)}$ intersect at $l = 1$ and at a value of l between 0 and 1.

this kind, given any of the pgf's we have considered. We shall prove this result for the case

$$f(l) = e^{(1+\alpha)(l-1)}$$

for which (primes denoting differentiation)

$$f'(l) = e^{(1+\alpha)(l-1)}(1+\alpha) \qquad f''(l) = e^{(1+\alpha)(l-1)}(1+\alpha)^2$$

both of which are necessarily positive. In following the proof, it will help to refer repeatedly to Figure 4.2.

The curves $Z = l$, $Z = f(l)$ will certainly intersect for l between 0 and 1 if (a) $f(l)$ lies *above* l when $l = 0$ and (b) $f(l)$ lies *below* l when l is near 1. Now (a) must be true since $f(0) = \pi_0$, which is greater than 0. To prove (b), we note that $f'(1) = 1 + \alpha$, which is greater than 1. Thus $Z = f(l)$ is *steeper* than $Z = l$ at the point $l = 1$, where these two curves meet, which is possible only if (b) is correct (otherwise, at that point, either the slopes would be equal or $Z = f(l)$ would be less steep than $Z = l$, as the reader will see by drawing a few diagrams). Hence the curves intersect when $0 < l < 1$.

We must show that our curves intersect only once when $0 < l < 1$; to do this, we show that $Z = f(l)$ will have the same general shape as in Figure 4.2. The characteristic features of this shape are that the slope of the curve is *positive and increasing*; that is, both the slope, $f'(l)$, and the rate of change of the slope, $f''(l)$, are positive for all l within our range $0 \leq l \leq 1$. We have already noted that both of these are correct.

The argument is easily extended to pgf's in general (apart from some very special cases, which do not occur in genetics).

It is clear that the solution $l = 1$ is irrelevant. For, on genetical grounds, l_t cannot decrease with time; l_t, starting at value 0 in generation 0, will increase over successive generations until it reaches the *smallest* possible limiting value, that is, the *smallest* solution of $l = f(l)$ (this may be proved more formally). Thus l_t never approaches 1.

We see then that if we can find a *non-unit* solution of $l = f(l)$, this solution must be the probability of ultimate loss of A. Let $l = 1 - y$; then y is the probability of ultimate fixation of A. We seek the *non-zero* solution of

$$(1 - y) = f(1 - y)$$

In the case $N_e = N$, we have

$$(1 - y) = e^{-(1+\alpha)y}$$

and so

$$\log(1 - y) = -y(1 + \alpha)$$

Let $\log(1 + \alpha) = x$. Then

$$\log(1 - y) = -ye^x$$

Expanding in logarithmic and in exponential series, respectively, we have

$$-y - \tfrac{1}{2}y^2 - \tfrac{1}{3}y^3 - \tfrac{1}{4}y^4 - \tfrac{1}{5}y^5 - \cdots = -y\left(1 + x + \frac{x^2}{2!} + \frac{x^3}{3!} + \frac{x^4}{4!} + \cdots\right)$$

Then $y = 0$ (irrelevant) or

$$\tfrac{1}{2}y + \tfrac{1}{3}y^2 + \tfrac{1}{4}y^3 + \tfrac{1}{5}y^4 + \cdots = x + \tfrac{1}{2}x^2 + \tfrac{1}{6}x^3 + \tfrac{1}{24}x^4 + \cdots$$

NEW MUTATIONS: BRANCHING PROCESSES

Following Fisher (1930a), we shall express y as

$$y = Ax + Bx^2 + Cx^3 + \cdots$$

where $A, B, C, \ldots$ are constants to be determined. Inserting this expression into the equation immediately preceding, we have, after some very simple algebra,

$$\tfrac{1}{2}Ax + (\tfrac{1}{2}B + \tfrac{1}{3}A^2)x^2 + (\tfrac{1}{2}C + \tfrac{2}{3}AB + \tfrac{1}{4}A^3)x^3 + \cdots = x + \tfrac{1}{2}x^2 + \tfrac{1}{6}x^3 + \cdots$$

Equating coefficients of powers of x then gives

$$\tfrac{1}{2}A = 1 \qquad A = 2$$

$$\tfrac{1}{2}B + \tfrac{1}{3}A^2 = \tfrac{1}{2} \qquad \tfrac{1}{2}B = \tfrac{1}{2} - \tfrac{4}{3} \qquad B = -\tfrac{5}{3}$$

$$\tfrac{1}{2}C + \tfrac{2}{3}AB + \tfrac{1}{4}A^3 = \tfrac{1}{6} \qquad \tfrac{1}{2}C = \tfrac{1}{6} + \tfrac{20}{9} - \tfrac{8}{4} = \tfrac{7}{18} \qquad C = \tfrac{7}{9}$$

Using this method, with more terms included, we can find y to any desired accuracy. The expression

$$y = 2x - \tfrac{5}{3}x^2 + \tfrac{7}{9}x^3 - \tfrac{131}{540}x^4 + \tfrac{95}{1620}x^5 - \tfrac{771}{68040}x^6$$

will be accurate enough for almost all practical purposes.

Since $x = \log(1 + \alpha) = \alpha - \tfrac{1}{2}\alpha^2 + \tfrac{1}{3}\alpha^3 - \cdots$ then when α is small, so that terms in $\alpha^2, \alpha^3, \ldots$ may be neglected, we have the close approximation

$$y = 2\alpha$$

given by Haldane (1927).

The equation

$$(1 - y) = \left(\frac{a}{1 - b(1 - y)}\right)^{a(1+\alpha)/b}$$

$(a + b = 1)$, appropriate when $N_e < N$, may be solved in a similar manner, although terms become rather complicated for higher powers of x. The expression

$$y = 2ax - \tfrac{5}{3}a(2a - 1)x^2 + \tfrac{1}{9}a(46a^2 - 46a + 7)x^3$$
$$- \tfrac{1}{540}a(4024a^3 - 6036a^2 + 2274a - 131)x^4$$

is, however, very accurate for $\alpha \lesssim 0.6$ and fairly accurate for larger α up to 1. If very high accuracy is desired, our equation can be solved numerically. Some values for y are given in Table 4.5.

When α is small, our formula reduces to

$$y = 2a\alpha = 2\alpha N_e/N$$

(from Table 4.5, this simplified formula gives a good approximation to y when $\alpha \lesssim 0.05$). Thus, when the selective advantage is small, the probability of fixation is quite low, especially when N_e is substantially less than N. Even

Table 4.5 Probability y that an advantageous allele, appearing once in generation 0, is ultimately fixed; and number n of independent mutations required for overall probability of fixation to equal 0.95; α = selective advantage, N_e = effective population size, N = actual population size.

α	N_e/N	y	$\log(1-y)$	n
0.50	1.0	0.5828	−0.8742	3.4
	0.5	0.3888	−0.4923	6.1
	0.1	0.1049	−0.1108	27.0
0.10	1.0	0.1761	−0.1937	15.5
	0.5	0.0951	−0.0999	30.0
	0.1	0.0203	−0.0205	146.1
0.05	1.0	0.0937	−0.0984	30.4
	0.5	0.0488	−0.0500	59.9
	0.1	0.0101	−0.0101	296.6
0.01	1.0	0.0197	−0.0199	150.5
	0.5	0.0100	−0.0100	299.6
	0.1	0.0020	−0.0020	1497.9

when α is large, fixation is by no means certain. We ask, then: 'How many independent mutations to A are required for fixation of A to be a near-certainty (say, to have probability 0.95)?'

If n such mutations occur, the probability that all n are lost equals $(1-y)^n$ given independent propagation. Thus the probability that A is fixed equals

$$1 - (1-y)^n$$

For this to equal some chosen probability P,

$$(1 - y)^n = 1 - P$$

whence, on taking logarithms of both sides

$$n = \frac{\log(1 - P)}{\log(1 - y)}$$

When $P = 0.95$, $\log(1 - P) = -2.9957$, yielding the values for n given in Table 4.5.

In practice, it is unlikely that all n advantageous mutants will have the same DNA sequence; thus advantageous 'allele' A will normally be a class of advantageous alleles, individual members of which may differ in selective advantage α and hence probability of survival y. Suppose there are k distinct types of advantageous mutant at our locus. Let the ith type have advantage α_i and probability of survival y_i; let the probability that, when an advantageous mutant appears, this mutant is of type i be u_i ($\sum_i u_i = 1$). Then if n mutants appear, the probability that a_1 are of type 1, a_2 are of type 2, ..., a_k are of type k will be

$$\frac{n!}{a_1! a_2! \cdots a_k!} u_1^{a_1} u_2^{a_2} \cdots u_k^{a_k}$$

in which case the probability that all n are lost is

$$(1 - y_1)^{a_1}(1 - y_2)^{a_2} \cdots (1 - y_k)^{a_k}$$

Multiplying these two expressions together and then summing over all $a_1, a_2, \ldots, a_k$ such that $a_1 + a_2 + \cdots + a_k = n$, we obtain (by a standard mathematical result)

$$[u_1(1 - y_1) + u_2(1 - y_2) + \cdots + u_k(1 - y_k)]^n$$
$$= [1 - (u_1 y_1 + u_2 y_2 + \cdots + u_k y_k)]^n$$

Thus the probability that A is fixed equals

$$1 - (1 - \bar{y})^n$$

where $\bar{y}$ is the weighted mean of the y_i's, the weights being the u_i's. If all α_i's are *small*, $\bar{y}$ will be about the same as the y for a single mutant with α equal to the weighted mean of the α_i's.

Population subdivision

So far we have assumed one large population; in practice, the world population of the species will be divided into subpopulations. We consider the case where selective advantages are the same in all localities – see Pollak (1966) for a more general treatment. Suppose, in the first instance, that subpopulation sizes are constant and, while not necessarily equal, are all large, so that in any subpopulation A's propagate independently when A is rare locally. If A is nowhere common, all A's in the world will propagate independently. Thus all localities are equivalent as regards propagation and the probability of survival of any single A mutant and its descendants, if any, will be quite independent of where these are found. For example, this probability will be the same whether all descendants stay in the same locality or some migrate elsewhere. As usual, the survival of a line of descent stemming from a single A mutant will normally be decided fairly soon after that mutant first appears. In fact, the whole of our branching process theory for advantageous mutants will continue to apply and the probability of survival is calculated exactly as before. Now consider the probability of fixation. If no subpopulation is completely isolated from all others, a mutant that spreads *anywhere* will eventually spread *everywhere*; A's will be supplied to all localities by direct or indirect migration and, given sufficient time, enough A's will reach every locality to ensure local fixation (of course, with very restricted migration, it will take a long time for this to happen). Thus our probability of survival y is also the probability of fixation *in the whole species*. This being so, our n of Table 4.5 is the number of independent mutations to A required for fixation of A in the whole species to be a near-certainty ($P = 0.95$); further, it matters not at all where in the world individual mutations to A occur.

How many generations are required for n independent mutations to A to appear? If u is the rate of mutation to A and N is the world population size of the species, $2Nu$ new A mutants appear every generation. Hence our n mutants could appear quite rapidly; if so, the low probability of fixation for any *single* mutant is of little consequence (Fisher 1930b).

Our assumption of fixed, large local population sizes is not in fact necessary. Our conclusions would still follow even if local population sizes were small and changed by migration or subpopulation fusion or splitting, provided the haploid model held in every locality (Maruyama 1977); given that no subpopulation was completely isolated, branching process methods would still give the correct world probability of fixation if the *world* population size was large.

However, other changes in local population size (e.g. those due to changes in climate) need special consideration. As an extreme case, consider the situation (mentioned in Ch. 1) where local populations frequently become extinct and are replaced by individuals from elsewhere (Maruyama & Kimura 1980). Suppose a mutant becomes established in some subpopulation; this subpopulation could well become extinct before mutants have emigrated in sufficient quantity to ensure establishment in any other locality. Thus our statement that a mutant that spreads anywhere will eventually spread everywhere no longer holds. A combination of low migration rate and high rate of subpopulation extinction and replacement would very substantially reduce the probability of fixation of a single mutant. While N can still be taken as the world population size, the number of independent mutations for fixation of the mutant to be near-certain would be very large. Of course, this combination might be very unusual.

Cyclical variation in population size

Finally, we discuss, in outline, the effect of cyclical variation in population size; for simplicity, we consider an undivided population. Suppose the cycle extends over k generations; let the population size in the ith generation of the cycle be N_i (for $i = 1, 2, \ldots, k$) with $N_{i+k} = N_i$. Suppose further that progeny distributions are Poisson. We seek y_i, the probability that a mutant that first appears in generation i of the cycle is ultimately fixed.

We recall that, when the population size is fixed, the probability y of ultimate fixation is given by

$$-\log(1 - y) = (1 + \alpha)y$$

Ewens (1967a) has shown that, in the case of cyclical fluctuation, the corresponding equation is

$$-\log(1 - y_i) = (1 + \alpha)(N_{i+1}/N_i)y_{i+1}$$

with $y_{i+k} = y_i$; α is defined as before.

When $\alpha \leqslant 0.05$, the following solution will be appropriate unless fluctuations in population size are very extreme or k is quite large. Let N^* be the harmonic mean of the N_i's defined by

$$\frac{1}{N^*} = \frac{1}{k} \sum_{i=1}^{k} \frac{1}{N_i}$$

CYCLICAL VARIATION IN POPULATION SIZE

Then, approximately

$$y_i = 2\alpha \frac{N^*}{N_i} - \alpha^2 \frac{N^*}{N_i}\left[(k+1) + \frac{2(N^*)^2}{3k}\sum_{i=1}^{k}\frac{1}{N_i^2}\right.$$
$$\left. - \frac{2N^*}{k}\left(\frac{1}{N_{i+1}} + \frac{2}{N_{i+2}} + \cdots + \frac{k-1}{N_{i+k-1}}\right)\right]$$

When fluctuations are modest and k moderate, with $\alpha \leqslant 0.05$, the simple formula

$$y_i = 2\alpha N^*/N_i$$

will be adequate, as shown by Ewens, who argues further as follows. Let the probability that the mutant first appears in generation i be u_i. The overall probability of fixation is, then, the weighted mean $\bar{y}$ of the y_i's:

$$\sum_{i=1}^{k} u_i y_i$$

Then taking

$$u_i = N_i \bigg/ \sum_{i=1}^{k} N_i$$

we find at once, using the simple formula for y_i,

$$\bar{y} = 2\alpha N^*/\bar{N}$$

where $\bar{N}$ is the *arithmetic* mean of the N_i's, that is $\Sigma_{i=1}^{k} N_i/k$. Since (note 9) N^* is necessarily less than $\bar{N}$ (unless all the N_i's are equal), $\bar{y}$ must be *less* than the corresponding y ($= 2\alpha$) for fixed population size, although with moderate fluctuations the effect is not very marked. For example, the four-generation cycle with population sizes

$$N, \quad 2N, \quad 4N, \quad 2N$$

gives $N^*/\bar{N} = 0.79$. However, when fluctuations are very marked, the reduction in probability can be substantial. Consider, for example, the eight-generation cycle

$$N, \quad 10N, \quad 100N, \quad 1000N, \quad 100N, \quad 1000N, \quad 10N, \quad 1000N$$

with $\alpha = 0.05$. Values of y_i can be found using our formula for y_i in terms of α and α^2 or, more accurately, by a numerical method given by Ewens. They are

$$0.4955, 0.0652, 0.0064, 0.0006, 0.0058, 0.0006, 0.0531, 0.0005$$

giving

$$\bar{y} = 0.0014$$

which is very much lower than the corresponding probability 0.0937 for a large population of constant size (see Table 4.5).

We now sum up. We have discussed three factors that can substantially reduce the probability of fixation of a newly arisen advantageous mutant: (a) large variation in parental fecundity; (b) frequent subpopulation extinction and replacement, accompanied by low migration rates; and (c) very marked fluctuations in population size. If the combined effect of these factors gives a probability of fixation, following n independent mutations,

$$1 - \left(1 - 2\alpha \frac{N_e}{N}\right)^n \simeq (1 - e^{2\alpha n N_e/N})$$

for an appropriately defined N_e and we equate this to 0.95, we have approximately (taking $\log 0.05 = -3$)

$$n = 3N/(2\alpha N_e)$$

With mutation rate u, about $2Nug$ independent mutants will appear over the course of g generations. Thus $n = 2Nug$, whence

$$g = 3/(4N_e \alpha u)$$

Hence, if N_e is really large, g will be quite small and the delay in the advance of even a mildly advantageous mutant due to low mutation rate and chance loss of individual mutants will be trivial, as argued by Fisher. On the other hand, estimates of N_e come out surprisingly small. Suppose, for example, that $N_e = 3.3$ million, as estimated for *Drosophila melanogaster* by Kreitman (1983), $\alpha = 0.01$ and $u = 10^{-8}$; then $g = 2273$. It is clear, then, that if these low estimates of N_e are reliable (a point that is very uncertain), the delay will be substantial in cases where αu is really small. In fact, the average delay is about $1/(2\alpha N_e/N)$ independent mutations corresponding to $1/(4N_e \alpha u)$ generations. Unfortunately, lack of data prohibits detailed discussion of this crucial problem.

A note on epistasis and linkage

It may surprise the reader that we have discussed individual loci in isolation from others and have made no mention of epistasis or linkage. Beginners often suppose that the genetic background in the individual in which a mutation first appears is necessarily all-important for the probability of fixation of that mutant. To see if this is so, we consider, in a very intuitive way, a case discussed rigorously by Ewens (1967c).

Suppose an advantageous mutant B arises in a large random-mating population of bb individuals, this population being polymorphic at another locus A/a on the same chromosome as B/b. Let W be the average viability of bb at the time when the mutation occurs; let the frequencies of alleles A, a at that time be q, p respectively. To use our preceding theory, we would naturally define the average viability of Bb as

$$W^* = q^2(\text{viability of } AABb) + 2pq(\text{viability of } AaBb) + p^2(\text{viability of } aabb)$$

and write $(1 + \alpha) = W^*/W$ to give the probability of fixation as

$$2[(W^*/W) - 1]$$

provided $(W^*/W) - 1$ was small.

However, this would, at first sight, appear to be wrong. If B arises in a chromosome carrying A and no recombination has yet occurred, a Bb individual must have an AB chromosome, so that the only possible genotypes for Bb at this early stage are

$$AB/Ab \quad \text{and} \quad AB/ab$$

with relative frequency $q:p$ ($q + p = 1$), so that the average viability of Bb will be

$$W_1 = q(\text{viability of } AABb) + p(\text{viability of } AaBb)$$

Similarly, if B arises in a chromosome carrying a the early average viability of Bb will be

$$W_2 = q(\text{viability of } AaBb) + p(\text{viability of } aaBb)$$

Now, if $W_1 > W$ but $W_2 < W$ (or $W_2 > W$ but $W_1 < W$) so that B is (on average) advantageous or disadvantageous, depending on whether B arose in a chromosome containing A or containing a, our earlier branching process theory for one locus is obviously inappropriate. However, if $W_1 > W$ and $W_2 > W$ so that B is advantageous irrespective of the chromosome in which it arose, our one-locus theory works surprisingly well, even though W_1 and W_2 may well be unequal.

Consider first the case where our two loci recombine freely (or almost so). Then the association between B and A (or between B and a) is dissolved very rapidly so that the average viability of Bb approaches W^* very rapidly in either case. Thus we expect the probability of fixation to be close to

$$2[(W^*/W) - 1]$$

provided $(W^*/W) - 1$ is small, as is indeed the case.

Now consider very tight linkage. With selective advantages small, the probability of fixation will presumably be

$$2[(W_1/W) - 1] \qquad \text{if } B \text{ arose in an } A \text{ chromosome}$$

$$2[(W_2/W) - 1] \qquad \text{if } B \text{ arose in an } a \text{ chromosome}$$

Again, these intuitive formulae are correct, as shown by Ewens, who makes the following important point. While it is true that these formulae differ, neither is very relevant on its own. We must allow for the fact that, since A has frequency q and a has frequency p, B will arise in an A chromosome with probability q and in an a chromosome with probability p. Hence the overall probability of fixation of B is

$$2q[(W_1/W) - 1] + 2p[(W_2/W) - 1]$$

But it follows directly from the definitions of W_1, W_2, and W^* that

$$qW_1 + pW_2 = W^*$$

so that, with very tight linkage, our probability of fixation reduces to

$$2[(W^*/W) - 1]$$

the same formula as for very loose linkage.

To sum up: if a mutant is unconditionally advantageous, its probability of fixation is affected very little by epistasis or linkage and is well approximated by the probability given by single-locus theory (we have discussed only the case where selective advantages are small, but our conclusion appears to hold generally). However, if the mutant is only conditionally advantageous, the intensity of linkage becomes very important. For this and other problems soluble by branching process methods, see Ewens (1967c, 1969), Schaffer (1970) and Haigh & Maynard Smith (1972).

Spread of a fungal pathogen

Our discussion may have seemed, at times, rather abstract; it may help, therefore, if we mention a specific problem. Suppose we have a cultivated species susceptible to some haploid fungal pathogen and breed a new variety resistant to that pathogen. Eventually, the pathogen will mutate to a virulent form. It is often supposed (although not proved) that a non-virulent strain of pathogen produces a protein recognized by the host as foreign, the acquisition of virulence being a change leading either to failure to produce the protein or to production of an altered form that is no longer recognized (Ellingboe 1982). From data on loss of antigenic activity in fungi (Fincham *et al.* 1979), it seems unlikely that the rate of mutation to virulence would be smaller than 10^{-7}. Further, *established* strains of pathogen spread very rapidly. One might suppose, then, that by the time a new resistant variety had completed trials, a corresponding virulent strain of pathogen would already have become well established. In practice, of course, this is not always so; the spread of virulent strains is often delayed, allowing a useful commercial life for a new resistant variety. The problem, then, is to explain this delay; we shall make some suggestions that could be tested empirically.

It is easy to adapt our branching process theory to the situation envisaged (Gale 1987). We define a fungal individual as a mass of mycelium derived from a single spore; by a mature individual we mean an individual that has reached reproductive age. We define the fitness of a genotype as the mean number of mature progeny individuals per mature individual of that genotype. Note that, since the non-virulent fails completely on the resistant variety, it will be the *absolute* fitness of the virulent genotype that is relevant to its survival, in contrast to situations considered earlier, where relative fitness was appropriate. Finally, we note that opportunities for recombination will be absent or infrequent; even if the pathogen can reproduce sexually, many generations could pass in which reproduction was purely asexual. As a first approximation, we shall ignore recombination, if any, and treat the complete genotype of a virulent strain as if it were a single locus. We write the fitness of the ith virulent strain arising as $(1 + \alpha_i)$ and we use the branching process theory developed earlier (substituting fitness for viability).

How large is α_i likely to be? Arguing along Fisherian lines, we suggest that the initial alteration of the recognized protein, although conferring virulence, could be very crippling to the pathogen in other ways, leading perhaps to a very reduced growth or sporulation rate. Only after a fairly lengthy evolution of the virulent strain would the rapid rate of spread found in established virulents be attained (Parlevliet 1981). Now we know that the survival of a mutant is normally decided soon after that mutant first appears; hence it is the fitness of the *primordial* virulent rather than the fitness of the established virulent that is relevant to survival. Hence α_i, although presumably normally positive, could usually be quite small (if, owing to interaction of the mutant allele with the rest of the genotype, α_i were negative, loss of the ith virulent strain would be certain).

Further, all the factors that we have mentioned as delaying the establishment of an advantageous allele are likely to be operating, perhaps in a very extreme form. Possibly environmentally produced variation in fecundity is particularly marked. Certainly, very large fluctuations in population size occur in these pathogens. Finally, consider subpopulation extinction and replacement. When the crop is withdrawn at the end of the season, the pathogen may become extinct locally, to be replaced at a later stage from spores produced elsewhere.

It follows, then, that the probability of survival of a single mutant to virulence could be quite small. Correspondingly, our n, now defined as the number of independent mutations to virulence, each of which gives rise to a mature virulent individual, necessary for fixation of virulence to be near-certain, could be very large.

Of course, even when the number of mutations is large enough for the success of the virulent to be assured, it will take some time for the virulent to spread. This time could be much affected by our postulated section of modifiers that raise the fitness of the successful virulent.

Notes and exercises

1 Suppose one A is present in generation t. Let the pgf for the number of A's in generation $(t + 1)$ be

$$f(x) = \pi_0 + \pi_1 x + \pi_2 x^2 + \pi_3 x^3 + \cdots$$

Now suppose two, independently propagating, A's are present in t. We show that the pgf for r, the number of A's in $(t + 1)$, is

$$[f(x)]^2$$

Write the contributions of the two A's in t to the number of A's in $(t + 1)$ as (u, v), where u is the contribution of one A, v the contribution of the other A; $r = u + v$. Then any of the following will give r A's in $t + 1$:

$$(r, 0), \quad (r - 1, 1), \quad (r - 2, 2), \quad (r - 3, 3), \quad \ldots, \quad (0, r)$$

NOTES AND EXERCISES

Thus the probability that r A's appear in $(t+1)$ is

$$\pi_r\pi_0 + \pi_{r-1}\pi_1 + \pi_{r-2}\pi_2 + \pi_{r-3}\pi_3 + \cdots + \pi_0\pi_r$$

which is just the coefficient of x^r in

$$(\pi_0 + \pi_1 x + \pi_2 x^2 + \pi_3 x^3 + \cdots)(\pi_0 + \pi_1 x + \pi_2 x^2 + \pi_3 x^3 + \cdots)$$

$$= [f(x)]^2$$

(any reader having difficulty in seeing this should choose a small value of r, say 3, and note how the coefficient of x^r arises when the expressions in parentheses are multiplied out).

The argument easily extends to the case where s A's are present in t; in this general case the pgf is

$$[f(x)]^s$$

2 Show that $f_{t+1}(x) = f_t(f(x))$ for all cases of independent propagation.
Proof. Suppose one A is present in a given generation t. Write $f(x)$ for the pgf for the number of A's one generation later. From note 1, the corresponding pgf when s A's are present in t is

$$[f(x)]^s$$

Hence, if $p_r(t+1|s)$ is the probability that r A's are present in $(t+1)$, conditional on s A's being present in t,

$$\sum_r p_r(t+1|s)x^r = [f(x)]^s$$

since each of these is the pgf in $(t+1)$ when s A's are present in t.

Writing, as usual, $p_r(t+1)$ for the unconditional probability that r A's are present in $(t+1)$, we have

$$p_r(t+1) = \sum_s p_s(t)p_r(t+1|s)$$

$$f_{t+1}(x) = \sum_r \sum_s p_s(t)p_r(t+1|s)x^r$$

145

Summing over r gives $[f(x)]^s$ (see above), so that

$$f_{t+1}(x) = \sum_s p_s(t)[f(x)]^s$$

$$= f_t(f(x))$$

Note that this is true irrespective of the number of A alleles in generation 0.

3 If just one A is present in generation 0

$$f_1(x) = f(x)$$

Using this and $f_{t+1}(x) = f_t(f(x))$ prove by induction that

$$f_t(x) = f(f_{t-1}(x))$$

Proof. Note first that in generation 0, $\pi_1 = 1$ and $\pi_r = 0$ unless $r = 1$, so that $f_0(x) = x$. Thus when $t = 1$, the relationship to be proved reduces to $f_1(x) = f(x)$, which is certainly true.

Now suppose that for some t

$$f_t(x) = f(f_{t-1}(x))$$

Then

$$f_{t+1}(x) = f_t(f(x)) = f(f_{t-1}(f(x))) = f(f_t(x))$$

Hence, if the relationship to be proved is true for any given value of t, it must be true for (that value of t) + 1. As it is true for $t = 1$, it must therefore hold for all $t \geq 1$.

Note that the relationship does not hold if more than one A is present initially. With s A's present in generation 0

$$f_0(x) = x^s \qquad f_1(x) = [f(x)]^s = f_0(f(x))$$

which does not equal

$$f(f_0(x)) = f(x^s)$$

unless $s = 1$. Nevertheless, the case $s > 1$ initially is easily discussed, using the approach we have given in the main text.

NOTES AND EXERCISES

4 Consider a single A and let π_r be the probability that it has r descendants one generation later. Then we have (primes denoting differentiation)

$$f(x) = \pi_0 + \pi_1 x + \pi_2 x^2 + \pi_3 x^3 + \pi_4 x^4 + \cdots$$

$$f'(x) = \pi_1 + 2\pi_2 x + 3\pi_3 x^2 + 4\pi_4 x^3 + \cdots$$

$$f''(x) = 1 \times 2\pi_2 + 2 \times 3\pi_3 x + 3 \times 4\pi_4 x^2 + \cdots$$

If we put $x = 1$ after differentiation, we see that

$$f'(1) = \pi_1 + 2\pi_2 + 3\pi_3 + 4\pi_4 + \cdots$$

$$f''(1) = 1 \times 2\pi_2 + 2 \times 3\pi_3 + 3 \times 4\pi_4 + \cdots$$

so that, by definition of Er, E$[r(r-1)]$

$$\mathrm{E}r = f'(1) \qquad \mathrm{E}[r(r-1)] = f''(1)$$

Thus the mean number of descendants is $f'(1)$ and the variance of the number of descendants being

$$\mathrm{E}r^2 - (\mathrm{E}r)^2 = \mathrm{E}[r(r-1)] + \mathrm{E}r - (\mathrm{E}r)^2$$

must be

$$f''(1) + f'(1) - [f'(1)]^2$$

Suppose then that

$$f(x) = \left(\frac{a}{1 - bx}\right)^{a/b} \qquad \text{where } a + b = 1$$

Show that

$$f'(x) = a^{1/b}(1 - bx)^{-1/b} \qquad f''(x) = a^{1/b}(1 - bx)^{-(1/b) - 1}$$

and hence that the mean number of descendants is 1 and the variance of the number of descendants is $1/a$.

5 Given the pgf

$$f(x) = \left(\frac{0.8}{1 - 0.2x}\right)^4$$

NEW MUTATIONS: BRANCHING PROCESSES

calculate the probability that a neutral allele, appearing R times in generation 0, is still present in generation t, for $R = 1$ to 5, $t = 1$ to 5.

Answer

| | \multicolumn{5}{c}{R} |
t	1	2	3	4	5
1	0.5904	0.8322	0.9313	0.9719	0.9884
2	0.4234	0.6676	0.8083	0.8895	0.9363
3	0.3314	0.5529	0.7011	0.8001	0.8663
4	0.2726	0.4710	0.6152	0.7201	0.7964
5	0.2318	0.4099	0.5467	0.6518	0.7325

6 Given the pgf for our conservation scheme

$$f(x) = \tfrac{1}{4} + \tfrac{1}{2}x + \tfrac{1}{4}x^2$$

carry out calculations analogous to those in note 5.

Answer

| | \multicolumn{5}{c}{R} |
t	1	2	3	4	5
1	0.7500	0.9375	0.9844	0.9961	0.9990
2	0.6094	0.8474	0.9404	0.9767	0.9909
3	0.5165	0.7663	0.8870	0.9454	0.9736
4	0.4498	0.6973	0.8335	0.9084	0.9496
5	0.3992	0.6391	0.7832	0.8697	0.9218

In the case $N = 24$, $R = 1$, exact solutions (to four decimal places), obtained with considerable labour, for $t = 1$ to 4, are 0.7500, 0.6094, 0.5167, 0.4505 – clearly our simple approach works very well. On the other hand, diffusion methods give particularly inaccurate results for this scheme.

Compare results with those obtained in note 5. The latter, with $N_e/N \simeq 0.8$, is fairly realistic for a scheme in which plants are raised on an experimental field, allowed to open-pollinate and seed bulked.

NOTES AND EXERCISES

Returning to our conservation scheme, suppose $2N = 100$. Consider an allele with frequency 0.01 in the source seed. Then, from the binomial distribution, we have:

R	1	2	3	4	5
probability	0.3697	0.1849	0.0610	0.0149	0.0029

and the probability that our allele is still present in generation 5 is about

$$0.3697 \times 0.3992 + 0.1849 \times 0.6391 + \cdots + 0.0029 \times 0.9218 = 0.33$$

7 A rigorous proof of Kolmogorov's formula for the probability that A is still present when t is large is given in Harris (1963). Informally, we argue as follows. Let $f(x)$ be some arbitrary pgf. We have

$$l_{t+1} = f(l_t)$$

or, writing

$$l_t = 1 - 1/v_t$$

$$1 - \frac{1}{v_{t+1}} = f\left(1 - \frac{1}{v_t}\right) = f(1) - \frac{1}{v_t} f'(1) + \frac{1}{2v_t^2} f''(1) - \cdots$$

From note 4, $f'(1) = 1$ and the variance of the number of A descendants, $V(g)$ say, is

$$f''(1) + f'(1) - [f'(1)]^2 = f''(1)$$

since $f'(1) = 1$. Further, for any pgf

$$f(1) = \pi_0 + \pi_1 + \pi_2 + \pi_3 + \cdots = 1$$

Hence

$$1 - \frac{1}{v_{t+1}} = 1 - \frac{1}{v_t} + \frac{V(g)}{2v_t^2} - \cdots$$

$$\frac{1}{v_{t+1}} = \frac{1}{v_t}\left(1 - \frac{V(g)}{2v_t} + \cdots\right)$$

and so

$$v_{t+1} = v_t\left(1 - \frac{V(g)}{2v_t} + \cdots\right)^{-1} = v_t\left(1 + \frac{V(g)}{2v_t} - \cdots\right)$$

Hence, approximately,

$$v_{t+1} = v_t + \tfrac{1}{2}V(g)$$

Thus v_t increases by about $\tfrac{1}{2}V(g)$ every generation. Since $v_0 = 1$, we have approximately

$$v_t = 1 + \tfrac{1}{2}tV(g)$$

or about

$$\tfrac{1}{2}tV(g)$$

when t is large.

Hence if one A is present initially, the probability that A is present in generation t, $1 - l_t$, is about $2/tV(g)$, that is, about $2N_e/tN$.

Of course, in a rigorous discussion, we should have to show that the terms omitted make relatively little contribution when t is large.

8 In conventional treatments of variance effective size when selection is present, the intensity of selection is supposed small and the treatment of effective size in the neutral case is used unmodified. If α is small, our $(1 + \alpha)/a$ will be close to $1/a$ and the relationship $a = N_e/N$ follows from our discussion of the neutral case. For general α, we argue as follows. Let p be the frequency of A in generation t, p^* the resulting frequency of A in $(t + 1)$, calculated deterministically, and p' the actual frequency of A in $t + 1$. When the Wright-Fisher model applies, we have

$$V(p') = p^*q^*/(2N)$$

and in the general case we define N_e by

$$V(p') = p^*q^*/(2N_e)$$

Thus the ratio of these variances is N/N_e by definition. Of course, since in all cases $V(2Np') = (2N)^2V(p')$, we can just as well take the ratio of variance of allele number rather than allele frequency. Our statement $a = N_e/N$, then, amounts only to a definition of N_e, in cases where α is large; however, it gives a better feel for the situation to work in terms of N_e/N rather than the unfamiliar parameter a.

9 Given k numbers $N_1, N_2, \ldots, N_k$, prove that

$$\bar{N}\text{(the arithmetic mean)} \geq N^*\text{(the harmonic mean)}$$

with equality only when the N_i's are all equal.

Proof. For a set of k paired numbers $(x_1, y_1), (x_2, y_2), \ldots, (x_k, y_k)$ we have Cauchy's inequality

$$(\Sigma x_i y_i)^2 \leq \Sigma x_i^2 \Sigma y_i^2$$

with equality only when

$$x_i y_j = x_j y_i$$

for all i, j (this inequality is used to prove that the square of the correlation coefficient cannot exceed unity). Putting $x_i = \sqrt{N_i}$, $y_i = 1/\sqrt{N_i}$, so that $\Sigma x_i y_i = k$, gives

$$k^2 \leq \Sigma N_i \Sigma (1/N_i)$$

$$\frac{k}{\Sigma N_i} \leq \frac{1}{k} \Sigma \frac{1}{N_i}$$

that is

$$\frac{1}{\bar{N}} \leq \frac{1}{N^*}$$

so

$$\bar{N} \geq N^*$$

For equality, we must have $\sqrt{N_i}/\sqrt{N_j} = \sqrt{N_j}/\sqrt{N_i}$, i.e. $N_i = N_j$ for all i, j.

5
Probability of fixation: the more general case

> You who can calculate the course of a biased bowl,
> shall I come near the jack?
> What twist can counter the force
> that holds back
> woods I roll?
>
> Basil Bunting, *Briggflatts*

Fundamental equation

In the last chapter, we derived the probability of fixation for an advantageous allele, on the assumption that individual advantageous alleles were propagating independently. Naturally, we would like to drop this assumption; we could then hope to obtain the probability of fixation given an arbitrary initial frequency p_0 for the advantageous allele, an arbitrary population size N and an arbitrary degree of dominance in viability. However, such a completely general treatment is not available. While discarding the assumption of independent propagation, we shall have to impose some other restriction; for example, we shall often suppose that the selective advantage is small. For the moment, however, our treatment will be general.

Let p_0 be the frequency of the advantageous allele A in generation 0. Then (Feller 1957)

probability of ultimate fixation given frequency p_0 in generation 0

$= \sum_{x}$ [(probability of going from frequency p_0 in generation 0 to frequency x in generation)

× (probability of ultimate fixation given that we go to frequency x in generation 1)]

UNIQUENESS OF THE SOLUTION

Now the probability of *ultimate* fixation, for any *given* frequency of A, will not depend on the generation for which that frequency is given. If, then, we write $u(p_0)$ for the probability on the left-hand side of the expression above, we may write $u(x)$ for the second probability on the right-hand side. Let g_{p_0x} be the probability of going from frequency p_0 in one generation to frequency x one generation later. Our expression may then be written

$$u(p_0) = \sum_x g_{p_0x} u(x)$$

We shall call this the *fundamental equation* for fixation probability.

Note, however, that this equation would still hold if we *redefined* u to be the probability of ultimate loss, or even the probability of ultimate absorption, rather than the probability of fixation. We can, however, distinguish these by stating what happens when p_0 takes its 'boundary' (i.e. most extreme) values, 0 and 1. If $p_0 = 0$, fixation is impossible; if $p_0 = 1$, fixation has already happened and is thus certain. Hence when (as throughout this chapter) u is the probability of ultimate fixation, we supplement our fundamental equation with the 'boundary conditions'

$$u(0) = 0 \qquad \text{and} \qquad u(1) = 1$$

Suppose that g_{p_0x} (which, of course, will depend on the model, which in turn takes account of the pattern of selection) has been found for all p_0 and x. Then, quite remarkably, our fundamental equation, taken in conjunction with our boundary conditions, is sufficient for us to calculate the probability of fixation for any given p_0. In other words, our fundamental equation, when supplemented by the boundary conditions, has a *unique solution* (once p_0 is given), as we now prove.

Uniqueness of the solution

For simplicity, we write the fundamental equation in terms of allele number. Thus, if p_0 is, say, $i/(2N)$, we write u_i as short for $u(p_0)$; similarly g_{ij} (or $g_{i,j}$ when necessary to avoid confusion) stands for $g_{i/(2N), j/(2N)}$. Then when $p_0 = 1/(2N)$ we have

$$u_1 = g_{10}u_0 + g_{11}u_1 + g_{12}u_2 + \cdots + g_{1, 2N-1}u_{2N-1} + g_{1, 2N}u_{2N}$$

On inserting the boundary conditions $u_0 = 0$, $u_{2N} = 1$ we find

$$u_1 = g_{11}u_1 + g_{12}u_2 + \cdots + g_{1, 2N-1}u_{2N-1} + g_{1, 2N}$$

with similar equations for $u_2, u_3, \ldots, u_{2N-1}$ on the left-hand side. It may help any readers who have difficulty with matrix algebra to write out the equation for u_2 and that for u_3 as we have done for u_1. We have, then, a set of simultaneous equations, which (from the usual rules for matrix multiplication and addition) may be written in the form

$$\begin{pmatrix} u_1 \\ u_2 \\ u_3 \\ \vdots \\ u_{2N-1} \end{pmatrix} = \begin{pmatrix} g_{11} & g_{12} & g_{13} & \cdots & g_{1,2N-1} \\ g_{21} & g_{22} & g_{23} & \cdots & g_{2,2N-1} \\ g_{31} & g_{32} & g_{33} & \cdots & g_{3,2N-1} \\ \vdots & \vdots & \vdots & \cdots & \vdots \\ g_{2N-1,1} & g_{2N-1,2} & g_{2N-1,3} & \cdots & g_{2N-1,2N-1} \end{pmatrix} \begin{pmatrix} u_1 \\ u_2 \\ u_3 \\ \vdots \\ u_{2N-1} \end{pmatrix}$$

$$+ \begin{pmatrix} g_{1,2N} \\ g_{2,2N} \\ g_{3,2N} \\ \vdots \\ g_{2N-1,2N} \end{pmatrix}$$

or

$$\mathbf{u} = \mathbf{Q}\mathbf{u} + \mathbf{R}$$

where $\mathbf{u}, \mathbf{Q}, \mathbf{R}$ are the matrices shown. Thus

$$\mathbf{u} - \mathbf{Q}\mathbf{u} = \mathbf{R}$$

or, writing $\mathbf{I}$ for the unit matrix,

$$(\mathbf{I} - \mathbf{Q})\mathbf{u} = \mathbf{R}$$

From standard theory of equations, we know that, provided $(\mathbf{I} - \mathbf{Q})$ has an inverse, or (equivalently) that $|\mathbf{I} - \mathbf{Q}|$, the determinant of $(\mathbf{I} - \mathbf{Q})$, is not zero, our simultaneous equations have unique solutions

$$\mathbf{u} = (\mathbf{I} - \mathbf{Q})^{-1}\mathbf{R}$$

Now g_{ij}, the element in the ith row, jth column of $\mathbf{Q}$ is the probability of changing from frequency $i/(2N)$ to frequency $j/(2N)$ *one* generation later; in $\mathbf{Q}$, i and j take values from 1 to $(2N - 1)$, inclusive. Now consider the matrix $\mathbf{Q}^n$. It is easily shown (note 1) that the element in the ith row, jth column of $\mathbf{Q}^n$ is the probability of changing from frequency $i/(2N)$ to frequency $j/(2N)$ n generations later; as before i, j go from 1 to $(2N - 1)$. But ultimately absorption is certain, so that in the limit, as $n \to \infty$, frequency $j/(2N)$ has probability zero for all these j. Thus

$$\lim_{n \to \infty} \mathbf{Q}^n = \mathbf{0}$$

a matrix with all elements equal to zero.

This being so, it is easy to prove (Kemeny & Snell 1960) that $|\mathbf{I} - \mathbf{Q}| \neq 0$. Consider the identity

$$(\mathbf{I} - \mathbf{Q})(\mathbf{I} + \mathbf{Q} + \mathbf{Q}^2 + \cdots + \mathbf{Q}^{n-1}) = \mathbf{I} - \mathbf{Q}^n$$

which is verified by multiplying out the left-hand side. Remembering the rule that 'the determinant of a product is the product of the determinants', we see that

$$|\mathbf{I} - \mathbf{Q}||\mathbf{I} + \mathbf{Q} + \mathbf{Q}^2 + \cdots + \mathbf{Q}^{n-1}| = |\mathbf{I} - \mathbf{Q}^n|$$

If $|\mathbf{I} - \mathbf{Q}|$ were zero, the right-hand side would be zero for *all n*. This is impossible. For $|\mathbf{I}| = 1$ and $\lim_{n \to \infty} \mathbf{Q}^n = \mathbf{0}$, so that, for sufficiently large n, the right-hand side will be close to unity and thus cannot be zero. Hence $(\mathbf{I} - \mathbf{Q})$ has an inverse and the fundamental equation, supplemented by the boundary conditions, has a unique solution.

While *in principle* the equation

$$\mathbf{u} = (\mathbf{I} - \mathbf{Q})^{-1}\mathbf{R}$$

supplies $u(p_0)$ for any given p_0, this equation can be used for finding $u(p_0)$ *in practice* only if we can find some way of inverting $(\mathbf{I} - \mathbf{Q})$. Of course, this can be done on the computer in cases where $(\mathbf{I} - \mathbf{Q})$ is not too large. In general, however, we shall have to proceed differently. We begin with Moran's model.

Probability of fixation for Moran's model

The reader will recall that in this model the population, which is haploid and has size $2N$, changes by a series of birth-death events and that, at any such event, the number of A alleles changes by *one at most*. If, then, we write g_{ij} for the probability of changing from frequency $i/(2N)$ to frequency $j/(2N)$ as a result of a single event, we have

$$g_{i,i+1} + g_{i,i-1} + g_{ii} = 1$$

PROBABILITY OF FIXATION: GENERAL CASE

With a few changes in wording, the argument set out earlier in this chapter continues to apply. Our fundamental equation, with 'generation' replaced by 'event' becomes

$$u_i = g_{i,i+1}u_{i+1} + g_{i,i-1}u_{i-1} + g_{ii}u_i \qquad (i = 1, 2, \ldots, 2N - 1)$$

On substituting $g_{ii} = 1 - g_{i,i+1} - g_{i,i-1}$ we find

$$g_{i,i+1}u_{i+1} - (g_{i,i+1} + g_{i,i-1})u_i + g_{i,i-1}u_{i-1} = 0$$

Since A has a selective advantage over a, we write

$$g_{i,i+1} = (1 + \alpha)g_{i,i-1}$$

with $\alpha > 0$ (for a careful discussion of this point see Moran (1962, p. 119)). Hence, on cancelling $g_{i,i-1}$,

$$(1 + \alpha)u_{i+1} - (2 + \alpha)u_i + u_{i-1} = 0$$

This is a *difference equation*, connecting values of u for three successive values of the subscript. We shall not discuss the theory of these equations (see e.g. Spiegel (1971) for a very lucid introductory account), which first appeared in the genetical literature in connection with the theory of inbreeding (Robbins 1917, 1918a, b) but will merely note a standard result. Suppose we have a difference equation of the form

$$au_{i+1} + bu_i + cu_{i-1} = 0 \qquad (a, b, c = \text{constants})$$

Write

$$a\lambda^{i+1} + b\lambda^i + c\lambda^{i-1} = 0$$

giving, on dividing through by λ^{i-1}, the quadratic equation

$$a\lambda^2 + b\lambda + c = 0$$

If this equation has two distinct solutions, say λ_1 and λ_2, all solutions of our difference equation are covered by the formula

$$u_i = A\lambda_1^i + B\lambda_2^i$$

where A and B are constants. If the quadratic equation has only one distinct solution λ ('equal roots'), we have

$$u_i = (A + Bi)\lambda^i$$

A and B again being constants. In our case

$$(1 + \alpha)\lambda^2 - (2 + \alpha)\lambda + 1 = 0$$

giving

$$\lambda = \frac{(2 + \alpha) \pm \sqrt{(4 + 4\alpha + \alpha^2 - 4 - 4\alpha)}}{2(1 + \alpha)} = \frac{2 + \alpha \pm \alpha}{2(1 + \alpha)}$$

$$= 1 \text{ or } (1 + \alpha)^{-1}$$

Therefore

$$u_i = A(1)^i + B[(1 + \alpha)^{-1}]^i = A + B(1 + \alpha)^{-i}$$

To find A, B we appeal to the boundary conditions. Remembering that u_i is short for $u(i/(2N))$, we have $u_0 = 0$, $u_{2N} = 1$. Hence

$$0 = A + B \qquad B = -A$$

and

$$1 = A + B(1 + \alpha)^{-2N} = A[1 - (1 + \alpha)^{-2N}]$$

$$A = \frac{1}{1 - (1 + \alpha)^{-2N}}$$

whence

$$u_i = \frac{1 - (1 + \alpha)^{-i}}{1 - (1 + \alpha)^{-2N}}$$

or, since u_i is short for $u(i/(2N))$, so that i is the initial number of A alleles,

$$u(p_0) = \frac{1 - (1 + \alpha)^{-2Np_0}}{1 - (1 + \alpha)^{-2N}}$$

(Moran 1960a, 1962) for initial frequency of A equal to p_0. Our proof is essentially the same as that given by Moran; an equivalent approach would be to invert $\mathbf{I} - \mathbf{Q}$ algebraically.

Note that our result is valid whatever the value of α. For later comparison with results for other models, however, we note that when α is small we have the close approximations

$$(1 + \alpha)^{-2Np_0} = e^{-2N\alpha p_0} \qquad (1 + \alpha)^{-2N} = e^{-2N\alpha}$$

giving

$$u(p_0) = \frac{1 - e^{-2N\alpha p_0}}{1 - e^{-2N\alpha}}$$

when α is small.

We shall mention a quite different method for finding $u(p_0)$, also due to Moran (1960a, 1962), since the approach ('Moran's method') is sometimes very helpful; in fact, we used this method in Chapter 3 to find the probability of fixation of a neutral allele (the method was given, for that special case, by Owen in 1956 in unpublished lectures). The idea is to find a quantity whose mean remains unchanged, irrespective of changes in allele frequency. Write j for the number of A alleles present at any time t. We shall show that

$$\mathop{E}_{j} (1 + \alpha)^{-j}$$

where E represents the mean, is the same for all t.

Write i for the number of A alleles present before the last birth–death event preceding t. Thus j can only be $i + 1$, $i - 1$ or i. Suppose in the first instance that the value of i is known. Then (by definition of mean) the mean of $(1 + \alpha)^{-j}$, conditional on this knowledge of i, written $E[(1 + \alpha)^{-j} | i]$, will be

$$(1 + \alpha)^{-(i+1)} g_{i,i+1} + (1 + \alpha)^{-(i-1)} g_{i,i-1} + (1 + \alpha)^{-i}(1 - g_{i,i+1} - g_{i,i-1})$$

which, on substituting $g_{i,i+1} = (1 + \alpha) g_{i,i-1}$, becomes

$$(1 + \alpha)^{-i}$$

Now, except at the very start, the value of i will not be known; i will be a random variable. We allow for this, in the usual way, by averaging our conditional mean over all i. Hence

$$\mathop{E}_{j} (1 + \alpha)^{-j} = \mathop{E}_{i} (1 + \alpha)^{-i}$$

these being unconditional means. We see that

$$E(1 + \alpha)^{-(\text{the number of } A\text{'s})}$$

does not change at any birth-death event and thus at all times, including ultimate time, equals its initial value

$$(1 + \alpha)^{-2Np_0}$$

Ultimately, A is fixed (number of A's is $2N$) with probability $u(p_0)$ or lost (number of A's is 0) with probability $1 - u(p_0)$. Hence, writing k for the ultimate number of A's, we have

$$E(1 + \alpha)^{-k} = (1 + \alpha)^{-2N} u(p_0) + (1 + \alpha)^{-0}[1 - u(p_0)]$$

$$= 1 - u(p_0)[1 - (1 + \alpha)^{-2N}]$$

Since this must equal $(1 + \alpha)^{-2Np_0}$, we find

$$u(p_0) = \frac{1 - (1 + \alpha)^{-2Np_0}}{1 - (1 + \alpha)^{-2N}}$$

as before. Notice that this is the same formula as given for the problem in notes to Chapter 1 (Q/P in that case being analogous to $g_{i,i-1}/g_{i,i+1} = (1 + \alpha)^{-1}$ here). In view of this the reader may care to return to our comments on that problem.

It is obviously very pleasing that, for Moran's model, we have a formula for $u(p_0)$ that is valid for all α. On the other hand, the model represents a very special biological situation, so that the reader may feel doubtful whether the conclusions we can draw from our formula will apply to populations in general. We shall, therefore, postpone further discussion until we have considered some different models, to which we now turn.

Moran's method for the Wright–Fisher haploid model

As already stated, a formula for $u(p_0)$ is not available for models in general; the best we can usually obtain are approximate formulae, reliable for small α. However, in the case of genic selection, we can, for any α, obtain limits between which $u(p_0)$ must lie.

Let p be the known frequency of A in any generation t. The reader will recall that, under genic selection, the corresponding frequency in generation

PROBABILITY OF FIXATION: GENERAL CASE

$(t + 1)$, calculated deterministically, is

$$p^* = \frac{p(1 + \alpha)}{1 + \alpha p}$$

We assume for the moment the Wright-Fisher model; that is, we suppose that the probability that r A alleles appear in $(t + 1)$ is

$$\frac{(2N)!}{r!(2N - r)!} (q^*)^{2N-r}(p^*)^r$$

where $q^* = 1 - p^*$.

Suppose then that p has known value. Let p' be the frequency of A in $(t + 1)$. In an attempt to use Moran's method, we consider (Moran 1960b) $e^{-4N\theta p'}$, where θ is a positive number, independent of p, to be determined; as usual, we write

$$E(e^{-4N\theta p'}|p)$$

for the mean, conditional on p having its known value.

If, for some appropriate θ, it were true that

$$E(e^{-4N\theta p'}|p) = e^{-4N\theta p}$$

whatever the value of p, we could find $u(p_0)$ as in the last section.

Now, by definition of mean

$$E(e^{-4N\theta p'}|p) = E(e^{-2\theta r}|p)$$
$$= \sum_{r=0}^{2N} e^{-2\theta r} \frac{(2N)!}{r!(2N - r)!} (q^*)^{2N-r}(p^*)^r$$
$$= \sum_{r=0}^{2N} \frac{(2N)!}{r!(2N - r)!} (1 - p^*)^{2N-r}(p^*e^{-2\theta})^r$$
$$= (1 - p^* + p^*e^{-2\theta})^{2N}$$

by the binomial theorem. For this to equal

$$e^{-4N\theta p} = (e^{-2\theta p})^{2N}$$

we must have

$$1 - p^* + p^*e^{-2\theta} = e^{-2\theta p}$$

whatever the value of p. This is impossible, since we want θ to be independent of p. For example, suppose $\alpha = 1$. Let $p = 0.5$ and substitute x for $e^{-\theta}$, x^2 for $e^{-2\theta}$. By simple algebra, we find that $x = 1$ or 0.5, so that θ ($= -\log x$) is either 0 (irrelevant) or $-\log 0.5 = 0.693\,147\,18$. With this value of θ, our equation is satisfied when $p = 0.5$ but not, for example, when $p = 0.75$, as the reader will easily verify.

Moran overcomes this difficulty in a very ingenious way. Suppose we 'cheat' by replacing p^* by π^*, with π^* chosen so that our equation does hold for all p. That is, we define π^* by the relationship

$$1 - \pi^* + \pi^* e^{-2\theta} = e^{-2\theta p}$$

so that

$$\pi^* = \frac{1 - e^{-2\theta p}}{1 - e^{-2\theta}}$$

and replace p^* by π^* every generation. Then, allowing for the fact that p is a random variable, we would have

$$\mathop{E}_{p'}(e^{-4N\theta p'}) = \mathop{E}_{p}(e^{-4N\theta p})$$

so that, at all times

$$E(e^{-4N\theta(\text{allele frequency})})$$

equals its initial value $e^{-4N\theta p_0}$. We would have, then, at ultimate time,

$$e^{-4N\theta} u(p_0) + e^0 [1 - u(p_0)] = e^{-4N\theta p_0}$$

whence

$$u(p_0) = \frac{1 - e^{-4N\theta p_0}}{1 - e^{-4N\theta}}$$

We shall refer to this as u_{fudge}.

This approach may seem at first sight contrived and irrelevant; that this is not necessarily so may be seen by considering some properties of u_{fudge}. First, u_{fudge} gives the correct fixation probability when $p_0 = 0$ and when $p_0 = 1$ (0 and 1, respectively). Now consider the case when p_0 lies between 0 and 1. If we expand numerator and denominator in exponential series and let $\theta \to 0$, we find that the limit of u_{fudge} as $\theta \to 0$ is p_0. The limit of u_{fudge} as $\theta \to \infty$ is 1.

PROBABILITY OF FIXATION: GENERAL CASE

Also, u_{fudge} is a continuous function of θ when θ is positive. Thus u_{fudge} takes all possible values between p_0 and 1 as θ increases. Now the true probability of fixation, u_{true}, say, also lies between p_0 (the fixation probability for a neutral allele) and 1. Hence, whatever the value of u_{true} in any particular case, it must be possible to find a value of θ that makes u_{fudge} equal to u_{true} for that case. The difficulty is that this θ will depend on p_0 and N as well as on α. Nevertheless, we hope to find that, in all cases, this θ lies within some *range* whose end-points depend on α *only*.

To see if this is so, we suppose that p_0 lies between 0 and 1 and assess (in a slightly more intuitive way than Moran) the effect of our apparent malpractice. Clearly, if π^* exceeded p^*, generation after generation, the tendency for A to rise in frequency owing to natural selection would be persistently exaggerated. Hence the calculated probability of fixation would exceed the correct probability u_{true}. Now, by direct substitution, we find that π^* equals p^* when p equals 0 or 1. If, however, for *all* other values of p

(a) $\pi^* > p^*$, then $u_{\text{fudge}} > u_{\text{true}}$;
(b) $\pi^* < p^*$, then $u_{\text{fudge}} < u_{\text{true}}$.

By simple algebra

$$p^* - \pi^* = \frac{p(1+\alpha)}{1+\alpha p} - \frac{1-e^{-2\theta p}}{1-e^{-2\theta}}$$

$$= \frac{p - 1 - pe^{-2\theta} - \alpha p e^{-2\theta} + e^{-2\theta p} + \alpha p e^{-2\theta p}}{(1+\alpha p)(1-e^{-2\theta})}$$

Since θ is positive, $e^{-2\theta}$ will lie between 0 and 1, so that the denominator is necessarily positive. If, then, we write

$$f(p) = p - 1 - pe^{-2\theta} - \alpha p e^{-2\theta} + e^{-2\theta p} + \alpha p e^{-2\theta p}$$

we see that if $f(p) > 0$, $p^* > \pi^*$, and if $f(p) < 0$, $p^* < \pi^*$.

To obtain the sign of $f(p)$, we consider the sign of $f''(p)$, the rate of change of the slope of $f(p)$. If, as in Figure 5.1, the slope of $f(p)$ always increases with p over the range $0 \leqslant p \leqslant 1$, then $f(p)$ must be negative throughout this range, apart from the end-points when $f(p) = 0$. Generally

(a) if $f''(p) > 0$ over $0 \leqslant p \leqslant 1$, then $f(p) < 0$ over $0 < p < 1$, so that $u_{\text{fudge}} > u_{\text{true}}$;
(b) if $f''(p) < 0$ over $0 \leqslant p \leqslant 1$, then $f(p) > 0$ over $0 < p < 1$, so that $u_{\text{fudge}} < u_{\text{true}}$.

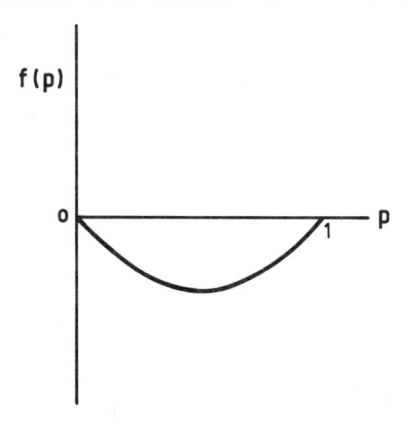

[Figure: plot of f(p) vs p, U-shaped curve with f(0)=f(1)=0]

Figure 5.1 Plot of $f(p)$ against p, with $f(0) = f(1) = 0$, in the case where the slope of $f(p)$ continually increases as p goes from zero to unity, being successively relatively large and negative, small and negative, zero, small and positive, relatively large and positive.

By elementary calculus

$$f''(p) = 4\theta e^{-2\theta p}(\theta - \alpha + \theta\alpha p)$$

The terms outside the parentheses are necessarily positive. Hence

(a) if $\theta > \alpha$, $u_{\text{fudge}} > u_{\text{true}}$;
(b) if $\theta < \alpha/(1 + \alpha)$, $\theta(1 + \alpha) < \alpha$, $\theta(1 + \alpha p) < \alpha$ (since $0 \leqslant p \leqslant 1$), $u_{\text{fudge}} < u_{\text{true}}$.

Thus, while there cannot be an exact expression, in terms of α alone, for the value of θ that makes $u_{\text{fudge}} = u_{\text{true}}$ in all cases, we have shown that this θ never exceeds α and is never less than $\alpha/(1 + \alpha)$. Thus in all cases

$$\alpha/(1 + \alpha) \leqslant \theta \leqslant \alpha$$

and, recalling that

$$u_{\text{fudge}} = \frac{1 - e^{-4N\theta p_0}}{1 - e^{-4N\theta}}$$

we have, for the true probability of fixation $u(p_0)$,

$$\frac{1 - e^{-4N\alpha' p_0}}{1 - e^{-4N\alpha'}} \leqslant u(p_0) \leqslant \frac{1 - e^{-4N\alpha p_0}}{1 - e^{-4N\alpha}}$$

where $\alpha' = \alpha/(1 + \alpha)$.

PROBABILITY OF FIXATION: GENERAL CASE

If α is small, these limits for $u(p_0)$ are very close to one another. For example, when $N = 1000$ and $p_0 = 1/(2N)$, the limits are 0.0196, 0.0198 when $\alpha = 0.01$, and 0.0908, 0.0952 when $\alpha = 0.05$. Generally, for α up to about 0.05, the limits are within about 6% of one another. Thus, at least for $\alpha \lesssim 0.05$, it will usually be sufficient to use the approximation

$$u(p_0) = \frac{1 - e^{-4N\alpha p_0}}{1 - e^{-4N\alpha}}$$

first obtained by Kimura (1957) using a general approximate method described below. If very great accuracy is desired, one may use a very accurate but unfortunately rather complicated formula derived by Ewens (1964a); we shall not give details, but merely note that his procedure is closely related to Moran's method. Taking $\theta = \alpha$, he writes (in our notation)

$$E(e^{-4N\alpha p'}|p) = e^{-4N\alpha p} + (\text{error})$$

He obtains a close approximation to this error and then uses an ingenious method to correct Kimura's formula to allow for this error. It is interesting to note that this error does not involve terms in α or α^2. It follows at once that if we substitute Kimura's formula into the fundamental equation, the discrepancy between the substituted left- and right-hand sides of this equation will not involve terms in α or α^2, again indicating that Kimura's formula gives very accurate results when α is small. Ewens (1963b) has compared these results with exact results in the case $2N = 12$; agreement was excellent for α up to 0.1, a rather larger value of α than might have been expected from our earlier arguments.

In the special case $p_0 = 1/(2N)$, Kimura's formula becomes

$$u(p_0) = \frac{1 - e^{-2\alpha}}{1 - e^{-4N\alpha}}$$

Expanding the numerator in exponential series and neglecting terms in α^2, $\alpha^3, \ldots$, we have the close approximation when α is small

$$u(p_0) = \frac{2\alpha}{1 - e^{-4N\alpha}}$$

first obtained by Fisher (1930a) and Wright (1931).

When α is large, Moran's limits for $u(p_0)$ are, of course, still valid but differ rather substantially from one another and are thus not as helpful as in the case where α is small. Further, Kimura's formula is not, in general, valid when

α is large. But this is not a problem. For with large α and N anything but very small indeed, $N\alpha'$ (and, *a fortiori*, $N\alpha$) will exceed 2. Hence $e^{-4N\alpha'}$ and $e^{-4N\alpha}$ will each be less than

$$e^{-8} = 0.0003$$

so that the denominators in Moran's limits may each be taken as unity; of course, if N were sufficiently large, this would also be justified even for small α. Let $R\ (=2Np_0)$ be the initial *number* of A alleles. Then when $N\alpha'$ exceeds 2, the probability of fixation will (near-enough) lie between limits

$$1 - e^{-2\alpha'R} \quad \text{and} \quad 1 - e^{-2\alpha R}$$

which are virtually identical when α is small and depend only on the selective advantage and initial *number* of A. Thus when $N\alpha' > 2$ and α is small, the probability of fixation is independent of the population size; *a fortiori* this independence holds when $N\alpha' > 2$ and α is large. Hence we may take the population size to be infinite if we wish. Thus, if $N\alpha' > 2$, the probability of fixation will be given, very accurately, by the branching process approach. The reader will easily check, for any given case, that the probability so determined lies within the limits just given. For example, we showed (Table 4.5) that with $R = 1$, $\alpha = 0.5$, the probability of fixation, given by the branching process approach, is 0.5828, which does indeed lie within limits 0.6321 and 0.4866; generally, as here, the true probability lies closer to the upper than to the lower limit.

Population subdivision

Suppose now that the population, of adult size N ($2N$ in the case of haploids), is divided into subpopulations, none of which is completely isolated from the others. Subpopulations may have any size and may fuse, split and change their genetic composition by migration. We suppose genic selection in every subpopulation, with α the same throughout, and N *constant over generations*. Note that absorption is certain; the *whole population* must end up either all A or all a. This would not necessarily be so if, say, one group of subpopulations were completely isolated from the rest; thus 'complete isolation' is quite distinct from 'no isolation but very low migration rates', with quite different consequences for the prospects of fixation in the whole population, as we shall see.

We shall show that the probability of fixation in the whole population is the same as that in an undivided population of size N (or $2N$ for haploids). Maruyama gave two proofs of this remarkable result. He considered his later proof (Maruyama 1977) superior; while the present author accepts this, we shall concentrate on the earlier proof (Maruyama 1970c), which is very simple and gives the essence of the matter.

Suppose that, at some time t, Y_i is the number of A alleles and $2N_i$ the total number of alleles in the ith subpopulation. The corresponding numbers for the whole population will be written Y and $2N$, that is

$$Y = \sum_i Y_i \qquad N = \sum_i N_i$$

We suppose that selection and sampling of gametes occurs independently in each subpopulation. Then writing Y'_i, N'_i, Y' for the quantities in generation $(t+1)$ corresponding to Y_i, N_i, Y in generation t, we have the close approximation

$$E(e^{-2\alpha Y'_i}) = E(e^{-2\alpha Y_i})$$

Further, since sampling of gametes occurs independently, we have, using the statistical rule for *independent* variables that 'the mean of a product is the product of the means',

$$\begin{aligned}
E(e^{-2\alpha Y'}) &= E(e^{-2\alpha \Sigma Y'_i}) \\
&= E(e^{-2\alpha Y'_1} e^{-2\alpha Y'_2} e^{-2\alpha Y'_3} \ldots) \\
&= (Ee^{-2\alpha Y'_1})(Ee^{-2\alpha Y'_2})(Ee^{-2\alpha Y'_3}) \ldots \\
&\simeq (Ee^{-2\alpha Y_1})(Ee^{-2\alpha Y_2})(Ee^{-2\alpha Y_3}) \ldots \\
&= E(e^{-2\alpha Y_1} e^{-2\alpha Y_2} e^{-2\alpha Y_3} \ldots) \\
&= E(e^{-2\alpha Y})
\end{aligned}$$

Writing then $Y' = 2Np'$, $Y = 2Np$ we have the familiar

$$E(e^{-4N\alpha p'}) = E(e^{-4N\alpha p})$$

This (approximate) relationship between allele frequencies *in the whole population* in two successive generations is not affected by factors, such as migration, that may change subpopulation, but not overall, frequencies. Thus all goes as in undivided populations in respect of probability of fixation, which is thus given approximately by Kimura's formula, with p_0 the initial frequency of A in the whole population.

This proof is not quite rigorous, since the approximation in the course of the proof might spoil the accuracy of the final result. Of course, we know that this is not a serious problem for an undivided population, but the situation is a little different here, since with very low migration rates the time until fixation will be very long; in principle, the error arising from using an approximate formula, every generation, over such a long period might perhaps be substantial. We shall, therefore, consider, very much in outline, Maruyama's alternative approach.

Consider two populations, one undivided and one subdivided, but identical with respect to α, N and p_0. Let U be the probability that A is fixed *by some stated time*. For the undivided population, U will depend solely on α, N, p_0 and the stated time; for the subdivided population, the population structure and pattern and intensity of migration will be relevant in addition to α, N, p_0 and the time. Thus the situation in the two populations would appear to be quite different. However, Maruyama shows that this difference can be completely summarized by the timescale of events in the two populations. If a given value of U is reached at time t, measured in generations, in the undivided population, the same value of U is reached at time τ, measured on the appropriate scale ('stochastic clock scale'), in the subdivided population. If we used a common scale, say years, for both populations, the value of U, at any *finite* time, could be very different for the two populations. However, the scale for the subdivided population is chosen so that as $t \to \infty$, $\tau \to \infty$. Thus at *ultimate* time, the distinction between t and τ disappears, so that the probability of *ultimate* fixation is the same for both populations.

We stress again that our whole discussion of subdivided populations has assumed genic selection and is therefore appropriate for haploids, for loci in diploids where there is little or no dominance in fitness or, in the case where there is dominance in fitness, when the advantageous allele is initially rare and all subpopulations are large. It is clear that if the probability of fixation is much affected by dominance, population subdivision must affect our probability to some extent, since the *overall* frequency of AA is affected by population subdivision, even though mating in every subpopulation is at random. Even more important is our assumption that the overall population size remains constant; obviously, that imposes fairly severe restraints on the manner in which the population structure can change.

The case $N_e < N$

We return now to an undivided population and consider the case where, owing to the effect of parental environment on reproductive output, N_e is less than N; naturally, we hope that Kimura's formula will continue to apply when α is small, provided we replace N by N_e. Moran (1961) has obtained limits for $u(p_0)$ for this case; unfortunately these limits are not as helpful as in the case $N_e = N$. We should, however, note the main features of this approach, which provides the only strictly rigorous analysis available of the problem.

We begin, as in Chapter 4, with the negative binomial distribution, which we suggested would be appropriate, in populations of infinite size, when fecundity depends on environment. If, given these conditions, s A's are present in a given generation, the probability that r A's are present one generation later is the coefficient of x^r in the expansion (in series) of

$$\left(\frac{a}{1-bx}\right)^{a(1+\alpha)s/b}$$

where a ($= N_e/N$ in this case) lies between 0 and 1, $b = 1 - a$. Moran modifies this probability generating function to allow for finite population size (N in diploids, $2N$ in haploids) and obtains the pgf

$$\frac{\Gamma(A+B)}{\Gamma(A)\Gamma(B)} \int_0^1 (1 - y + yx)^{2N} y^{A-1} (1-y)^{B-1} \, dy$$

here $A = a(1 + \alpha)s/b$, $B = a(2N - s)/b$ and y is a dummy variable; as before, the probability that r A's are present in the progeny generation is the coefficient of x^r. We shall call the distribution corresponding to this pgf the 'modified negative binomial distribution'. It may be shown that, for this distribution,

$$N_e/N = a + b/(2N)$$

when $\alpha = 0$; strictly, this formula should be modified to allow for selection, but the modification proves trivial when α is small. It may be shown that as $N_e \to N$, our modified negative binomial distribution tends to the usual binomial distribution and may thus be regarded as an extension of the latter to the case $N_e < N$.

From a consideration of the moments of this modified distribution, Moran obtained limits for the probability of fixation $u(p_0)$. When N is sufficiently large and α sufficiently small (comparable in magnitude with $1/N$), both limits prove very close to Kimura's formula

$$\frac{1 - e^{-4N_e\alpha p_0}}{1 - e^{-4N_e\alpha}}$$

While the derivation of this formula that we have outlined is rigorous, the conditions on N and α are rather restrictive. It seems, however, that in practice, just as in the case $N_e = N$, this formula is a good approximation to the true $u(p_0)$, provided only that α does not exceed a few per cent. For example, it follows at once from Moran's paper that if we substitute the formula in the fundamental equation, with N_e/N given as $a + b/(2N)$, or even using the completely accurate formula for N_e/N mentioned earlier, the discrepancy between the substituted left- and right-hand sides does not involve terms in α or α^2. We can also follow the Ewens approach: we take $2N = 12$, selected values of α up to 0.1 and compare values of $u(p_0)$ from Kimura's formula (taking $N_e/N = a + b/(2N)$) with exact values given by

$$\mathbf{u} = (\mathbf{I} - \mathbf{Q})^{-1}\mathbf{R}$$

where the elements of $\mathbf{Q}$ and $\mathbf{R}$ are taken from the modified negative binomial distribution. Agreement between the Kimura values and exact values is at least as good as in the case $N_e = N$ and indeed better when N_e/N is small. We may conclude, then, with some confidence, that Kimura's formula will be accurate enough for practical purposes when α is smallish (<0.05, say). For larger α, the branching process approach will be appropriate; we would expect also that this approach would give much the same answers as Kimura's formula when α is small but N_e large enough for $N_e\alpha$ to exceed 2, and this also seems to be true, as the reader will readily check for some special cases. Finally, population subdivision will not affect the probability of fixation, given the conditions stated earlier for the case $N_e = N$, provided also that N_e/N is the same in all subpopulations.

It is particularly remarkable that the probability of fixation for Moran's model is also given by Kimura's formula, when α is small. We showed earlier that, for Moran's model, when α is small,

$$u(p_0) = \frac{1 - e^{-2N\alpha p_0}}{1 - e^{-2N\alpha}}$$

which is just Kimura's formula with $N_e = \tfrac{1}{2}N$; note that this is the value of N_e for Moran's model that we postulated, rather tentatively, in Chapter 2.

Clearly, the Wright–Fisher model, Moran's model and the modified negative binomial model, when N_e is much smaller than N, represent very different biological situations. Yet the probability of fixation is given by the same formula in all three cases; indeed, this formula is also correct for the simple random-walk model that we gave in notes in Chapter 1. It seems clear that there are some leading properties of natural populations that are much the same for a wide range of models. Of course, we have not yet considered the effect of dominance in viability in any detail; however, we argued earlier that this effect would not be critical over quite a wide range of degrees of

PROBABILITY OF FIXATION: GENERAL CASE

dominance. It will be appropriate, therefore, to attempt some conclusions at this stage.

Effect of selection on fixation probability

We have shown that, when α is small, $u(p_0)$ for a population with $N_e < N$ is obtained by taking $u(p_0)$ for a Wright–Fisher population and replacing $N\alpha$ by

$$N_e \alpha = N(\alpha N_e/N)$$

at least for models considered so far. Effectively, then, the selective advantage is reduced by a factor N_e/N. More generally, we can define the *effective* selective advantage α_e; an advantage α_e in a Wright–Fisher population gives the same probability of fixation as an advantage α in a different type of population. Thus, when α is small

$$u(p_0) = \frac{1 - e^{-4N\alpha_e p_0}}{1 - e^{-4N\alpha_e}}$$

where $\alpha_e = \alpha N_e/N$. In fact (see e.g. Table 4.5) α_e is also less than α in cases where α is large, although the effect is not quite as marked as when α is small, α_e/α being rather more than N_e/N when α is large.

We consider in detail the case where A is an advantageous mutant, newly arisen by mutation, so that

$$p_0 = 1/(2N)$$

Some numerical values for $u(p_0)$ are given in Table 5.1.

Clearly a selective advantage, *even if quite small*, substantially increases the probability of fixation of a newly arisen mutant, especially when N, the

Table 5.1 Probability (per cent) that an advantageous mutant is ultimately fixed; α_e = effective selective advantage, N = (diploid) population size.

N	α_e			
	0	0.001	0.01	0.05
200	0.250 00	0.3628	1.9808	9.5163
2 000	0.025 00	0.1999	1.9801	9.5163
20 000	0.002 50	0.1998	1.9801	9.5163
200 000	0.000 25	0.1998	1.9801	9.5163

population size, is large (in view of results on population subdivision mentioned earlier, we would expect N to be at least moderate in practice – small values of N must surely be unusual). Thus, if $N = 200\,000$, $\alpha_e = 0.001$, the probability of fixation is nearly 800 times as large as the corresponding probability, $1/(2N)$, for a neutral mutant. More generally, when $\alpha < 0.01$ but $N\alpha_e > 1$, we have the close approximations

$$1 - e^{-2\alpha_e} = 1 - (1 - 2\alpha_e) = 2\alpha_e \qquad 1 - e^{-4N\alpha_e} = 1$$

so that $u(p_0)$, for $p_0 = 1/(2N)$, will be close to $2\alpha_e$ and the probability of fixation, relative to the corresponding probability for a neutral, will be about

$$2\alpha_e \div 1/(2N) = 4N\alpha_e$$

Thus even with N as small as 10 000 and α_e as small as 0.001, an advantageous mutant is about 40 times as likely to be fixed as a neutral. Similar results are obtained when p_0, while still small, exceeds $1/(2N)$. For example, with $N = 10\,000$, $\alpha_e = 0.001$, $p_0 = 0.01$, our advantageous allele is about 33 times as likely to be fixed as a neutral with the same initial frequency.

Only if $4N\alpha_e$ is really quite small (less than about 0.25) will the advantageous allele be effectively neutral. We have, in this case, the close approximation

$$1 - e^{-4N\alpha_e} = 1 - (1 - 4N\alpha_e) = 4N\alpha_e \qquad 1 - e^{-4N\alpha_e p_0} = 4N\alpha_e p_0$$

giving

$$u(p_0) = p_0$$

as for a neutral allele.

The consequences for the neutral theory of molecular evolution are readily apparent. Consider first the case where the selective advantage, if any, is apparent from the very start (i.e. immediately after mutation). Now, although most mutants, even if advantageous, are lost by drift, what matter in evolution are the mutants that are actually fixed; hence we just consider the latter. The chance of fixation, giving $4N\alpha_e$ large, is so much greater for an advantageous mutant than for a neutral that even if a fairly large number of neutral mutants arise for every advantageous mutant that arises, the advantageous allele is still much more likely to be fixed than any of the neutrals taken *en bloc*, in contradiction to the neutralist scheme.

To avoid this difficulty, the neutralist supposes that the overwhelming majority of non-harmful mutants are neutral or almost neutral. Further, he will naturally stress those factors which might keep $4N\alpha_e$ small. Thus one may plausibly suppose that α is quite small for most molecular changes and this will give a very small value of α_e if we postulate a sufficiently small N_e/N. Clearly, the neutral theory is very much easier to accept if one assumes that effective population sizes are typically small or moderate – see also our discussion of recurrent mutation in Chapter 4. In fact, this assumption, although not always explicitly stated, appears to underlie most of Kimura's thinking on the neutral theory. This explains how Kimura, while accepting Fisher's mathematics, does not accept his conclusions; Fisher thought in terms of very large populations and appears to have assumed that N_e/N would not be substantially less than unity. In the present author's opinion, no conclusive evidence on values of N_e has, as yet, been produced. It remains an open question, therefore, whether the neutralist is justified in invoking such modest values of N_e.

> *Glendower* I can call spirits from the vasty deep.
> *Hotspur* Why, so can I, or so can any man:
> But will they come when you do call for them?
>
> William Shakespeare, *Henry IV, Part One*

The assumption of moderate N_e has a further very interesting consequence. It is often suggested that alleles that spread under natural selection may previously have been disadvantageous and maintained at low frequency by the opposing effects of recurrent mutation and selection, a change in environment being responsible for a disadvantageous allele becoming advantageous (presumably only a fairly small proportion of the possible disadvantageous alleles at a locus would ever change in this way). This scheme is not very plausible if N_e is modest, since in that case the allele, when disadvantageous, could well have been absent from the population for most of the time, owing to drift.

Our discussion has been based on the haploid model. While we have given intuitive arguments suggesting that conclusions derived from this model apply over a wide range of situations, we really need to justify our intuitions by a quantitative argument. To do this, we consider a general method for finding $u(p_0)$ valid when selection is mild (or absent), due to Kimura (1957, 1962).

Kimura's method for finding the probability of fixation

We begin with the fundamental equation

$$u(p_0) = \sum_x g_{p_0 x} u(x)$$

By definition of mean, the right-hand side is the mean, over all x, of $u(x)$, so that

$$u(p_0) = \underset{x}{E}\, u(x)$$

Let δp_0 be the change in allele frequency in going from generation 0 to generation 1. Then $x = p_0 + \delta p_0$, where δp_0 is a random variable. For given p_0, averaging over all $(p_0 + \delta p_0)$ on the right-hand side is the same as averaging that side over all δp_0, whence

$$u(p_0) = \underset{\delta p_0}{E}\, u(p_0 + \delta p_0)$$

Kimura's idea is to approximate this equation and obtain $u(p_0)$ from this approximate version. The boundary conditions

$$u(0) = 0 \qquad u(1) = 1$$

remain as before.

Now, the allele frequency is a discontinuous variable, changing in discrete steps of size $1/(2N)$, but if N is large, we may treat the allele frequency as a continuous variable. Expanding in Taylor series, we have (primes denoting differentiation)

$$u(p_0) = \underset{\delta p_0}{E} \left(u(p_0) + (\delta p_0)u'(p_0) + \frac{(\delta p_0)^2}{2!} u''(p_0) + \frac{(\delta p_0)^3}{3!} u'''(p_0) + \cdots \right)$$

Now $u(p_0)$ and the derivatives $u'(p_0)$, $u''(p_0)$, $u'''(p_0)$,... are not functions of δp_0 and may be taken as constants when averaging. Also the mean of a sum is the sum of the corresponding means. Hence

$$u(p_0) = u(p_0) + [E(\delta p_0)]u'(p_0) + \left(\frac{E(\delta p_0)^2}{2!}\right)u''(p_0) + \left(\frac{E(\delta p_0)^3}{3!}\right)u'''(p_0) + \cdots$$

where means are taken over all values of δp_0.

We may assume (see Ch. 2) that $E(\delta p_0)^r$ is negligible when $r > 2$ (provided N is not too small). Then, to a close approximation

$$[E(\delta p_0)]u'(p_0) + \tfrac{1}{2}[E(\delta p_0)^2]u''(p_0) = 0$$

Of course, $E(\delta p_0)$ will depend on the pattern and intensity of selection, if any. Unfortunately, the equation, as it stands, is intractable, except in the

PROBABILITY OF FIXATION: GENERAL CASE

neutral case. To proceed further, when selection is acting, we shall have to approximate both $E(\delta p_0)$ and $E(\delta p_0)^2$. In order to obtain useful approximations, we assume that selection is mild. Consider then $E(\delta p_0)$. For brevity, we write p_0 as p. Then given:

	AA	Aa	aa
frequency	p^2	$2pq$	q^2
viability	$1+s$	$1+sh$	1

the frequency of A, one generation later, calculated deterministically, will be

$$p^* = \frac{p^2(1+s) + pq(1+sh)}{1 + sp^2 + 2pqsh}$$

whence

$$p^* - p = \frac{p^2 + sp^2 + pq + pqsh - p - sp^3 - 2p^2qsh}{1 + sp^2 + 2pqsh}$$

Now

$$p^2 + pq - p = p^2 + p - p^2 - p = 0$$

so that s appears in all terms in the numerator. Further,

$$(1 + sp^2 + 2pqsh)^{-1} = 1 + \text{(terms in } s, s^2, s^3, \ldots)$$

If selection is mild (s small), we may neglect terms in $s^2, s^3, \ldots$, so that, to a close approximation,

$$p^* - p = sp^2 + pqsh - sp^3 - 2p^2qsh$$
$$= sp^2q + pqsh(1 - 2p)$$
$$= spq(p + h - 2ph)$$
$$= spq[h + (1 - 2h)p]$$

We recall that the change in allele frequency over one generation, calculated deterministically, equals the mean change in allele frequency over that generation. Hence, with mild selection, we have approximately

$$E(\delta p_0) = sp_0q_0[h + (1 - 2h)p_0]$$

Now consider $E(\delta p_0)^2$. Given mild selection, we approximate this by the corresponding formula when selection is absent. In that case, $E(\delta p_0) = 0$, so that

$$E(\delta p_0)^2 = V(\delta p_0)$$

the variance of δp_0. Since adding a constant does not affect the variance

$$V(\delta p_0) = V(p_0 + \delta p_0) = p_0 q_0/(2N_e)$$

Analogous considerations apply in more complicated cases (for example, when viabilities fluctuate – see below). We write M and V for the appropriate approximations for $E(\delta p_0)$ and $E(\delta p_0)^2$, respectively, for the case in question. Then, writing u as short for $u(p_0)$, we have in all cases

$$M \frac{du}{dp_0} + \tfrac{1}{2} V \frac{d^2 u}{dp_0^2} = 0$$

or

$$\frac{d^2 u}{dp_0^2} + \frac{2M}{V} \frac{du}{dp_0} = 0$$

Our discussion has, of course, been very loose in this section. A very much more careful discussion than ours is given in Karlin & Taylor (1981, Ch. 15, see especially pp. 176-80); this should convince the sceptical reader that our procedure is not nearly as haphazard as it might seem. Ultimately, our whole approach can be justified by a comparison of the results obtained with exact results available in special cases.

To find $u(p_0)$, we introduce the celebrated function

$$G(p_0) = e^{-\int (2M/V) dp_0}$$

Let

$$H(p_0) = 1/G(p_0) = e^{\int (2M/V) dp_0}$$

Then

$$\frac{dH(p_0)}{dp_0} = \frac{2M}{V} H(p_0)$$

from the usual 'function of a function' rule, taken in conjunction with the fact that differentiation is the reverse of integration.

Multiplying our equation

$$\frac{d^2u}{dp_0^2} + \frac{2M}{V}\frac{du}{dp_0} = 0$$

by $H(p_0)$ we have

$$H(p_0)\frac{d^2u}{dp_0^2} + \frac{2M}{V}H(p_0)\frac{du}{dp_0} = 0$$

and so

$$\frac{d}{dp_0}\left(H(p_0)\frac{du}{dp_0}\right) = 0$$

from the usual rule for differentiation of a product. Hence

$$H(p_0)\,du/dp_0 = A$$

where A is a constant to be determined later. Thus

$$du/dp_0 = A/H(p_0) = AG(p_0)$$

so that one further integration will supply u. It will be convenient to express u in terms of *definite* integrals, with p_0 as one of the limits of integration. To avoid p_0 doing double duty as a limit of integration and as the variable in the integrand, we replace p_0 in the last equation by the dummy variable x. Then

$$du/dx = AG(x)$$

where

$$G(x) = e^{-\int (2M/V)dx}$$

with p_0 replaced by x in the formulae for M and for V. (For any reader who finds this strange, we note that, in a definite integral, the variable in the integrand may be represented by any letter we please; thus $\int_0^1 2x\,dx$ is the same as $\int_0^1 2y\,dy$, as the reader will readily verify.)

KIMURA'S METHOD FOR PROBABILITY OF FIXATION

On integrating with respect to x from $x = 0$ to $x = p_0$ we have

$$u(p_0) - u(0) = A \int_0^{p_0} G(x)\, dx$$

But $u(0) = 0$; hence

$$u(p_0) = A \int_0^{p_0} G(x)\, dx$$

Also $u(1) = 1$, so that

$$1 = A \int_0^1 G(x)\, dx$$

whence, finally

$$u(p_0) = \frac{\int_0^{p_0} G(x)\, dx}{\int_0^1 G(x)\, dx}$$

(Kimura 1962). Thus for viabilities:

AA	Aa	aa
$1 + s$	$1 + sh$	1

we have

$$M = sx(1 - x)[h + (1 - 2h)x]$$
$$V = x(1 - x)/(2N_e)$$
$$2M/V = 4N_e s[h + (1 - 2h)x]$$
$$\int (2M/V)\, dx = 4N_e shx + 2N_e s(1 - 2h)x^2$$
$$G(x) = e^{-4N_e shx - 2N_e s(1 - 2h)x^2}$$
$$u(p_0) = \frac{\int_0^{p_0} e^{-4N_e shx - 2N_e s(1 - 2h)x^2}\, dx}{\int_0^1 e^{-4N_e shx - 2N_e s(1 - 2h)x^2}\, dx}$$

In general, this will have to be evaluated numerically on the computer.

Despite all our approximations, this formula proves very accurate for values of s up to 0.1, even when N is as small as 10 (Carr & Nassar 1970), a truly astonishing result.

In the case of genic selection, we have (ignoring terms in α^2) $1 + sh = 1 + \alpha$, $1 + s = 1 + 2\alpha$, whence $s = 2\alpha$, $sh = \alpha$, $h = \frac{1}{2}$ and

$$u(p_0) = \frac{\int_0^{p_0} e^{-4N_e\alpha x}\,dx}{\int_0^1 e^{-4N_e\alpha x}\,dx} = \frac{1 - e^{-4N_e\alpha p_0}}{1 - e^{-4N_e\alpha}}$$

as usual. We shall refer to this as 'the haploid formula' and our general formula for $u(p_0)$ as the 'the diploid formula'.

We may now return to our suggestion that, provided N_e is large and the advantageous allele is not recessive or near-recessive in viability, $u(p_0)$ may be calculated from the haploid formula, even when h is not close to $\frac{1}{2}$.

We note first that recessivity in viability is synonymous with $h = 0$. Let us suppose then that h lies in the range 0.2 to 1. We take $p_0 = 1/(2N)$ and compute $u(p_0)$ from the diploid formula for many different values of s, h and N_e. In every case, we compare our result with the $u(p_0)$ from the haploid formula for the same N_e and with $\alpha = sh$. For $N_e s > 90$, agreement proves excellent in all cases; for $N_e s > 45$, agreement is quite good (sometimes excellent).

For smaller $N_e s$, results from the haploid model are sometimes inaccurate, but never out by more than one-third of the true value. It follows that, despite the inaccuracy (in some cases) when $N_e s$ is small, the conclusions that we obtained from the haploid formula are *qualitatively* correct; we may continue to maintain that a selective advantage, even if small, has a large effect on probability of fixation, especially when N_e is large. Indeed this is still true to some extent when $h < 0.2$, despite the fact that the haploid formula is often inaccurate in such cases (as h approaches zero, we need ever larger values of N_e for the two formulae to agree). Consider, for example, the case $h = 0$ (advantageous allele completely recessive in viability). Then

$$u(p_0) = \frac{\int_0^{p_0} e^{-2N_e s x^2}\,dx}{\int_0^1 e^{-2N_e s x^2}\,dx}$$

Consider first the numerator, with $p_0 = 1/(2N)$. The integrand, which has value unity when $x = 0$, will depart only very marginally from unity over the range of integration 0 to $1/(2N)$. Thus the numerator will be very close to

$$\int_0^{1/(2N)} 1\,dx = 1/(2N)$$

Thus
$$u(p_0) = 1/(2NI)$$
where
$$I = \int_0^1 e^{-2N_e sx^2}\, dx$$

This integral is easily evaluated from tables of the normal distribution. Let
$$c = 2\sqrt{(N_e s)}x$$
so that
$$\tfrac{1}{2}c^2 = 2N_e sx^2 \qquad \frac{dx}{dc} = \frac{1}{2\sqrt{(N_e s)}}$$

Then, by the usual integration by substitution,
$$I = \int_0^{2\sqrt{(N_e s)}} e^{-c^2/2} \frac{1}{2\sqrt{(N_e s)}}\, dc$$
$$= \frac{\sqrt{(2\pi)}}{2\sqrt{(N_e s)}} \frac{1}{\sqrt{(2\pi)}} \int_0^{2\sqrt{(N_e s)}} e^{-c^2/2}\, dc5 \frac{\sqrt{(2\pi)}}{2\sqrt{(N_e s)}} \frac{P}{2}$$

where
$$P = \frac{1}{\sqrt{(2\pi)}} \int_{-2\sqrt{(N_e s)}}^{2\sqrt{(N_e s)}} e^{-c^2/2}\, dc$$

by the symmetry of the normal curve. Hence the probability of fixation of a newly arisen advantageous recessive mutant is

$$[\sqrt{(2N_e s)}/N\sqrt{\pi}]/P$$

If $N_e s$ exceeds about 1.66, P will cover almost the whole area under the normal curve and we may take $P = 1$, thus giving the formula obtained by Kimura (1957) for large $N_e s$.

The probability of fixation, relative to that of a newly arisen neutral mutant, is therefore

$$2\sqrt{(2N_e s)}/\sqrt{\pi} = 1.6\sqrt{(N_e s)}$$

In practice, this formula is not misleading for $N_e s$ down to 1. While the effect of mild selection and large N_e on probability of fixation is not as marked as in the cases we considered earlier, the effect is still apparent.

For small $N_e s$, we may evaluate P by expanding the integrand in exponential series, retaining just the first few terms, and integrating term by term. Then

$$P = \frac{4\sqrt{(N_e s)}}{\sqrt{(2\pi)}} [1 - \tfrac{2}{3} N_e s + \tfrac{2}{5}(N_e s)^2 - \tfrac{4}{21}(N_e s)^3 + \cdots]$$

$$u\left(\frac{1}{2N}\right) = \frac{1}{2N} [1 - \tfrac{2}{3} N_e s + \tfrac{2}{5}(N_e s)^2 - \tfrac{4}{21}(N_e s)^3 + \cdots]^{-1}$$

$$= \frac{1}{2N} [1 + \tfrac{2}{3} N_e s + \tfrac{2}{45}(N_e s)^2 - \tfrac{44}{945}(N_e s)^3 + \cdots]$$

The simpler formula

$$\frac{1}{2N} (1 + \tfrac{2}{3} N_e s)$$

works rather well for $N_e s$ up to 1. Thus $N_e s$ would have to be quite small for the allele to be effectively neutral.

Population subdivision reconsidered

As mentioned earlier, population subdivision will affect the probability of ultimate fixation in some cases. Suppose, for simplicity, that the population is composed of n subpopulations, each of size N and effective size N_e. Let u be the probability of fixation of an allele of initial frequency $1/(2N)$ in a population of size N, as given by our diploid formula. Slatkin (1981) has shown that, when migration rates are very low, the probability that an advantageous allele, of initial frequency (in the whole population) $1/(2nN)$, is ultimately fixed in the whole population equals approximately

$$u \, \frac{1 - e^{-2N_e s}}{1 - e^{-2nN_e s}}$$

We designate this U_1; we write U_2 for the corresponding probability for an allele of initial frequency $1/(2nN)$ in an undivided population size nN, effective size nN_e. For genic selection ($s = 2\alpha$), u will be given by the haploid formula, whence

$$U_1 = \frac{1 - e^{-2(N_e/N)\alpha}}{1 - e^{-4N_e\alpha}} \cdot \frac{1 - e^{-4N_e\alpha}}{1 - e^{-4nN_e\alpha}} = \frac{1 - e^{-2(N_e/N)\alpha}}{1 - e^{-4nN_e\alpha}} = U_2$$

Thus $U_1 = U_2$ in this case, in agreement with Maruyama's results described earlier. Otherwise U_1 and U_2 will differ, the true probability of fixation lying somewhere between U_1 and U_2. Consider now the case $h > 0.2$. If the local effective size N_e is sufficiently large to make $N_e s$ large, u will be given accurately by the haploid formula and U_1 will be very close to U_2; hence population subdivision does not affect probability of fixation, as we surmised in the previous chapter. Otherwise, the haploid formula may give only a rough idea of the correct u, so that U_1 and U_2 can be distinctly different; however, the difference will not be so marked as to upset our conclusions on the effect of selection on probability of fixation.

The case in which h is small is best dealt with separately. As $h \to 0$, the value of N_e required to make the diploid and haploid formulae agree increases without limit, so that the preceding argument is not appropriate. For simplicity, we consider the case $h = 0$, $N_e = N$. The effect of subdivision will be greater when n, the number of subpopulations, is large. For example, if the local population size N is large, sufficiently large to make Ns large, we have approximately

$$\frac{1 - e^{-2Ns}}{1 - e^{-2nNs}} = 1 \qquad U_1 = u = 1.128\sqrt{\left(\frac{s}{2N}\right)} \qquad U_2 = 1.128\sqrt{\left(\frac{s}{2nN}\right)}$$

so that, with very low migration rates, subdivision increases the overall probability of fixation by a factor $\sqrt{n}$. If Ns is small, but nNs large, then approximately

$$1 - e^{-2Ns} = 2Ns \qquad 1 - e^{-2nNs} = 1$$

$$u = \frac{1}{2N}(1 + \tfrac{2}{3}Ns) \qquad U_1 = s(1 + \tfrac{2}{3}Ns)$$

with U_2 as before, giving

$$U_1/U_2 = 1.254\sqrt{(nNs)}(1 + \tfrac{2}{3}Ns)$$

Thus, as expected, subdivision increases the probability of overall fixation, rather substantially if n is large, in either case. On the other hand, for a neutral mutant

$$U_1 = U_2 = 1/(2nN)$$

in all cases.

Generally, then, the probability of fixation of an advantageous recessive (or near-recessive) mutant will, in some circumstances, be substantially greater than the corresponding probability for a neutral mutant, especially in cases where migration is very low and subdivision very extreme.

Fluctuating viabilities

An objection might be raised to the whole argument given in this chapter. We have assumed that relative viabilities remain fixed. In practice, viabilities may fluctuate over successive generations. If so, will our conclusions remain valid?

Unfortunately, the whole topic is difficult. Indeed, as Gillespie (1973) and Jensen (1973) first showed, most of the early work on fluctuating viabilities is wrong, giving rise to very misleading conclusions. Modern treatments, for infinite populations, are given by Hartl & Cook (1973, 1976), Cook & Hartl (1975), Karlin & Lieberman (1974) and Levikson & Karlin (1975), and for finite populations by Karlin & Levikson (1974), Avery (1977) and Narain & Pollak (1977).

We shall summarize results on finite populations. We stress that all investigations have supposed (effectively) very mild selection in every generation. It is not clear (at least to the present author) to what extent the main results would apply in other cases.

We suppose that, in every generation, viabilities are drawn at random, *independently from one generation to another*, from a joint distribution of viabilities for the different genotypes. In this distribution, which is the same for all generations, the viabilities of different genotypes can be correlated or independent. Then, from the investigations carried out so far, the following tentative generalizations can be made.

First, variances and covariances of viabilities can be just as important as means in determining the probability of fixation of a new mutant. Consider the case where variances of different genotypes are equal (and correlations not very close to unity). Then the probability of fixation rises with increase in variance, virtually independently of differences, if any, in genotypic means. Hence the distinction between the case where genotypes differ in mean viability and the 'quasi-neutral' case, where all means are identical, is, to some extent, obscured; 'variance in selection expression reduces and mitigates the mean effects of selection differentials' (Karlin & Levikson 1974).

Possibly a neutralist might feel encouraged by this, although the consequences for evolution would not be those of the 'standard' neutral theory (Takahata 1981). Indeed, Kimura (1983, p. 51) has stressed the role of physiological homeostasis in buffering organisms, especially higher organisms, against environmental fluctuation and appears to regard fluctuation of fitness as fairly unusual.

On the other hand, the selectionist might argue that a combination of differing mean viabilities but identical variances in viability would be unlikely. It has been shown that differences in variance will, to some extent at least, affect the probability of fixation; the large the variance of a genotype (relative to the variance of other genotypes), the lower the fitness of that genotype. Given our very restricted knowledge of actual fitness values and their possible fluctuations, we shall not pursue the problem further at this stage.

Rate of evolution

We now return to the rate of nucleotide substitution in evolution, which we discussed qualitatively in Chapter 1. Let S be a set of base pairs. For example, S could consist of all the base pairs in the genome, in a specific gene or in a specific coding sequence or even all the base pairs at the first position of every triplet in that sequence. Even when S does not represent a specific gene, we shall speak of S as mutating whenever a mutation occurs somewhere in S. Provided S is large, it can mutate in many different ways; we may assume then, that whenever a mutant form of S appears, that particular mutant is a novel form of S, which was not present in the species immediately before mutation. We shall further suppose that no subpopulation, or group of subpopulations, is completely isolated from the rest of the species.

Given these conditions, we shall find the rate of evolution of S under neutral theory. We define k, the rate of mutant substitution in evolution, as the long-term average of the number of mutants in S substituted per generation (i.e. one mutant is substituted, on average, every $1/k$ generations). Let v_T be the overall mutation rate, per generation, for the set and f_0 the proportion of such mutants that are neutral, so that the 'neutral mutation rate' for S is $v_T f_0$. If, in a given generation, the world population size of the species is N (or $2N$ for a haploid species), we expect $2Nv_T f_0$ new neutral mutants in S in that generation. A fraction $1/(2N)$ of these neutral mutants are (on average) eventually fixed in the species. Hence

$$k = 2Nv_T f_0 \times 1/(2N) = v_T f_0$$

Thus, in the neutral case, the rate of evolution of S equals the neutral mutation rate (Kimura 1968a). The same will apply if we follow the evolution of S over a line of evolutionary descent, which may cover several intermediate species.

If an event occurs, on average, once in every g generations, the mean interval between successive events equals g. Hence the average interval between successive fixations equals $1/k$ generations. This must be distinguished from the mean fixation time, which is the average number of generations for a newly arisen, ultimately successful, mutant to be fixed (compare the average time between the reader's meals with the average time he takes to eat a meal).

In practice, we often wish to compare rates of substitution for different sets. Rates estimated from actual data, therefore, are usually given as mean number of base-pair substitutions per base-pair site per unit time, for the set under consideration. In the older work, in which evolution was studied at the protein level, results were given as mean number of amino-acid substitutions per amino-acid site per unit time. Details of these calculations and extensive discussion of results are given in Kimura (1983), Li et al. (1985) and Nei (1987).

For any given set, we would expect the overall mutation rate per generation to be roughly the same in different lines of descent (but see note 5). If the same applies to f_0, the rate of evolution of a given set should be roughly the same in different lines of descent, provided the unit of time is a generation. It has long been clear, however, from extensive analysis of protein evolution, that a (roughly) constant rate is found only if time is measured in years rather than generations (see e.g. Wilson et al. 1977). It is not clear, therefore, to what extent the constant rate supports the neutral theory. On the other hand, the neutralist is not obliged to accept a constant f_0 for a given set (Ohta 1977, Kimura 1979). We noted earlier that a mildly advantageous mutant will behave more and more like a neutral mutant as the effective population size is decreased; similarly, a mildly disadvantageous mutant will approach 'effective neutrality' as N_e is diminished. Thus *other things being equal*, the value of f_0, for any given S, will be greater for less abundant species; such species, then, should show a more rapid rate of evolution *per generation* than shown by abundant species. However, less abundant species tend to have a relatively long generation time, counteracting the more rapid evolution per generation. Given the appropriate assumptions, the outcome will be a constant rate per year, rather than per generation (Kimura 1979); we stress the assumption that other things really are equal, which, no doubt, the selectionist would contest. Thus Kimura supposes that the distribution of actual (rather than effective) mild selective disadvantages among mutants at different sites within a given gene remains constant over long evolutionary periods. Further, this distribution is written in terms of two parameters, one of which must be much the same for different loci if the argument is to work. The reader wishing to pursue the matter is referred to the summary given by Kimura (1983, pp. 240-8).

On the other hand, the constant rate per year is an even more serious problem for the selectionist. The discrepancy between rate of evolution at the morphological and the protein level obviously suggests a difference in the underlying mode of evolution. Further, it does not help to suppose a (roughly) constant 'challenge', giving a (roughly) constant distribution of actual mild selective advantages among mutants at a locus. For in that case effective advantage would decline with diminishing N_e and less abundant species would show a *slower* rate of evolution per generation (and an even slower rate per year) than abundant species.

Now consider differences between sets in rate of evolution; here comparisons are made over the same line of descent, so that problems over the time until do not arise. As mentioned earlier, rates are highest for pseudogenes, rather lower for synonymous mutations and introns, and decidedly lower for non-synonymous changes in most coding sequences, especially where the function of the corresponding protein would be disrupted by almost any amino-acid substitution. It is agreed that these differences reflect in part differences between sets in the fraction of mutants that are disadvantageous and thus eliminated by natural selection; this fraction will be greatest for those sets for which base-pair substitutions have a marked effect on biochemical function, since large changes will usually be disadvantageous. To the neutralist, almost all non-disadvantageous mutants are neutral; thus, the lower the intensity of negative selection, the more rapid the rate of evolution. To the selectionist, the lower the intensity of negative selection, the more likely are advantageous mutants to appear; for sets with small negative selection are sets with mutants of small effect and the less marked the effect of a substitution, the more likely is that effect to be advantageous. However, the less marked the effect, the smaller (presumably) is the associated selective advantage and hence the probability of fixation. Hence we ask: 'Does selectionist theory given the same prediction as neutral theory, namely that the lower the intensity of negative selection, the more rapid the rate of evolution?'

Since we are arguing on the basis of standard selection theory, we shall suppose that N is large. It will be reasonable, therefore, to restrict our attention to advantageous mutants that have sufficient expression in the heterozygote to give probability of fixation $2\alpha N_e/N$. Let O be an optimum genome and A some other genome. The closeness of A to O can be described in terms of the many characteristics (considered as metrical characters) that make O superior. Following Fisher (1930b) we represent this situation geometrically. Each characteristic is represented by a single dimension, so that for a full treatment multidimensional geometry is required. However, in

PROBABILITY OF FIXATION: GENERAL CASE

the first instance, we may think in terms of two or three characteristics (dimensions). Suppose that A starts at distance a from O; mutation in A now gives a new genome A' at a distance r from A. If r exceeds $2a$, A' is further from O than was A and the mutation is certainly disadvantageous (we need not consider this further). If r is very small, the probability that the mutant is advantageous is fairly obviously $\frac{1}{2}$, as we confirm below. We wish to find the probability that the mutant is advantageous, expressed in terms of r and a, in the general case $0 \leq r/(2a) \leq 1$, when n, the number of dimensions, is chosen arbitrarily. Fisher (1930b) gave answers for the cases $n = 3$ and n large; Kimura (1983) gives a solution for general n. The derivation of these results requires rather advanced mathematics, but we shall make the result for general n plausible by considering the special case $n = 2$.

In Figure 5.2a, A is at distance a from O. To achieve a move of distance r in a random direction, OA is produced to Y, where AY = r, and AY rotated

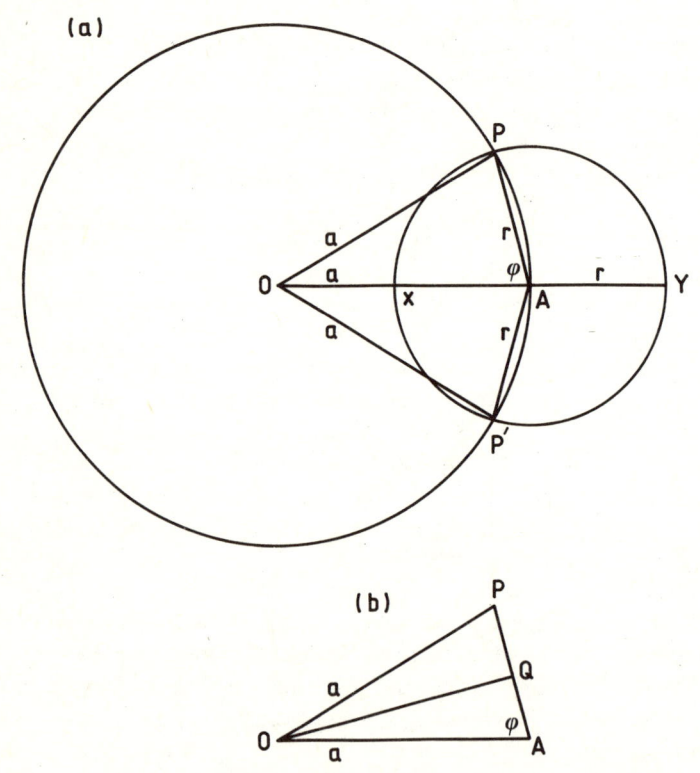

Figure 5.2 (a) Genome A, starting at distance a from the optimum genome O, moves by mutation a distance r in a randomly chosen direction. The mutant will be advantageous if the resulting genome lies on the arc PXP'. (b) Triangle AOP redrawn; OQ is the bisector of the angle AOP.

anticlockwise through an angle θ, chosen at random from a uniform distribution.

If the new genome lies on the arc PXP' and is thus closer to O than was A, the mutant is advantageous. Thus the probability, p say, that the mutant is advantageous equals the length of PXP' divided by $2\pi r$, the length of the circumference of the smaller circle.

Triangles OAP, OAP' are congruent (SSS), so that $\angle$ OAP = $\angle$ OAP' = ϕ, say. Then with ϕ measured in radians, PXP' has length $2\phi r$, so that

$$p = \phi/\pi$$

We wish to express this in terms of r and a. Consider Figure 5.2b, in which triangle AOP is redrawn; the bisector of $\angle$ AOP meets AP at Q. Then triangles AOQ, POQ are congruent (SAS) whence AQ = QP = $\tfrac{1}{2}r$ and also $\angle$ OQA is a right angle. Hence

$$\cos \phi = \tfrac{1}{2}r/a = r/d$$

where $d = 2a$. To obtain p in standard form, we note that

$$\cot \phi = \frac{\cos \phi}{\sin \phi} = \frac{\cos \phi}{\sqrt{(1 - \cos^2 \phi)}} = \sqrt{\left(\frac{r^2/d^2}{1 - r^2/d^2}\right)} = T \quad \text{say}$$

so

$$p = (1/\pi)\cot^{-1} T$$

Now, for a dummy variable y,

$$\frac{d}{dy} \cot^{-1} y = -\frac{1}{1 + y^2}$$

Also

$$\cot^{-1} \infty = 0$$

Hence

$$p = \frac{1}{\pi} \int_\infty^T \left(-\frac{1}{1 + y^2}\right) dy = \int_T^\infty \frac{1}{\pi(1 + y^2)} dy$$

PROBABILITY OF FIXATION: GENERAL CASE

Thus, when $n = 2$, the probability that the mutant is advantageous is just the probability of exceeding T in a *Cauchy distribution*, that is, a distribution with probability density function

$$1/\pi(1 + y^2)$$

Now the Cauchy distribution is a special case of the familiar Student's t-distribution, when the number of degrees of freedom equals 1. The result for general n is not, therefore, altogether surprising. Let T be defined as

$$\sqrt{\left(\frac{(n-1)r^2/d^2}{1-r^2/d^2}\right)}$$

(reducing to the previous definition when $n = 2$). It can be shown that the probability that the mutant is advantageous is just the probability of exceeding T in Student's t-distribution; the degrees of freedom being $(n-1)$.

> Oh Julius Caesar, thou art mighty yet!
> Thy spirit walks abroad!

Kimura gives this result in a different but equivalent form. As $r \to d$, $T \to \infty$ and the probability p that the mutant is advantageous tends to 0. If r is very small, T is close to zero and p is close to half the area under the t-curve, that is 0.5, as we surmised earlier.

Now consider the case in which n is very large. Then p will be very small unless $(r/d)^2$ is very small. Since we shall be interested only in cases where p is at least moderate, we may approximate $1 - r^2/d^2$ by 1. Also $(n - 1)$ will be close to n when n is large. Then, at least when n is very large, we may approximate T by

$$x = r\sqrt{(n)}/d$$

Further, the t-distribution will be very close to the normal distribution under these circumstances. We have then

$$p = \int_x^\infty \frac{1}{\sqrt{(2\pi)}} e^{-y^2/2} \, dy$$

as given by Fisher. At first sight, we would not expect this to be a good approximation when n is modest in size. In fact, the use of x rather then T neatly compensates for the error introduced by approximating the t-distibution by the normal distribution under these circumstances, except for values of p too small to be relevant here. As Kimura shows, Fisher's formula is good enough for practical purposes when $n \geq 9$.

Kimura assumes that the selective advantage α is proportional to x. Then the probability of fixation is proportional to $2xN_e/N$. In practice, n will be large, since adaptation usually involves conformity in a large number of respects. Hence, for mutants with effect r, the rate of evolution will be *proportional to*

$$2x \int_x^\infty \frac{1}{\sqrt{(2\pi)}} e^{-y^2/2} \, dy$$

As x increases, this expression rises from zero to a maximum when $x = 0.7518$ and falls steadily thereafter. Hence, with n and d fixed, the mutants with very small values of r do *not* have the most rapid rate of evolution.

The argument is very interesting but could be challenged. One might propose, for example, that x exceeds 0.7518 for all sets studied so far (Clarke, personal communication). Futher, we have supposed, for the two-dimensional case, that points representing genomes of equal fitness lie on the circumference of a circle; in n dimensions, such points would lie on the surface of a hypersphere. Although, in the present context, we are free to apply any mathematical transformation we wish to our metrical characters, it seems unlikely that any transformation would exist that would give rise to hyperspheres of equal fitness, in cases where all characteristics making up the genome were considered. Thus the degree of approximation in Kimura's result is uncertain.

Notes and exercises

1 *The matrix* $\mathbf{Q}^n$. Let f_i be the probability that, at some stage, the allele frequency is $i/(2N)$. Let

$$\mathbf{f} = (f_1 \quad f_2 \quad f_3 \quad \cdots \quad f_{2N-1})$$

Suppose that, in generation n, $\mathbf{f}$ takes value $\mathbf{F}_n$. Then, with $\mathbf{Q}$ defined as in the main text,

$$\mathbf{F}_{n+1} = \mathbf{F}_n \mathbf{Q}$$

directly from the definitions of these matrices. Then

$$\mathbf{F}_1 = \mathbf{F}_0 \mathbf{Q} \qquad \mathbf{F}_2 = \mathbf{F}_1 \mathbf{Q} = \mathbf{F}_0 \mathbf{Q}^2 \qquad \mathbf{F}_3 = \mathbf{F}_2 \mathbf{Q} = \mathbf{F}_0 \mathbf{Q}^3$$

and generally

$$\mathbf{F}_n = \mathbf{F}_0 \mathbf{Q}^n$$

Thus the element in the ith row, jth column of $\mathbf{Q}^n$ must be the probability of changing from frequency $i/(2N)$ in generation 0 to frequency $j/(2N)$ in generation n.

2 In our notes to Chapter 1, we gave, but did not derive, a formula for the probability that our man eventually reaches home. It is interesting that this formula can be derived by Moran's method (Moran 1968).

We require a quantity whose *mean* never changes. Show that, if $P = Q$, a suitable quantity is the number of paces from the lake and that, if $P \neq Q$,

$$(Q/P)^{\text{number of paces from lake}}$$

will be suitable. Hence derive our formula for the probability of eventually reaching home (you can assume that the man cannot wander back and forth along the street indefinitely but must eventually end up either at home or in the lake, which follows from the fact that home or lake can be reached in a finite number of paces).

3 *Moran's model.* Let $u(p_0)$ be the probability that allele A, conveying selective advantage α, at initial frequency p_0 in a population of size $2N$, is ultimately fixed. We have shown that

$$u(p_0) = \frac{1 - (1 + \alpha)^{-2Np_0}}{1 - (1 + \alpha)^{-2N}}$$

this result being exact. As a partial check on the accuracy of Kimura's method for finding probability of fixation, we shall derive $u(p_0)$ by that method.

We think in terms of events rather than generations. We shall write probabilities for the three possible outcomes of any birth–death event in a convenient abbreviated form. We have, for any such event:

Change in frequency of A	Probability
$+1/(2N)$	$g(1 + \alpha)$
$-1/(2N)$	g
0	$1 - 2g - \alpha g$

for an appropriately chosen g. The reader will easily verify that

$$M = \text{mean change in frequency of } A = \frac{1}{2N}\alpha g$$

$$MS = \text{mean (change in frequency of } A)^2 = \left(\frac{1}{2N}\right)^2 (2+\alpha)g$$

so that the variance of the change in frequency equals

$$\left(\frac{1}{2N}\right)^2 [(2+\alpha)g - \alpha^2 g^2]$$

Writing u as short for $u(p_0)$, we have, as usual,

$$M \frac{du}{dp_0} + \tfrac{1}{2}V \frac{d^2 u}{dp_0^2} = 0$$

From the derivation of this formula, the correct expression for V is our MS given above. However, Kimura, as we have seen, uses, as approximation for V, the variance of the change in frequency for a *neutral* allele, which in the present case is seen (on putting $\alpha = 0$) to be $2g/(2N)^2$.

Show that, if the exact V is used,

$$u(p_0) = \frac{1 - (e^{-2\alpha/(2+\alpha)})^{2Np_0}}{1 - (e^{-2\alpha/(2+\alpha)})^{2N}}$$

whereas, if the approximate V is used,

$$u(p_0) = \frac{1 - (e^{-\alpha})^{2Np_0}}{1 - (e^{-\alpha})^{2N}}$$

In these solutions, the exact $(1+\alpha)^{-1}$ is replaced by

$$e^{-2\alpha/(2+\alpha)} \quad \text{and} \quad e^{-\alpha}$$

respectively.
    Now

$$(1+\alpha)^{-1} = 1 - \alpha + \alpha^2 - \alpha^3 + \alpha^4 - \cdots$$
$$e^{-2\alpha/(2+\alpha)} = 1 - \alpha + \alpha^2 - \tfrac{11}{12}\alpha^3 + \tfrac{19}{24}\alpha^4 - \cdots$$
$$e^{-\alpha} = 1 - \alpha + \tfrac{1}{2}\alpha^2 - \tfrac{1}{6}\alpha^3 + \tfrac{1}{24}\alpha^4 - \cdots$$

Obviously, one obtains a better approximation using the exact V; on the other hand, if α is small, the difference between these expressions is relatively slight (e.g. if $\alpha = 0.1$, they have value 0.9091, 0.9092, 0.9048, respectively).

4   While the results in note 3 are encouraging, they offer no guarantee that Kimura's method will always work for the Wright–Fisher model when α is small. Consider genic selection. With the usual Kimura approximations for M and V, we obtain

$$\frac{d^2u}{dp_0^2} + 4N\alpha \frac{du}{dp_0} = 0$$

whereas using exact expressions for M and V gives

$$\frac{d^2u}{dp_0^2} + 4N\alpha \frac{1 + \alpha p_0}{1 + \alpha + 2N\alpha^2 p_0 q_0} \frac{du}{dp_0} = 0$$

($q_0 = 1 - p_0$). It would *appear*, then, that 'α small' is not a sufficient condition for Kimura's method to supply an accurate value of $u(p_0)$. It would *seem* that we also need '$N\alpha^2$ small', as Fisher (1930a) noted in connection with his own derivation of $u(p_0)$ in the case $p_0 = 1/(2N)$.

It is here that Moran's limits for $u(p_0)$, in the case of genic selection, are so helpful, since they make it clear that, *in spite of the argument we have given*, 'α small' is, after all, sufficient.

Slightly less accurately: when α, but not necessarily $N\alpha^2$, is small, we have the close approximation

$$\frac{d^2u}{dp_0^2} + \frac{4N\alpha}{1 + 2N\alpha^2 p_0 q_0} \frac{du}{dp_0} = 0$$

This gives

$$u(p_0) = \int_0^{p_0} g(x)\, dx \bigg/ \int_0^1 g(x)\, dx$$

where

$$g(x) = \left(\frac{1 - 2x/(a+1)}{1 + 2x/(a-1)} \right)^{2/\alpha a} \qquad a = \sqrt{\left(\frac{2 + N\alpha^2}{N\alpha^2} \right)}$$

Kimura's formula has $g(x)$ replaced by $e^{-4N\alpha x}$.

It is easily shown that $g(x)$ will be close to $e^{-4N\alpha x}$ provided $(a-1) \gg 2x$. This condition holds for all x if $N\alpha^2$ is small. Otherwise, it still holds when x is sufficiently small; for larger x, $g(x)$ and $e^{-4N\alpha x}$ will both be very small and differ by very little in the *absolute* sense. Thus in all cases, values for $u(p_0)$ will be close to those given by Kimura's formula. Thus the latter is very accurate, provided α is small.

Presumably, the main point is as follows. Suppose $N\alpha^2$ is large ($\gg \alpha$); since α is small, this amounts to saying that $N\alpha$ is very large. If we choose p_0 small, say $k/(2N)$ where k is not too large, the term $2N\alpha^2 p_0 q_0$ is still quite small and may be ignored when finding $u(p_0)$.

On the other hand, if we choose p_0 at all large, greater than $2/(N\alpha)$, say, the true probability of fixation is very close to unity. Although, in some cases, the coefficient $4N\alpha$ of du/dp_0 in Kimura's equation will be much too large (owing to the neglect of $2N\alpha^2 p_0 q_0$), the sole effect will be to give a probability of fixation very very close, rather than very close, to unity.

Thus, even when $2N\alpha^2 p_0 q_0$ is large, its effects on probability of fixation is of no practical importance.

An analogous apparent difficulty arises when calculating mean sojourn times when $N\alpha^2$ is large. A slightly modified version of our argument can be used in that case to suggest that standard results will be valid, provided only that α is small.

5 Our belief that the mutation rate per locus per generation is much the same in different eucaryotic species is based, very largely, on the study of rates of mutation to harmful mutants. Many of these arise through the insertion of a transposable element. Possibly, the mutations of evolutionary importance (especially, of course, base-pair changes) arise at constant rate per year, as suggested by Kimura (1986).

6

Some notes on continuous approximations

The diffusion equation approach has been used extensively in the study of gene frequency change because of its extreme usefulness. But it is an approximation based on rather intuitive arguments. Therefore, to investigate the conditions under which such an approximation may be valid, is an important task for mathematicians.

M. Kimura (1964)

The problem stated

In the previous chapter, we derived the probability of fixation of an allele A by treating the allele frequency as a continuous variable. We noted that this approach gave a very accurate result, provided that the intensity of natural selection was not too great. In the present chapter, we consider some problems that arise when allele frequency, changing in discrete generations, is treated as a continuous variable changing in continuous time. We shall be especially concerned with the reliability of results obtained using these continuous approximations. By considering quite a simple biological problem, it will be shown that the continuous approach can, in some circumstances, yield distinctly inaccurate results. This should not, however, be interpreted as a general criticism of this approach, which, as we have seen, often works very well. Unfortunately, it is often difficult to identify in advance whether the approach will lead to an accurate solution to a given problem – see, for example, our brief discussion in Chapter 4. Frequently, the best that can be done is to check results obtained by the continuous approach against exact results obtainable in special cases – we shall give an example of this. Our main point is that results obtained from diffusion methods cannot be accepted uncritically and that every formula has, of necessity, to be checked individually. Of course, with the advent of computers, such checking has been carried out very extensively.

Probability of reaching a given allele frequency

For brevity, we shall, in most of this chapter, write the initial frequency of A as p (rather than p_0). We ask: 'What is the probability, say, f_{px}, that the frequency of A ever reaches some given value x?' For simplicity, we consider neutral alleles, although similar arguments will apply when natural selection is acting.

Consider first Moran's model. At any birth–death event, the number of A alleles either stays the same or changes by *one*, giving in the latter case a change in frequency of $1/(2N)$, positive or negative. It follows that for the frequency of A to change, in the course of time, from, say, value y to value z, this frequency must at some time pass through *all possible values between y and z*. For example, if y is less than z, then to reach z from y the frequency must necessarily take values

$$y + \frac{1}{2N}, \quad y + \frac{2}{2N}, \quad \ldots, \quad z - \frac{2}{2N}, \quad z - \frac{1}{2N}$$

in due course. As a special case, an allele with initial frequency p that is ultimately fixed must visit (at least once) all possible frequencies between p and unity (the actual pattern of such visits could, in any individual case, be very elaborate, involving many returns to frequencies already visited, but these complexities need not be considered in the present context).

This being so, we may use the following indirect but simple argument. Suppose first that p is less than x. Then

probability that A is ultimately fixed when its initial frequency is p

= probability that the frequency of A ever reaches x

× probability that A is ultimately fixed when its initial frequency is x

We know that the probability on the left is p and that the second probability on the right is x; hence

$$f_{px} = p/x$$

in any case where p is less than x. Note that when $p = x$, our formula gives the correct solution $f_{px} = 1$.

When p is greater than x, the argument just given is not applicable, since an allele changing in frequency from p to 1 is not obliged to visit frequency x.

195

However, an allele with initial frequency p that is ultimately *lost* must visit all possible frequencies between p and *zero*. Hence if p is greater than x

probability that A is ultimately lost when its initial frequency is p

$\quad$ = probability that the frequency of A ever reaches x

$\quad\quad \times$ probability that A is ultimately lost when its initial frequency is x

This gives

$$f_{px} = (1-p)/(1-x)$$

in any case where p is greater than x. Here, as before, our formula gives the correct solution $f_{px} = 1$ in the special case $p = x$.

We now leave Moran's model and consider a model in which both allele frequency and time are approximated by continuous variables. If, for any given set-up, we plot allele frequency against time, we obtain a curve which, though usually very erratic, will necessarily be continuous; we say that the allele frequency follows a continuous trajectory through time. Then by common sense (or by a standard mathematical result) an allele frequency changing in time from value y to value z must necessarily visit *every* value lying between y and z. There will be an infinity (in strict terminology a 'continuous infinity') of such intermediate values, in contrast to the finite number of intermediate values in Moran's model, but the arguments advanced earlier for Moran's model will clearly still apply and we have at once for our continuous model

$$f_{px} = p/x \quad\quad\quad \text{when } p < x$$
$$\phantom{f_{px}} = (1-p)/(1-x) \quad\quad \text{when } p > x$$

with either formula giving $f_{px} = 1$ when $p = x$.

We shall note in passing some interesting properties of f_{px}, considered as a function of the continuous variable x. Now, although the formula for f_{px} changes abruptly when $x = p$, both formulae give the same value for f_{px} at that point. We see then that f_{px} is a continuous function of x throughout the range $0 \leqslant x \leqslant 1$. On the other hand, while f_{px} may be differentiated with respect to x for all values of x *other* than $x = p$, no meaning can be given to the symbol

$$df_{px}/dx$$

in the special case $x = p$. To see this, start with $x = p$ and try to differentiate f_{px} from first principles. To achieve consistency with the notation of our schooldays, write y for f_{px}. If, in the usual calculus notation, we take δx positive, we have $(x + \delta x) > p$, so that

$$\delta y = \frac{p}{x + \delta x} - \frac{p}{x} = \frac{-p\delta x}{x(x + \delta x)} = -\frac{\delta x}{p + \delta x}$$

since we are starting at $x = p$. On the other hand, if we take δx negative, we have $(x + \delta x) < p$, so that

$$\delta y = \frac{1-p}{1 - x - \delta x} - \frac{1-p}{1-x} = \frac{(1-p)\delta x}{(1-x)(1 - x - \delta x)} = \frac{\delta x}{1 - p - \delta x}$$

Thus

$$\delta y/\delta x = -1/(p + \delta x) \quad \text{if } \delta x \text{ is positive}$$
$$= 1/(1 - p - \delta x) \quad \text{if } \delta x \text{ is negative}$$

If then we let $\delta x \to 0$, we get quite *different* limits, namely $-1/p$ and $1/(1-p)$ in the two cases. Clearly, 'our old friend', *the limit of $\delta y/\delta x$ as $\delta x \to 0$, does not exist*, since we get different answers for this limit if, assuming it exists, we attempt to calculate it by two valid methods.

Formally, we say that f_{px} is a continuous function of x (throughout the range of the latter) but is 'not differentiable' with respect to x in the special case $x = p$. We may surmise (correctly) that we shall meet functions of this kind elsewhere in our subject. Let $\bar{s}_{px}$ be the mean sojourn time at frequency x, given initial frequency p. We proved in Chapter 3 that

$$\bar{s}_{px}/\bar{s}_{xx} = f_{px}$$

where f_{px} is defined as earlier in this chapter. In view of this close relationship between $\bar{s}_{px}$ and f_{px}, it will hardly surprise us to learn that $\bar{s}_{px}$ is a continuous function of x, but not differentiable when $x = p$, as was first shown by Ewens (1963a). It is hoped that the preceding discussion will have shown that such functions arise quite naturally when we use continuous approximations.

Leaving this digression, we now consider the value of f_{px}, given the Wright–Fisher model. On this model, the allele frequency may 'jump' over several frequencies in the course of a single change in frequency. Hence, in order to pass from frequency y to frequency z, it is *not* necessary to visit all possible frequencies between y and z. Thus none of our preceding discussion is appropriate. We conclude that, whereas the continuous approach gives the correct value of f_{px} for Moran's model, it cannot do so for the Wright–Fisher model.

SOME NOTES ON CONTINUOUS APPROXIMATIONS

The evolutionist will be particularly interested in the case where A is a newly arisen mutant, so we shall discuss this case in detail, With $p = 1/(2N)$, p/x is most simply written as $1/i$, where $i = 2Nx$ is the number of A alleles corresponding to frequency x. Table 6.1 gives the exact value of f_{px} and corresponding values of $1/i$, for several values of i in a diploid population of size 25. It is immediately apparent that the approximation $1/i$, while giving a very rough idea of the value of f_{px}, can be quite inaccurate (the relative error is always large, but the absolute error is probably of greater importance; however, even the absolute error is disturbingly large when i is small).

We can pursue the matter further and thus obtain an improved approximation for f_{px}, but only at the cost of a rather intricate argument. We try to decide whether the error in f_{px} ($= \bar{s}_{px}/\bar{s}_{xx}$) is associated in the main with an error in $\bar{s}_{px}$ or with an error in $\bar{s}_{xx}$ (of course, it could be both, but we try one as a possibility). In fact, it seems clear that the continous approach will not supply an accurate value for $\bar{s}_{xx}$ in the Wright–Fisher model. To see this, we turn to our modified process of Chapter 3. We recall that in this process we consider only those mutants which are eventually fixed; as before, we use the symbol * to indicate that we are dealing with the modified process, other symbols having the same meaning as above. Now we showed in Chapter 3 that

$$\bar{s}_{xx} = \bar{s}_{xx}^*$$

Table 6.1 Probability that the number of A alleles ever reaches i, given that A is a newly arisen neutral mutant in a Wright–Fisher diploid population of size 25.

i	Exact value	Approximate values $1/i$	Improved
1	1.00	1.00	1.00
2	0.38	0.50	0.38
3	0.26	0.33	0.25
4	0.19	0.25	0.18
5	0.15	0.20	0.14
6	0.12	0.17	0.12
7	0.10	0.14	0.10
8	0.09	0.13	0.09
9	0.08	0.11	0.08
10	0.07	0.10	0.07

so we consider the latter. Now in the continuous (and in Moran's) model, an allele with initial frequency p less than x must, en route to fixation, visit x. Thus when considering the mean sojourn time at frequency x in the modified process, no special significance need be attached to the case where the initial frequency of A is x, rather than less than x. In fact, since

$$\bar{s}^*_{px}/\bar{s}^*_{xx} = f^*_{px}$$

and $f^*_{px} = 1$ if p is less than or equal to x, we have

$$\bar{s}^*_{px} = \bar{s}^*_{xx}$$

provided $p \leqslant x$. But none of this is true for the Wright–Fisher model, where a visit to the starting frequency is obligatory by definition, whereas a visit to any given frequency other than p or 1 is optional. Hence when considering the mean sojourn time at frequency x, we must give a special status to the case where x is the initial value; it is just this special status that is destroyed in the continuous model. We can conclude with fair confidence, then, that the value of $\bar{s}^*_{xx}$ and hence of $\bar{s}_{xx}$ will be significantly larger in the Wright–Fisher model than the value given in the corresponding continuous approximation. It follows that if we can adjust our formula for f_{px} to allow for this error in $\bar{s}_{xx}$, a substantial gain in accuracy will result. Although we have used rather intuitive arguments, the conclusion is indeed correct. We shall not give details but merely note that it follows from the work of Pollak & Arnold (1975) that, when i exceeds 1, we shall obtain an improved value for f_{px} by multiplying our approximate value $1/i$ by

$$\frac{2\{1 - [(2N)!/i!(2N-i)!][1 - i/(2N)]^{2N-i}[i/(2N)]^i\}}{3 - 4\{[(2N)!/i!(2N-i)!][1 - i/(2N)]^{2N-i}[i/(2N)]^i\}}$$

A glance at the last column of Table 6.1 will at once convince the reader that the gain in accuracy is substantial.

Perhaps the main point to emerge from this lengthy discussion is that the use of continuous approximations sometimes introduces features, such as the lack of special status for the initial frequency in the case described above, that are intrinsic properties of the continuous approximation but not necessarily of the model being approximated. When, therefore, one obtains a result that seems unnatural, it is well worth considering whether it is the continuous representation that is responsible. Of course, the ultimate test is not whether the results are intuitively reasonable, but whether they are in good agreement with results deriving from the model that is being approximated. A general theory of such approximations has not, however, been achieved, as is hardly surprising in view of the complexities that we have described (and others that we have not!). Although algebraic checks of agreement are sometimes possible, a comparison of numerical results in special cases is often the only method of checking available.

Events at the boundaries

So far we have discussed some effects of continuous approximations but have deliberately avoided mentioning some other features essential to the use of diffusion methods (hence our use of the term 'continuous model' rather than 'diffusion model'). If, however, we wish to discuss the way in which the diffusion model formulae represent events at the 'boundaries' (i.e. at allele frequency 0 and 1), we must say just a little more about diffusion methods. The rigorous treatment of events at the boundaries (Feller 1952) is well beyond the scope of this book; we shall, therefore, treat the matter *very* informally. We shall investigate what happens at the boundaries in diffusion models and compare our results with those obtained from the Wright–Fisher model (the latter and Moran's model are essentially the same as far as boundaries are concerned). To facilitate the comparison, we borrow some descriptive terms from diffusion theory when discussing the Wright–Fisher model.

The crux of the matter lies in the representation of the variance of the allele frequency. Suppose that the frequency of A in some generation is x; then the variance of the frequency of A one generation later is

$$x^*(1 - x^*)/2N$$

where x^* is the value of the allele frequency that would be obtained in this later generation if drift were absent. However, it is universal practice to approximate the formula just given by

$$x(1 - x)/2N$$

Such an approximation is justified provided evolution is slow, since in such a case x^* will differ very little from x (of course, in the neutral case x^* will equal x exactly, if mutation is absent). In fact, in the rigorous treatment of diffusion models, the theory is formulated in such a way that the difference between our true and approximate variance is, at most, comparable in magnitude to $1/(2N)^2$ and may therefore, like other terms of comparable magnitude, be neglected. However, the approximation has a remarkable consequence; when x equals 0 or 1, the approximate variance is zero, implying no drift. Thus in diffusion models, the behaviour of the allele frequency is *essentially deterministic* at the boundaries, i.e. when $x = 0$ and when $x = 1$.

In the absence of mutation, this is certainly correct. For if $x = 0$, $x^* = 0$, and if $x = 1$, $x^* = 1$. Thus our *exact* variance is zero when $x = 0$ or 1 and we really do have deterministic behaviour at the boundaries. More specifically, consider the boundary $x = 1$ (precisely the same considerations will apply at $x = 0$). In the Wright–Fisher model, there is a non-zero probability that the frequency of allele A reaches 1 in finite time. Once the frequency reaches 1, it stays there for good. A boundary of this kind is known as an *exit*; starting at any intermediate value, the allele frequency can reach such a boundary in finite time, but once there cannot leave it.

Now consider the diffusion model. We showed in the last chapter that, in the absence of mutation, there is a non-zero probability that the frequency of A will reach 1 and will show later that this can happen in finite time. Once the boundary $x = 1$ is reached, we have a deterministic process with mean change in frequency $x^* - x$ equal to zero; since the process is deterministic, the mean change is just the actual change, which is thus zero for all time. The same applies at the other boundary. Hence the boundaries are both exits and the two models agree.

However, things are quite different when mutation is acting. Suppose, for simplicity, that there are just two alleles A, a at a given locus. Let A mutate to a at rate u and a mutate to A at rate v. For simplicity, we suppose that these alleles are neutral (nothing essentially new is added, in the present context, if we bring in natural selection). Now, owing to the mutation, permanent fixation or loss of either allele is impossible but by chance one allele may be temporarily fixed. Consider allele A and concentrate on the boundary $x = 1$. In the Wright–Fisher model, there is a non-zero probability that the frequency of A reaches 1 in finite time. Having reached this boundary, the allele frequency can certainly leave it, owing to mutation. A boundary of this kind is known as *regular*; starting at any intermediate value, the allele frequency can reach such a boundary in finite time and move from the boundary to some intermediate value in finite time. Notice that the allele frequency may spend some time at frequency 1. If the frequency reaches 1 in some generation, the probability that it is still 1 a generation later is $(1 - u)^{2N}$, or about e^{-2Nu} since u is small. Hence if Nu is not too large, there is a good chance that A, having reached fixation, will remain fixed for some generations. Similar considerations apply at the other boundary. Thus both boundaries are regular and in addition there is the possibility of a temporary stay at either boundary.

Now consider the diffusion model. Suppose that the frequency of A equals 1. Even if we think in terms of generations, it is clear that, since the process is essentially deterministic, the frequency of A must fall to $(1 - u)$ one generation later. However, the situation is much exacerbated by the use of the continuous time approximation. Owing to this, we must 'spread' the mutation rate u over all the time making up a generation. For any short period of time δt comprising part of a generation, the mutation rate will be about $u\delta t$ in this representation. Thus, if the frequency of A is 1, the frequency then changes deterministically with mutation rate $u\delta t$, however small the value of δt. Hence, given the diffusion model, the frequency of A leaves 1 *immediately*. More precisely, if we think of a time, however small (say the time for which God's anger lasts - 1/58 888th part of an hour, according to the *Talmud*), the frequency of A will remain at 1 for less than that time.

Clearly, the diffusion model greatly exaggerates the strength of the forces driving the frequency of A away from the boundary. Now we showed above that this strength is more properly measured by the value of Nu; in the diffusion model, the strength will be considerably greater than this. It is not, therefore, altogether surprising that if Nu is sufficiently large, an allele that starts at an intermediate frequency is *forever unable to reach the boundary* $x = 1$. More specifically, by a slight modification of the procedure for finding probability of fixation given in Chapter 5, we can show that if $4Nu \geq 1$, the probability that A ever reaches frequency 1 is zero. A boundary of this kind, which has no counterpart in the Wright–Fisher or Moran models, is known as an *entrance*; starting at such a boundary, the allele frequency can reach an intermediate value in finite time, but it can never return to that boundary.

If $4Nu < 1$, the probability that A ever reaches frequency 1 proves to be greater than zero; it can also be shown that this frequency can be reached in finite time (and, of course, left in finite time). Thus $x = 1$ is a regular boundary in this case, as in the Wright–Fisher and Moran models. But, in contrast to the set-up for those models, the allele frequency, having reached 1, leaves that value instantly, as explained earlier. In that the allele frequency spends less than any arbitrarily small time at frequency 1, the probability that the allele frequency equals 1 must be zero, even when the boundary is regular.

The situation at the boundary $x = 0$ is entirely similar. If $4Nv \geq 1$, the boundary is an entrance; otherwise the boundary is regular, but once reached is left immediately.

We see then that, given two-way mutation, diffusion models will necessarily represent the allele frequency as lying *between* 0 and 1. It is remarkable that Wright's treatment of two-way mutation has always incorporated this feature; its necessity was first shown by Crow & Kimura (1956), using a different approach from that given above. In practice, of course, the allele frequency could well be 0 or 1 if $4Nv$ and $4Nu$ were small. It would appear that the diffusion approach would be useless for finding the probability that the allele frequency is 0, or is 1, in such cases. Fortunately, this is not so; these probabilities can be extracted by special devices (Wright 1931, Ewens 1965, 1979, Kimura 1968b), which we discuss in a later chapter.

Finally, we consider the case of multiple alleles. With, say, k alleles mutating *inter se* (Kimura 1968b), things come out much the same as in the two-allele case. Now consider the 'infinite allele' model of mutation. Here it is supposed that a given allele A mutates to all other possible alleles combined at rate u, say, but that the number of possible alleles is so large that back-mutation to A is negligibly small; thus if A is lost, it never reappears. The diffusion representation at the boundaries comes out as expected from our earlier discussion. The boundary $x = 0$ is an exit; $x = 1$ is an entrance if $4Nu \geq 1$, regular if $4Nu < 1$; in the latter case an allele frequency reaching 1 leaves this value instantaneously.

The problem of representing the initial frequency

We conclude these notes on continuous approximations by discussing the following problem. Suppose we are given the initial allele frequency. It will be simpler here to return to our standard notation and write p for the allele frequency and p_0 for the given initial frequency. Then initially (i.e. in generation 0) $p = p_0$ with probability one and p equals any other value with probability zero. As long as we think of p as a discontinuous variable, changing in steps of $1/(2N)$, the probability distribution just described presents no problems. Suppose, however, we wish to use continuous approximations; then for consistency, we must treat p as a continuous variable, even in generation 0. How are we to represent our probability distribution in generation 0 in this case?

Ad hoc method

The following procedure is plausible and attractive. We write the distribution of p in generation 0 as a normal distribution with mean p_0 and *very small variance*. Thus we could write the variance, σ^2, in quite arbitrary fashion as $1/(2n)$, n being a dummy variable, and suppose that n is very large. This gives us the probability density function

$$\psi(p) = \frac{1}{\sigma\sqrt{(2\pi)}} e^{-(p-p_0)^2/2\sigma^2}$$

$$= \sqrt{\left(\frac{n}{\pi}\right)} e^{-n(p-p_0)^2}$$

with n very large. Then, as we are dealing with a probability distribution

$$\int_{-\infty}^{\infty} \psi(p)\, dp = 1 \tag{6.1}$$

and since the variance is very small, we have to a good approximation

$$\psi(p) = 0 \tag{6.2}$$

unless p is very close to its mean p_0. Notice that properties 6.1 and 6.2, taken together, are all that we require to show that our procedure is reasonable, since they make it clear that we are dealing with a continuous probability distribution with probability concentrated in the neighbourhood of p_0.

A third property of this distribution is slightly less obvious. Let $g(p)$ be any function of p. Then, by definition of mean, the mean value of $g(p)$, given our distribution of p, is

$$Eg(p) = \int_{-\infty}^{\infty} g(p)\psi(p)\,dp$$

But, with very high probability, p is very close to its mean p_0, so that the mean of $g(p)$ will be very close to $g(p_0)$. Hence, to a good approximation

$$\int_{-\infty}^{\infty} g(p)\psi(p)\,dp = g(p_0) \tag{6.3}$$

Objections to our *ad hoc* method

Clearly, our procedure gives a representation in which the allele frequency in generation 0 is close to its given value p_0. One might, however, object to our procedure on several reasonable grounds.

First, the allele frequency, in our representation, does not have to equal p_0 exactly, but only to be very close to p_0 with very high probability; thus our procedure is approximate. Secondly, our procedure does not seem very practicable, since we have given no numerical value to n, but have merely stated that n is very large. Thirdly, and most seriously, our whole procedure seems very arbitrary. The normal distribution has been chosen purely on the basis of familiarity, but with a little ingenuity one can find many other distributions that would be just as appropriate. Consider, for example, the probability density function, shown in Figure 6.1,

$$\chi(p) = 0 \quad \text{when } p < p_0 - 1/(2n)$$
$$= n \quad \text{when } p_0 - 1/(2n) \leq p \leq p_0 + 1/(2n)$$
$$= 0 \quad \text{when } p > p_0 + 1/(2n)$$

Clearly (from the figure or otherwise) the area under the curve is unity and if n is chosen very large, p will be very close to p_0; in other words properties 6.1,

DIRAC'S DELTA FUNCTION

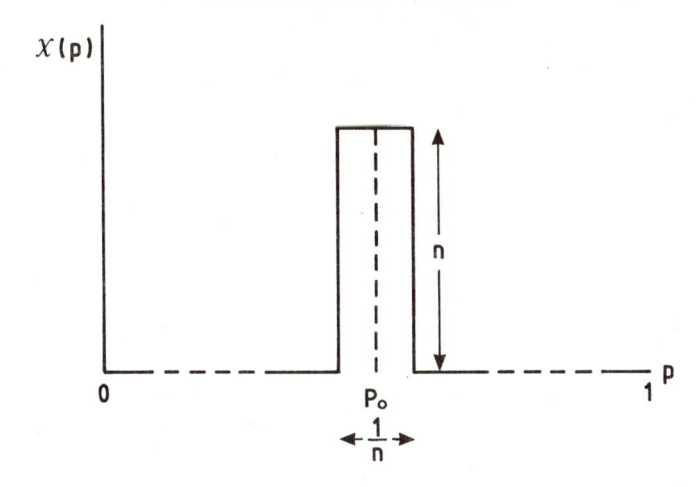

Figure 6.1 Alternative representation of distribution of allele frequency p in generation 0; n is supposed very large.

6.2 (and hence 6.3) apply to $\chi(p)$. Thus, since 6.1 and 6.2 apply, $\chi(p)$ is just as appropriate to our purpose as is $\psi(p)$.

It is easily seen what is wrong. All that we require is embodied in properties 6.1 and 6.2 (property 6.3 following automatically from these); our uncertainty on whether to use $\psi(p)$, $\chi(p)$, or any of the other functions mentioned, has arisen only because we have attempted to exhibit these properties in the context of one or other standard distribution. There is no need to do this. Rather, we should try to construct our distribution on the basis of 6.1 and 6.2 only, with 6.2 holding to as high a degree of accuracy as possible when $p \neq p_0$. On the other hand, since, after all, $\psi(p)$, $\chi(p)$ and other functions do seem to have some relevance to the problem, we naturally hope that this relevance will appear at a later stage.

Dirac's delta function

It is clear what we would like to do, provided we can do so without talking nonsense. We define a function, say

$$\delta(p - p_0)$$

having property 6.1 and for which property 6.2 holds, *exactly*, unless p is *exactly* equal to p_0. That is

$$\int_{-\infty}^{\infty} \delta(p - p_0) \, dp = 1 \qquad (6.1')$$

and

$$\delta(p - p_0) = 0 \qquad (6.2')$$

when $p \neq p_0$ (in view of this, $\delta(p - p_0)$ will certainly be zero outside the range 0 to 1 and it need cause no concern that in 6.1' the limits of integration are $-\infty$, ∞ rather than the more natural 0, 1). It follows at once that for any function $g(p)$

$$\int_{-\infty}^{\infty} g(p)\delta(p - p_0) \, dp = g(p_0) \qquad (6.3')$$

exactly. Thus multiplying $g(p)$ by $\delta(p - p_0)$ and integrating 'spotlights' the value of $g(p)$ when $p = p_0$.

Our 'delta' function $\delta(p - p_0)$ was introduced by the celebrated atomic physicist, Paul Dirac, whose main interest lay in the correct use of his function in applications in physics and who did not really explain the nature of this function, which at first sight seems bizarre and altogether impossible. Gamow (1966) relates:

> In the case of Dirac all the stories are really true.... At the question period after a Dirac lecture at the University of Toronto, somebody in the audience remarked: 'Professor Dirac, I do not understand how you derived the formula on the top left side of the blackboard.'
>
> 'This is not a question,' snapped Dirac, 'it is a statement. Next question, please.'

Certainly no ordinary function has the properties of Dirac's delta function. It is an example of a 'symbolic' (or 'generalized') function. The nature of these functions has been elucidated by Laurent Schwarz, who has shown that symbolic functions may be manipulated in the same way as ordinary functions. For example, in expressions involving Dirac's delta function, the usual rules for integration by substitution or by parts still apply. Further, any function which, in the limit, has properties 6.1' and 6.2' may be used to represent the delta function, provided the function chosen is taken at its limiting value. For example, we can regard the delta function as the limit, as $n \to \infty$, of either $\psi(p)$ or $\chi(p)$. In the limit, all such functions are equivalent and our difficulty over arbitrariness in choice of function disappears.

We have built our discussion around the problem of representing the distribution of allele frequency in generation 0 and this seems the easiest way in which to approach the delta function. In fact, this function first appears in population genetics as a solution to our problem (Kimura 1955a). However, the delta function also appears in other contexts (Kimura 1955b, Maruyama 1977). Suppose, for example, that we wish to represent the distribution of allele frequency, taken as a continuous variable, when the allele frequency is changing *deterministically* in continuous time. Suppose that the allele frequency at time t, calculated by the usual deterministic methods, is p_t. Then at time t, $p = p_t$ with probability one and p equals anything else with probability zero, so that, given that change is deterministic, the distribution of allele frequency at time t has probability density function

$$\delta(p - p_t)$$

With a view to coping with this and other formulae in which the delta function appears, we end our discussion by setting out a few procedures, which will be useful later.

Suppose we have the integral

$$\int_a^b g(p)\delta(p - p_0)\,dp$$

where a and b are arbitrary. If p_0 lies outside the range a to b, then p, ranging from a to b, never takes the value p_0; thus $\delta(p - p_0)$ is always zero and the integral is zero. On the other hand if p_0 lies inside the range a to b, it follows from our earlier arguments that the integral is $g(p_0)$. For example

$$\int_0^x \frac{1}{p(1-p)} \delta(p - p_0)\,dp$$

equals zero if $p_0 > x$ (and thus lies outside the range 0 to x) but is

$$1/[p_0(1 - p_0)]$$

if $p_0 < x$ (and thus lies inside the range 0 to x). Some further examples are given in the notes and exercises.

Sometimes we need to check expressions in which the delta function appears. Supposing, for example, it is asserted that

$$\delta(p - p_0) = \delta(p_0 - p)$$

To verify this and other expressions involving the delta function, we use a 'test function', say $\theta(p)$. A wide variety of functions would be suitable; the sole restrictions placed on $\theta(p)$, for it to qualify as a test function, are that $\theta(p)$ may be successively differentiated any number of times and that $\theta(p)$ is zero outside some finite range of p. We shall not attempt to justify these esoteric restrictions. We now multiply each side of the equation to be verified by $\theta(p)$ and integrate with respect to p over the limits $-\infty, \infty$. If we obtain the same number whichever side we start with, our original equation is correct. Thus

$$\int_{-\infty}^{\infty} \theta(p)\delta(p - p_0)\, dp = \theta(p_0)$$

the usual 'spotlight' result. Now consider

$$\int_{-\infty}^{\infty} \theta(p)\delta(p_0 - p)\, dp$$

Let $p = -q$, $p_0 = -q_0$ and use the usual results for integration by substitution. Our integral becomes

$$\int_{\infty}^{-\infty} \theta(-q)\delta(q - q_0)(-1)\, dq = \int_{-\infty}^{\infty} \theta(-q)\delta(q - q_0)\, dq$$
$$= \theta(-q_0) \qquad \text{(the spotlight result)}$$
$$= \theta(p_0)$$

Thus we obtain $\theta(p_0)$, irrespective of whether we started with $\delta(p - p_0)$ or $\delta(p_0 - p)$ and these expressions are therefore identical.

The reader will probably be relieved to discover that our discussion of the delta function should be adequate for following the literature of population genetics. Any reader, however, who is particularly interested in the delta function is referred to the many excellent books dealing with this function and related topics; the present author has found Mackie (1965) particularly helpful.

Notes and exercises

1 Consider our f_{px}, given by

$$f_{px} = p/x \qquad \text{when } p \leqslant x$$
$$= (1 - p)/(1 - x) \qquad \text{when } p \geqslant x$$

Show that f_{px}, considered as a function of p, is not differentiable with respect to p when $p = x$.

NOTES AND EXERCISES

2 Consider a case of one-way mutation: A mutates to a at rate u, with zero back-mutation rate. While A must be lost *eventually*, by chance the frequency of A could reach unity at an earlier stage. Do we obtain this result if we use diffusion methods?

Let p be the initial frequency of A and (following Karlin & Taylor 1981) let $y(p)$ be the probability that the frequency of A reaches value $b (b > 0)$ before eventual loss. Assume the Wright–Fisher model. Use Kimura's approach to show that (on writing y as short for $y(p)$)

$$\frac{d^2 y}{dp^2} - \frac{4Nu}{1-p}\frac{dy}{dp} = 0$$

Multiply through by $(1-p)^{4Nu}$ and hence show that

$$dy/dp = A(1-p)^{-4Nu}$$

where A is a constant. Thus obtain

$$y(p) = C(1-p)^{1-4Nu} + B \qquad \text{when } 4Nu \neq 1$$
$$= E \log(1-p) + D \qquad \text{when } 4Nu = 1$$

where B, C, D, E, are constants, easily evaluated from the conditions

$$y(0) = 0 \qquad y(b) = 1$$

Show that

$$y(p) = \frac{1 - (1-p)^{1-4Nu}}{1 - (1-b)^{1-4Nu}} \qquad \text{when } 4Nu \neq 1$$

$$= \frac{\log(1-p)}{\log(1-b)} \qquad \text{when } 4Nu = 1$$

Now let $b \to 1$. Then, provided $4Nu < 1$, so that $(1 - 4Nu)$ is positive,

$$y(p) \to 1 - (1-p)^{1-4Nu}$$

so that frequency unity is attainable (in finite time in fact, as can be shown by diffusion methods).

But, if $4Nu \geq 1$

$$y(p) \to 0 \qquad \text{(unless, of course, } p = 1\text{)}$$

SOME NOTES ON CONTINUOUS APPROXIMATIONS

Thus, according to the diffusion approach, frequency unity is a regular boundary only when $4Nu < 1$. Otherwise it is an entrance boundary. This result is clearly misleading.

3 When we say that, in the limit as $n \to \infty$, $\psi(p)$, $\chi(p)$ are equivalent, we mean, strictly speaking: given any test function $\theta(p)$

$$\lim_{n \to \infty} \int_{-\infty}^{\infty} \psi(p)\theta(p) \, dp = \lim_{n \to \infty} \int_{-\infty}^{\infty} \chi(p)\theta(p) \, dp$$

(in fact, both of these expressions equals $\theta(p_0)$).

4 A heuristic device when evaluating integrals involving the delta function is to put

$$\delta(x - a) = 0 \qquad \text{when } x \neq a$$
$$= 1/dx \qquad \text{when } x = a$$

5 For a little practice in using the delta function, evaluate

(a) $\int_{-\infty}^{\infty} f(x)\delta(x) \, dx$

(b) $\int_{-\infty}^{\infty} f(x)\delta(x - a) \, dx$

(c) $\int_{-\infty}^{\infty} f(x - a)\delta(x - a) \, dx$

Answers

(a) $f(0)$, (b) $f(a)$, (c) $f(0)$.

6 Use a test function $\theta(x)$ to verify that

(a) $f(x)\delta(x - a) = f(a)\delta(x - a)$
(b) $\delta(ax) = (1/a)\delta(x)$

(a is a positive constant).

7 We can use a test function $\theta(x)$ to define $\delta'(x)$, the derivative of $\delta(x)$. For, on a formal application of integration by parts, we have (recalling that our test function must equal 0 when x lies outside a finite range)

$$\int_{-\infty}^{\infty} \delta'(x)\theta(x) \, dx = -\int_{-\infty}^{\infty} \delta(x)\theta'(x) \, dx = -\theta'(0)$$

8 The 'function of a function' rule may validly be used when differentiating the delta function. Suppose, for example, that we wish to differentiate, with respect to t, the expression

$$\delta(x - y)$$

where y, but not x, is a function of t. Writing $\delta'(x - y)$ for the derivative of $\delta(x - y)$ with respect to $(x - y)$, we have

$$\frac{d}{dt} \delta(x - y) = \delta'(x - y)\left(-\frac{dy}{dt}\right)$$

9 Consider the equation (Kimura 1955b)

$$\frac{\partial \phi(x, t)}{\partial t} = -M(x) \frac{\partial \phi(x, t)}{\partial x} - M'(x)\phi(x, t)$$

Here x and t are independent variables; $M(x)$ is a function of x only and $M'(x)$ is the derivative of $M(x)$ with respect to x.

Also, x_t is a function of t only, x_t being defined by

$$dx_t/dt = M(x_t)$$

where $M(x_t)$ is obtained by taking $M(x)$ and replacing x by x_t.

We wish to show that our first equation has solution

$$\phi(x, t) = \delta(x - x_t)$$

Show that, if we substitute this proposed solution in our equation, we obtain

$$\delta'(x - x_t)M(x_t) = M(x)\delta'(x - x_t) + M'(x)\delta(x - x_t)$$

Use a test function $\theta(x)$ to show that the LHS of this equation really does equal the RHS (when integrating, use integration by parts when necessary and do not forget that $M(x_t)$ is not a function of x, whereas both $M(x)$ and $\theta(x)$ are!), thus verifying our solution. Kimura gives a proof that, if we supplement our first equation by the condition for $t = 0$,

$$\phi(x, 0) = \delta(x - x_0)$$

(where x_0 is a given constant), our solution is unique. Essentially, his argument amounts to showing that all solutions of our first equation are covered by the formula

$$M(x)\phi(x, t) = f\left(\int_x^{x_t} \frac{d\theta}{M(\theta)}\right)$$

where f is an arbitrary function, θ is a dummy variable and $M(\theta)$ is $M(x)$ with x replaced by θ. Then, by our condition for $t = 0$,

$$M(x)\delta(x - x_0) = f\left(\int_x^{x_0} \frac{d\theta}{M(\theta)}\right)$$

and to accommodate this, we shall have to write

$$f\left(\int_x^{x_t} \frac{d\theta}{M(\theta)}\right) = M(x)\delta(x - x_t)$$

giving

$$\phi(x, t) = \delta(x - x_t)$$

7

Mean sojourn, absorption and fixation times

> Men have fallen in love with statues and pictures. I find it far easier to imagine a man falling in love with a differential equation, and I am inclined to think that some mathematicians have done so.
> J. B. S. Haldane, *Scientific Calvinism*

Retrospect

The conservationist, wishing to preserve genetic variability in his material for as long as practically possible, will naturally ask: 'How many generations, on average, will it take for such variability to be lost?' More strictly, consider a population in which two or more alleles at a given locus coexist in generation 0. Let t be the first generation in which all members of the population are identically homozygous at the locus. Then at least two alleles will have coexisted in the population in generations $0, 1, 2, \ldots, (t-1)$, that is, for t generations in all; we stress that the initial generation is included. Then, as in Chapter 3, t is called the absorption time, which, owing to drift, is a random variable. If the identical conservation programme were carried out on identical initial material on different occasions, the value of t would vary from occasion to occasion. Our conservationist, then, wishes to know the mean absorption time, that is, the value of t, averaged over all possible occasions.

In contrast, the evolutionist will be primarily concerned with the mean time to fixation of *those alleles that are eventually fixed* (generation 0 counting, as in the case of mean absorption time). As in Chapter 3, we shall call this the mean fixation time and use the term 'modified process' to indicate that we are considering only a restricted class of alleles, in contrast to the 'unmodified process' in which all alleles are considered.

For either process, we defined the mean sojourn time at any frequency from $1/(2N)$ to $(2N-1)/2N$ as the mean number of generations (including generation 0) spent at that particular frequency by a specific allele, A say, prior to absorption or fixation. Let the initial frequency of A be $i/(2N)$. Write, as before, $\bar{s}_{ij}$ for the mean sojourn time of A at frequency $j/(2N)$ in the unmodified process and $\bar{s}_{ij}^*$ for the corresponding mean sojourn time in the modified process. Then the mean fixation time $\bar{s}_i^*$ is

$$\sum_{j=1}^{2N-1} \bar{s}_{ij}^*$$

For most of our discussion, we shall suppose that just two alleles are present at our locus in generation 0, so that the absorption time is the first generation in which one allele reaches frequency 0 or 1. Then the mean absorption time $\bar{s}_i$ is

$$\sum_{j=1}^{2N-1} \bar{s}_{ij}$$

This restriction to two alleles at a locus need not cause concern; once we have discussed this case, results for multiple alleles will seem quite natural.

We recall that it is unnecessary to calculate $\bar{s}_{ij}$, $\bar{s}_{ij}^*$ separately, since

$$\bar{s}_{ij}^* = \bar{s}_{ij} u_j / u_i$$

where u_j, u_i are the probabilities of ultimate fixation when A has initial frequency $j/(2N)$, $i/(2N)$ respectively, as calculated in Chapters 4 and 5. Thus all quantities of interest in the present context follow once $\bar{s}_{ij}$ has been determined for all i and j. Most of our discussion, therefore, will centre on finding $\bar{s}_{ij}$, although occasionally it will be convenient to find other quantities directly.

It will help to recall (from Ch. 3) that f_{ij}, the probability that A, starting at frequency $i/(2N)$ in the unmodified process, ever reaches frequency $j/(2N)$, is given by

$$f_{ij} = \bar{s}_{ij} / \bar{s}_{jj}$$

and (from Ch. 6) that for Moran's model

$$f_{ij} = u_i/u_j \qquad \text{when } i \leqslant j$$
$$= (1 - u_i)/(1 - u_j) \qquad \text{when } i \geqslant j$$

u_i, u_j being defined as above.

Fundamental equations

Given initial frequency of A equal to $i/(2N)$, we wish to find the mean sojourn time of A at frequency $r/(2N)$ in the unmodified process, for all values of r from 1 to $(2N-1)$ inclusive.

Imagine, just for the moment, that the frequency of A changed from known value $i/(2N)$ in generation 0 to *known* value $j/(2N)$ in generation 1. Suppose that we calculate mean sojourn times on the assumption, apparently irrelevant, that the initial frequency of A was $j/(2N)$. Then we have 'lost' the initial generation at frequency $i/(2N)$, but otherwise mean sojourn times will be the same as for initial frequency $i/(2N)$. Thus, when j is *known*

mean sojourn time at frequency $r/(2N)$, for initial frequency $i/(2N)$

= mean sojourn time at frequency $r/(2N)$ for initial frequency $j/(2N)$

+ one generation extra in the case $r/(2N)$ equal to $i/(2N)$

We may write this as

$$\bar{s}_{ir} = \bar{s}_{jr} + d_{ir}$$

where

$$d_{ir} = 1 \quad \text{when } r = i$$
$$= 0 \quad \text{otherwise}$$

In practice, j will not be known, but we can easily modify our results to allow for this. Let g_{ij} be the probability of going from frequency $i/(2N)$ in generation 0 to frequency $j/(2N)$ in generation 1. Then we have the fundamental equation

$$\bar{s}_{ir} = \sum_{j=0}^{2N} g_{ij}\bar{s}_{jr} + d_{ir}$$

There will be one such equation for every value of r from 1 to $(2N-1)$, giving $(2N-1)$ equations in all for every value of i. Note that, in the present context, sojourn times at frequencies 0 or 1 are meaningless; thus we shall never take r to be zero or $2N$. Hence, for initial number of A equal to 0 or $2N$, A can never reach frequency $r/(2N)$. Hence we have the boundary conditions

$$\bar{s}_{0r} = 0 \qquad \bar{s}_{2N,r} = 0$$

so that our fundamental equation becomes

$$\bar{s}_{ir} = \sum_{j=1}^{2N-1} g_{ij}\bar{s}_{jr} + d_{ir} \tag{7.1}$$

With both i and r taking any value from 1 to $(2N-1)$, we have then $(2N-1)^2$ simultaneous equations, which, as we shall show shortly, provide a unique solution for every $\bar{s}_{ir}$.

Summing both sides of our fundamental equation over all r from 1 to $(2N-1)$ and writing

$$\bar{s}_i = \sum_{r=1}^{2N-1} \bar{s}_{ir} \qquad \bar{s}_j = \sum_{r=1}^{2N-1} \bar{s}_{jr}$$

so that $\bar{s}_i$, $\bar{s}_j$ are the mean absorption times for initial frequencies $i/(2N)$, $j/(2N)$ respectively, we obtain

$$\bar{s}_i = \sum_{j=0}^{2N} g_{ij}\bar{s}_j + \sum_{r=1}^{2N-1} d_{ir}$$

the fundamental equation for mean absorption times. The boundary conditions are

$$\bar{s}_0 = 0 \qquad \bar{s}_{2N} = 0$$

so that

$$\bar{s}_i = \sum_{j=1}^{2N-1} g_{ij}\bar{s}_j + \sum_{r=1}^{2N-1} d_{ir}$$

When i does not equal 0 or $2N$, r, in ranging from 1 to $(2N-1)$, must equal i at some stage. Hence, since $d_{ir} = 1$ when $r = i$ but $d_{ir} = 0$ for all other values of r, then

$$\bar{s}_i = \sum_{j=1}^{2N-1} g_{ij}\bar{s}_j + 1 \tag{7.2}$$

Of course, this equation could have been deduced directly, without reference to sojourn times, the 1 appearing since the mean absorption time, starting at generation 0, is one generation longer than the mean absorption time starting at generation 1. However, our approach, although admittedly tortuous, will prove very helpful later.

Uniqueness of solutions

Our next step is to put Equations 7.1 and 7.2 in matrix form. Any reader finding this difficult is advised to consider the special case $2N = 4$ and write out 7.1 in full for all nine combinations of i and r ($i = 1, 2, 3; r = 1, 2, 3$). For example

$$\bar{s}_{11} = g_{11}\bar{s}_{11} + g_{12}\bar{s}_{21} + g_{13}\bar{s}_{31} + 1$$
$$\bar{s}_{12} = g_{11}\bar{s}_{12} + g_{12}\bar{s}_{22} + g_{13}\bar{s}_{32}$$

and so on. Let **S** be a matrix of $(2N - 1)$ rows, $(2N - 1)$ columns, with element $\bar{s}_{ij}$ in the ith row, jth column; correspondingly, let **Q** be the matrix, familiar from Chapter 5, with element g_{ij} in the ith row, jth column. In either case, i and j each go from 1 to $(2N - 1)$. Finally, let **I** be the unit matrix with $(2N - 1)$ rows, $(2N - 1)$ columns. Then Equation 7.1 becomes

$$\mathbf{S} = \mathbf{QS} + \mathbf{I}$$

and therefore

$$\mathbf{S} - \mathbf{QS} = (\mathbf{I} - \mathbf{Q})\mathbf{S} = \mathbf{I}$$

We know, from Chapter 5, that $(\mathbf{I} - \mathbf{Q})$ has an inverse, so that we have the unique solution

$$\mathbf{S} = (\mathbf{I} - \mathbf{Q})^{-1}\mathbf{I} = (\mathbf{I} - \mathbf{Q})^{-1}$$

Thus, provided N is not too large, exact values for the various $\bar{s}_{ij}$ can be found by inverting $(\mathbf{I} - \mathbf{Q})$ on the computer.

Two points follow which will be helpful in the sequel. First, since any matrix, **M** say, that has an inverse 'commutes with that inverse', that is

$$\mathbf{MM}^{-1} = \mathbf{M}^{-1}\mathbf{M}$$

we have

$$(\mathbf{I} - \mathbf{Q})\mathbf{S} = \mathbf{S}(\mathbf{I} - \mathbf{Q}) = \mathbf{S} - \mathbf{SQ}$$

MEAN SOJOURN, ABSORPTION AND FIXATION TIMES

so that 7.1 may also be written as

$$S = SQ + I$$

which is the same as

$$\bar{s}_{ir} = \sum_{j=1}^{2N-1} \bar{s}_{ij} g_{jr} + d_{ir} \tag{7.1'}$$

Secondly, mean sojourn times are closely connected to fixation probabilities. Let **u** be a single column matrix with $u(i/(2N))$, the probability of fixation for initial frequency $i/(2N)$, in its ith row and **R** be a single column matrix with ith row $g_{i,2N}$. Then, as we proved in Chapter 5,

$$\mathbf{u} = (\mathbf{I} - \mathbf{Q})^{-1}\mathbf{R}$$

Thus

$$\mathbf{SR} = \mathbf{u}$$

If we write out the first row of this ($i = 1$) and recall that, in the neutral case,

$$u(i/(2N)) = i/(2N)$$

we obtain, for that case,

$$\bar{s}_{11}g_{1,2N} + \bar{s}_{12}g_{2,2N} + \bar{s}_{13}g_{3,2N} + \cdots + \bar{s}_{1,2N-1}g_{2N-1,2N} = 1/(2N)$$

Similarly, by considering the probability of loss rather than fixation, we obtain

$$\bar{s}_{11}g_{10} + \bar{s}_{12}g_{20} + \bar{s}_{13}g_{30} + \cdots + \bar{s}_{1,2N-1}g_{2N-1,0} = 1 - 1/(2N)$$

Finally, we consider mean absorption times. Let **T** be a single column matrix with $\bar{s}_i$ in the ith row. Since $\bar{s}_i$ is the sum of $\bar{s}_{ij}$ over all j, we obtain the ith row of **T** by summing over elements in the ith row of **S**, which is achieved by multiplying **S** by a single column matrix, **1**, all the elements of which are unity. We have, then,

$$\mathbf{T} = \mathbf{S}\mathbf{1} = (\mathbf{I} - \mathbf{Q})^{-1}\mathbf{1}$$

which could have been deduced directly from 7.2, as the reader may care to verify.

Mean times for Moran's model

With very minor changes in wording, the preceding discussion will apply to Moran's model. In the first instance, we shall measure times in events rather than generations. Substituting event for generation throughout, we obtain the following analogue of 7.1

$$\bar{s}_{ir} = g_{i,i-1}\bar{s}_{i-1,r} + g_{ii}\bar{s}_{ir} + g_{i,i+1}\bar{s}_{i+1,r} + d_{ir}$$
$$= g_{i,i-1}\bar{s}_{i-1,r} + g_{i,i+1}\bar{s}_{i+1,r} + (1 - g_{i,i-1} - g_{i,i+1})\bar{s}_{ir} + d_{ir}$$

since

$$g_{i,i-1} + g_{ii} + g_{i,i+1} = 1$$

Now, as we noted earlier for Moran's model,

$$\bar{s}_{ir} = f_{ir}\bar{s}_{rr}$$
$$= (u_i/u_r)\bar{s}_{rr} \qquad \text{when } i \leq r$$
$$= [(1 - u_i)/(1 - u_r)]\bar{s}_{rr} \qquad \text{when } i \geq r$$

Since (from Ch. 5) we know u_i and u_r, it will be sufficient to find $\bar{s}_{rr}$. Putting i equal to r in our analogue of 7.1 we obtain

$$\bar{s}_{rr} = g_{r,r-1}\bar{s}_{r-1,r} + g_{r,r+1}\bar{s}_{r+1,r} + (1 - g_{r,r-1} - g_{r,r+1})\bar{s}_{rr} + 1$$

Therefore

$$g_{r,r-1}\bar{s}_{r-1,r} + g_{r,r+1}\bar{s}_{r+1,r} - (g_{r,r-1} + g_{r,r+1})\bar{s}_{rr} + 1 = 0$$

For simplicity, we consider the neutral case. Then

$$u_i = \frac{i}{2N} \qquad \frac{u_{r-1}}{u_r} = \frac{r-1}{r} \qquad \frac{1 - u_{r+1}}{1 - u_r} = \frac{2N - r - 1}{2N - r}$$

whence

$$\bar{s}_{r-1,r} = \frac{r-1}{r}\bar{s}_{rr} \qquad \bar{s}_{r+1,r} = \frac{2N - r - 1}{2N - r}\bar{s}_{rr}$$

Also (in the neutral case)

$$g_{r,r-1} = g_{r,r+1} = \frac{r}{2N} \frac{2N-r}{2N}$$

since, for a change of 1 in the number of A's, an A has to be chosen for one part of the birth–death event and an a chosen for the other part (see Ch. 2 for details). Thus

$$\bar{s}_{rr} \frac{r(2N-r)}{(2N)^2} \left(\frac{r-1}{r} + \frac{2N-r-1}{2N-r} - 2 \right) + 1 = 0$$

or

$$\bar{s}_{rr} \frac{r(2N-r)}{(2N)^2} \left(1 - \frac{1}{r} + 1 - \frac{1}{2N-r} - 2 \right) + 1 = 0$$

so that

$$\bar{s}_{rr} \frac{r(2N-r)}{(2N)^2} \left(-\frac{2N}{r(2N-r)} \right) + 1 = 0$$

i.e

$$\bar{s}_{rr} = 2N$$

events. Hence, in the neutral case (measuring time in events)

$$\bar{s}_{ir} = 2N(2N-i)/(2N-r) \quad \text{when } r \leqslant i$$
$$= 2Ni/r \quad \text{when } r \geqslant i$$

and

$$\bar{s}^*_{ir} = \bar{s}_{ir} r/i$$
$$= 2N(2N-i)r/(2N-r)i \quad \text{when } r \leqslant i$$
$$= 2N \quad \text{when } r \geqslant i$$

as given by Ewens (1979).

We shall be particularly interested in the case where A is a newly arisen neutral mutant. Then $i = 1$, so that $r \geqslant i$ always and the mean sojourn times at frequency $r/(2N)$ are $2N/r$ and $2N$ events respectively. In terms of generations ($2N$ events corresponding to one generation) these become $1/r$ and 1 respectively. It is particularly interesting that, in the modified process, A spends (on average) one generation at every frequency. Notice that either sojourn time, when measured in generations, is independent of the population size. This is quite remarkable, since necessarily $r \leqslant (2N - 1)$, so that possible values for r in any population must, of course, depend on the population size. We shall return to this point in the next section.

Mean absorption and fixation times are obtained by summing the corresponding mean sojourn times over all r from $r = 1$ to $(2N - 1)$. Since the summand changes form when $r = i$, we must take the sum in two parts. Thus the mean absorption time, measured in *generations*, is

$$\sum_{r=1}^{i} \frac{2N-i}{2N-r} + \sum_{r=i+1}^{2N-1} \frac{i}{r}$$

When $i = 1$, this becomes

$$\sum_{r=1}^{2N-1} \frac{1}{r}$$

To obtain an approximation for this, we note the standard mathematical result

$$\lim_{n \to \infty} (\tfrac{1}{1} + \tfrac{1}{2} + \tfrac{1}{3} + \cdots + 1/n - \log n) = 0.577\,215\,66 \cdots$$

so that for large n at least we have approximately

$$\sum_{r=1}^{n} 1/r = \log n + 0.5772$$

Thus when $i = 1$, the mean absorption time is approximately

$$\log(2N - 1) + 0.5772$$

generations. In practice, this formula will give a useful approximation unless N is very small. For example, when $N = 10$, the correct value is 3.5477 generations, whereas our approximate formula gives 3.5216 generations. Notice that the mean absorption time, when $i = 1$, increases very slowly with increasing N; for example, if $N = 10^6$, our formula gives 15.0859 generations.

On the other hand, when $i = 1$, the mean fixation time will be

$$\sum_{r=1}^{2N-1} 1 = (2N-1)$$

generations. Obviously, the two times will be very different when $i = 1$ and N is large. We stress this point, since these two quantities have occasionally been confused. We shall postpone further discussion until we have considered some other models.

A tentative approach for the Wright–Fisher model

Continuing with the neutral case and $i = 1$, we now consider mean times for the Wright–Fisher model. We begin with equation 7.1', with $i = 1$; for simplicity, we shall write b_r as short for $\bar{s}_{ir}$, giving

$$b_r = \sum_{j=1}^{2N-1} b_j g_{jr} + d_{1r} \tag{7.3}$$

where $d_{1r} = 0$ when $r \neq 1$, $d_{11} = 1$.

We suppose, until stated otherwise, that N is large, or at least fairly large. Then big changes in frequency of A in a single generation are very unlikely; that is, g_{jr} will be small unless r is fairly close to j. Hence the sum on the right-hand side of 7.3 will be dominated by terms for which j is fairly close to r.

Our previous experience suggests that results for the Wright–Fisher model are rather similar to those for Moran's model. Very tentatively, then, we consider a solution of the form

$$b_r = C/r$$

where C is a constant, and attempt to check this by substitution in 7.3. Of course, from our argument in Chapter 6 ('special status of the initial frequency' in Wright–Fisher model but not in Moran's model) we would not expect our solution to be correct when $r = 1$, so that the substitution $b_j = C/j$ would appear unjustified when $j = 1$. However, an error in the substituted b_1

will be immaterial when r is large; in such cases, substitution $b_j = C/j$ gives a right-hand side

$$\sum_{j=1}^{2N-1} \frac{C}{j} \frac{(2N)!}{r!(2N-r)!} \left(1 - \frac{j}{2N}\right)^{2N-r} \left(\frac{j}{2N}\right)^r$$

$$= C \frac{(2N)!}{r!(2N-r)!} \sum_{j=1}^{2N-1} \left(1 - \frac{j}{2N}\right)^{2N-r} \left(\frac{j}{2N}\right)^{r-1} \left(\frac{1}{2N}\right)$$

Notice that the summand is zero when $j = 0$, so that we may conveniently take the sum from $j = 0$ to $(2N - 1)$. Let $x = j/(2N)$, so that x ranges from 0 to $(1 - 1/(2N))$ in steps of $\delta x = 1/(2N)$. This gives us

$$C \frac{(2N)!}{r!(2N-r)!} \sum_{x=0}^{1-\delta x} (1-x)^{2N-r} x^{r-1} \delta x$$

Assuming that we can approximate the sum by the corresponding integral

$$\int_0^1 (1-x)^{2N-r} x^{r-1} \, dx = \int_0^1 (1-x)^{(2N-r+1)-1} x^{r-1} \, dx$$

and noting the standard mathematical result

$$\int_0^1 (1-x)^{m-1} x^{n-1} \, dx = \frac{(m-1)!(n-1)!}{(m+n-1)!}$$

(valid when $m > 0$, $n > 0$) we obtain

$$C \frac{(2N)!}{r!(2N-r)!} \frac{(2N-r)!(r-1)!}{(2N)!} = \frac{C}{r}$$

since (by definition of factorial) $r! = r(r-1)!$. Thus our formula C/r solves 7.3 when r is large, provided the approximation of the sum by the integral is valid. The relation between a sum and the corresponding integral is given by the *Euler-Maclaurin formula* (see e.g. Spiegel 1971). From this, it follows that (N being large) the approximation is excellent when r is large, except for values of r close to $2N$, where the approximation is rather less satisfactory, so that our formula $b_r = C/r$ is not quite correct in the latter case. This need not worry us overmuch, since our formula correctly gives b_r as very small in this case; thus the contribution of these b_r to the mean absorption time is very small, so that the error in b_r is of very minor importance. We shall take it then that $b_r = C/r$ when r is large. Our argument has presupposed that N is large,

MEAN SOJOURN, ABSORPTION AND FIXATION TIMES

but otherwise the precise value of N is (essentially) irrelevant and we may let $N \to \infty$, if we wish, when considering r large.

The consequence is surprising. Since N is large, we may certainly approximate the binomial by the Poisson distribution (amounting to letting $N \to \infty$) when r (and hence, for practical purposes, j) is small. For large r, however, substitution of the binomial by the Poisson would appear quite illegitimate; however, we try it and see what happens. Then on the right-hand side of 7.3 we have

$$\sum_{j=1}^{2N-1} \frac{C}{j} e^{-j} \frac{j^r}{r!} = \frac{C}{r!} \sum_{j=1}^{2N-1} e^{-j} j^{r-1} \delta j$$

where $\delta j = 1$ (a notation justified since j increases in steps of 1). The summand is zero when $j = 0$, so that we may conveniently suppose that j goes from 0 to $(2N - 1)$. Now let $N \to \infty$ and replace the sum by the corresponding integral

$$\int_0^\infty e^{-j} j^{r-1} \, dj$$

as can be justified by the Euler–Maclaurin formula when r is at all large. By a standard mathematical result, this integral has value $(r - 1)!$ so that the right-hand side comes out correctly as C/r. Thus the substitution of the binomial by the Poisson, however wrong at first sight when r is large, gives the correct solution!

It is intriguing to see what happens in any special case. Suppose, for example, $2N = 50$, $r = 40$. If we evaluate, term by term,

$$\frac{50!}{40!10!} \sum_{j=1}^{49} \left(1 - \frac{j}{50}\right)^{10} \left(\frac{j}{50}\right)^{39} \frac{1}{50} \quad \text{and} \quad \frac{1}{40!} \sum_{j=1}^{\infty} e^{-j} j^{39}$$

we find that, for given j, corresponding terms in these two series can be very different; yet either series sums to $1/40 = 0.025$; as the reader may readily confirm.

We now sum up this rather tentative discussion. It seems that, at least when N is large, we can obtain correct sojourn times by letting $N \to \infty$ and replacing the binomial distribution by the Poisson distribution. This, in fact, accords with the procedure of Fisher (1930a, b), who left it to the reader to see that his approach was correct. Possibly Fisher thought along the lines we have given, since he had decided, using diffusion methods, that b_r is proportional to $1/r$, at least when r is large. We have suggested that his procedure will yield very accurate results except when r is close to $2N$ and that the error in the latter case will scarcely affect the mean absorption time, especially when N is very large.

A generating function for mean sojourn times

Now let

$$\phi(x) = b_1 x + b_2 x^2 + b_3 x^3 + \cdots + b_{2N-1} x^{2N-1}$$

x being a dummy variable ($0 \leqslant x \leqslant 1$). Thus b_r is the coefficient of x^r, so that $\phi(x)$ is a generating function for mean sojourn times, analogous to the probability generating functions we encountered earlier. Note that, on putting $x = 1$, we have

$$\phi(1) = b_1 + b_2 + b_3 + \cdots + b_{2N-1}$$

so that $\phi(1)$ is the mean absorption time.

Our aim is to find an equation for $\phi(x)$ and deduce the values of b_r from this equation. Multiplying both sides of 7.3 by x^r and summing over r we have

$$\phi(x) = \sum_{r=1}^{2N-1} b_r x^r = \sum_{r=1}^{2N-1} \sum_{j=1}^{2N-1} b_j g_{jr} x^r + \sum_{r=1}^{2N-1} d_{1r} x^r$$

We shall evaluate the sums over r on the right-hand side. First, since d_{1r} is zero unless $r = 1$ and since $d_{11} = 1$

$$\sum_{r=1}^{2N-1} d_{ir} x^r = d_{11} x = x$$

Now consider

$$\sum_{r=1}^{2N-1} g_{jr} x^r$$

If r ran from 0 to $2N$, rather than from 1 to $(2N - 1)$, this expression would be (the definition of) the familiar probability generating function

$$\sum_{r=0}^{2N} \frac{(2N)!}{r!(2N-r)!} \left(1 - \frac{j}{2N}\right)^{2N-r} \left(\frac{j}{2N}\right)^r x^r = \left(1 - \frac{j}{2N} + \frac{jx}{2N}\right)^{2N}$$

$$= \left(1 + \frac{j}{2N}(x - 1)\right)^{2N}$$

Since this includes the unwanted cases $r = 0$ and $r = 2N$, we subtract the values of $g_{ir}x^r$ for these cases, namely

$$g_{j0} + g_{j, 2N}x^{2N}$$

in all, and our summation over r is complete. We have, then

$$\phi(x) = \sum_{j=1}^{2N-1} b_j\left(1 + \frac{j}{2N}(x-1)\right)^{2N} - \sum_{j=1}^{2N-1} b_j g_{j0} - \sum_{j=1}^{2N-1} b_j g_{j, 2N}x^{2N} + x$$

We pointed out earlier in this chapter that

$$\sum_{j=1}^{2N-1} b_j g_{j0} = 1 - \frac{1}{2N} \qquad \sum_{j=1}^{2N-1} b_j g_{j, 2N} = \frac{1}{2N}$$

so

$$\phi(x) = \sum_{j=1}^{2N-1} b_j\left(1 + \frac{j}{2N}(x-1)\right)^{2N} - 1 + \frac{1}{2N} - \frac{x^{2N}}{2N} + x$$

This expression is exact for all N, but intractable. To achieve some simplification, we follow our suggestions in the last section and let $N \to \infty$; then the terms $1/(2N)$, $x^{2N}/(2N) \to 0$. Further, from our argument in that section, we replace the probability generating function for the binomial distribution by the probability generating function for the Poisson distribution. This gives us (with $N \to \infty$)

$$\phi(x) = \sum_{j=1}^{2N-1} b_j e^{j(x-1)} + x - 1$$

Now

$$\sum_{j=1}^{2N-1} b_j e^{j(x-1)} = \sum_{j=1}^{2N-1} b_j(e^{x-1})^j = b_1 e^{x-1} + b_2(e^{x-1})^2 + \cdots + b_{2N-1}(e^{x-1})^{2N-1}$$

which (from the definition of $\phi(x)$) is just $\phi(x)$ with x replaced by e^{x-1}; we may write our sum then as $\phi(e^{x-1})$, so that

$$\phi(x) = \phi(e^{x-1}) + x - 1$$

This equation was obtained by Fisher (1930a, b), who used a different but equivalent approach. Either way, the argument amounts to an application of the theory of branching processes, since our Poisson distribution implies independent propagation. Now, it might be supposed that diffusion methods, in which the population is taken as finite, will *always* yield more accurate results than branching process methods, in which the population is supposed infinite. However, we showed in Chapter 4 that the branching process approach is *sometimes* superior; our present investigation provides another opportunity for comparing the accuracy of the two methods.

Calculating mean sojourn times for the Wright–Fisher model

We write our equation in the form

$$\phi(e^{x-1}) - \phi(x) = 1 - x$$

and consider Fisher's (1930a) beautiful solution. Let u_n be a function of a continuous variable n, with the following properties:

$$u_0 = 0 \qquad u_{n+1} = e^{u_n - 1}$$

We have encountered this function in Chapter 4; in that case, we wrote the function as l_t, where t was necessarily an integer. In the present context, n takes all values from 0 to ∞; we have changed notation to emphasize this difference. The relationships just given will supply u_n for any integral value of n; u_n for other n can be found by an appropriate interpolation procedure (Fisher 1930b), but it will be unnecessary to use this procedure here. Note that $u_n \to 1$ as $n \to \infty$.

From the usual 'function of a function' rule

$$\frac{du_{n+1}}{dn} = e^{u_n - 1} \frac{du_n}{dn}$$

and so

$$\log\left(\frac{du_{n+1}}{dn}\right) = u_n - 1 + \log\left(\frac{du_n}{dn}\right)$$

By elementary calculus

$$\frac{du_n}{dn} = 1 \div \frac{dn}{du_n}$$

Thus

$$\log\left(\frac{du_n}{dn}\right) = \log 1 - \log\left(\frac{dn}{du_n}\right) = -\log\left(\frac{dn}{du_n}\right)$$

We have, then,

$$1 - u_n = \log\left(\frac{dn}{du_{n+1}}\right) - \log\left(\frac{dn}{du_n}\right)$$

Now, in our equation for ϕ, put $x = u_n$. Then

$$e^{x-1} = e^{u_n - 1} = u_{n+1}$$

so giving

$$\phi(u_{n+1}) - \phi(u_n) = 1 - u_n$$
$$= \log\left(\frac{dn}{du_{n+1}}\right) - \log\left(\frac{dn}{du_n}\right)$$

Hence

$$\phi(u_1) - \phi(u_0) = \log\left(\frac{dn}{du_1}\right) - \log\left(\frac{dn}{du_0}\right)$$

$$\phi(u_2) - \phi(u_1) = \log\left(\frac{dn}{du_2}\right) - \log\left(\frac{dn}{du_1}\right)$$

$$\phi(u_3) - \phi(u_2) = \log\left(\frac{dn}{du_3}\right) - \log\left(\frac{dn}{du_2}\right)$$

and so on; here dn/du_0, dn/du_1, dn/du_2, ... are short for dn/du_n calculated at the values $u_0, u_1, u_2, \ldots$ respectively.

If we add these equations, almost all terms cancel and we obtain

$$\phi(u_n) - \phi(u_0) = \log\left(\frac{dn}{du_n}\right) - \log\left(\frac{dn}{du_0}\right)$$

Now $u_0 = 0$ and, by definition of $\phi(x)$, $\phi(0) = 0$. Therefore

$$\phi(u_n) = \log\left(\frac{dn}{du_n}\right) - \log\left(\frac{dn}{du_0}\right)$$

Thus we can find $\phi(u_n) = \phi(x)$ and hence all the mean sojourn times provided we can find an expression for n in terms of u_n. To do this, we write, as in Chapter 4,

$$u_n = 1 - 1/v_n \qquad v_n = 1/(1 - u_n)$$

and substitute in

$$u_{n+1} = e^{u_n - 1}$$

giving

$$1 - \frac{1}{v_{n+1}} = e^{-1/v_n} = 1 - \frac{1}{v_n} + \frac{1}{2v_n^2} - \frac{1}{6v_n^3} + \cdots$$

$$\frac{1}{v_{n+1}} = \frac{1}{v_n}\left(1 - \frac{1}{2v_n} + \frac{1}{6v_n^2} - \cdots\right)$$

$$v_{n+1} = v_n\left(1 - \frac{1}{2v_n} + \frac{1}{6v_n^2} - \cdots\right)^{-1}$$

$$= v_n\left(1 + \frac{1}{2v_n} + \frac{1}{12v_n^2} - \cdots\right)$$

$$= v_n + \tfrac{1}{2} + 1/(12v_n) - \cdots$$

We take as a first approximation

$$v_{n+1} = v_n + \tfrac{1}{2}$$

so that, roughly speaking, v_n increases by $\tfrac{1}{2}$ when n increases by 1. Further $v_0 = 1/(1 - u_0) = 1$, so that within the limits of our approximation

$$v_n = 1 + \tfrac{1}{2}n$$

Therefore

$$n = 2v_n - 2 = \frac{2}{1 - u_n} - 2$$

$$dn/du_n = 2/(1 - u_n)^2$$

and since $u_0 = 0$

$$dn/du_0 = 2$$

Hence, to a first approximation

$$\phi(u_n) = \log\left(\frac{2}{(1-u_n)^2}\right) - \log 2$$
$$= \log 2 - 2\log(1 - u_n) - \log 2$$
$$= -2\log(1 - u_n)$$

Therefore

$$\phi(x) = -2\log(1 - x) = \tfrac{2}{1}x + \tfrac{2}{2}x^2 + \tfrac{2}{3}x^3 + \cdots$$

Since, by definition of $\phi(x)$, b_r is the coefficient of x^r, $b_r = 2/r$ generations approximately. Note that our formula for $\phi(x)$ gives an approximation to b_r in an infinite population. While we have argued that this will also be an approximation for a large finite population, we must truncate the series just given for $\phi(x)$ at the term in x^{2N-1} in this finite case; hence it would not be valid to find the mean absorption time $\phi(1)$ by putting $x = 1$ in $-2\log(1 - x)$. Rather, our results show that the mean absorption time is approximately $\phi(1)$ for the truncated series, that is

$$\sum_{r=1}^{2N-1} 2/r = 2[\log(2N - 1) + 0.577\,215\,665]$$
$$= 2\log(2N - 1) + 1.154\,431\,33$$

generations.

To obtain better approximations, we need an improved formula for n in terms of u_n. By a series of neat manipulations (see note 1), Fisher obtained

$$n = \frac{2}{1 - u_n} + \tfrac{1}{3}\log(1 - u_n) - 2(1.014\,648\,607) + \frac{1 - u_n}{36} + \frac{(1 - u_n)^2}{540}$$
$$- \frac{(1 - u_n)^3}{54 \times 144} - \frac{71(1 - u_n)^4}{84 \times 72^2} - \frac{8759(1 - u_n)^5}{315 \times 720^2} + \frac{62(1 - u_n)^6}{81 \times 720^2}$$
$$+ \frac{1637(1 - u_n)^7}{504 \times 720^2} - \frac{20\,879\,093(1 - u_n)^8}{9504 \times 420 \times 720^2}$$

which is very accurate when $n \geqslant 5$ and only slightly less so for smaller n.

We find first the mean absorption time. Write n for n as determined by the full formula just given and n^* for n as determined by our earlier formula $2/(1 - u_n) - 2$; write ϕ, ϕ^* for the corresponding ϕ's:

$$\phi(u_n) = \log\left(\frac{dn}{du_n}\right) - \log\left(\frac{dn}{du_0}\right)$$

$$\phi^*(u_n) = \log\left(\frac{dn^*}{du_n}\right) - \log\left(\frac{dn^*}{du_0}\right)$$

with appropriate truncation to allow for the finite population size. $\phi(1)$ should provide a very accurate formula for the mean absorption time when N is large. As we have already found that

$$\phi^*(1) = 2\log(2N - 1) + 1.154\,431\,33$$

our aim is to correct this by calculating

$$\phi(1) - \phi^*(1)$$

Ignoring truncation for the moment, we can find dn/du_n and dn^*/du_n, from our formulae for n and n^*, by elementary calculus. It then follows at once that

$$\frac{dn}{du_n} \div \frac{dn^*}{du_n} = 1 \qquad \text{when } u_n = 1$$

Therefore

$$\log\left(\frac{dn}{du_n}\right) - \log\left(\frac{dn^*}{du_n}\right) = \log\left(\frac{dn}{du_n} \div \frac{dn^*}{du_n}\right) = \log 1 = 0 \qquad \text{when } u_n = 1$$

Strictly, this result should be corrected to allow for truncation, but the correction will be negligibly small when N is very large. We have, therefore,

$$\phi(1) - \phi^*(1) = -\log\left(\frac{dn}{du_0}\right) + \log\left(\frac{dn^*}{du_0}\right)$$

$$= -\log\left(\frac{dn}{du_0}\right) + \log 2$$

We could find dn/du_0 by finding dn/du_n from our full formula for n and then putting $u_n = u_0 = 0$. This gives

$$dn/du_0 = 1.637\,290\,97$$

This is slightly inaccurate, since our formula for n is not quite correct when n is very small, although very accurate when $n \geq 5$. From our earlier result

$$\frac{du_{n+1}}{dn} = e^{u_n - 1}\frac{du_n}{dn} = u_{n+1}\frac{du_n}{dn}$$

we have

$$\frac{dn}{du_n} = u_{n+1}\frac{dn}{du_{n+1}}$$

whence

$$\frac{dn}{du_0} = u_1\frac{dn}{du_1} = u_1 u_2 \frac{dn}{du_2} = \cdots = u_1 u_2 u_3 u_4 u_5 \frac{dn}{du_5}$$

This should supply a very accurate value of dn/du_0. Using this method, Fisher obtained

$$dn/du_0 = 1.636\,405\,56$$

$$-\log\left(\frac{dn}{du_0}\right) + \log 2 = 0.200\,645\,07$$

whence the mean absorption time, $\phi(1)$, equals $2\log(2N - 1) + 1.355\,076\,40$ generations subject to very minor rounding-off error. In fact, this formula is accurate to at least five decimal places when N is large (Ewens 1979). The conservationist, however, will be interested in cases where N is small. Since we have assumed, for most of our proof, that N is very large, we can scarcely hope that our formula will apply for small N. In Table 7.1, we compare some results from our formula with corresponding exact results. It will be seen that by 'small' we mean $N = 1$; for larger N our formula gives very reasonable results!

We recall that for Moran's model, the corresponding mean absorption time is $\log(2N - 1) + 0.5772$ generations, so that the term in $\log(2N - 1)$ is just half the corresponding term for the Wright–Fisher model. Now, for

Table 7.1 Mean absorption times when one A is present initially. Exact values and approximate values obtained from Fisher's formula.

Population size	Approximate time	Exact time
1	1.3551	2.0000
2	3.5523	3.6897
3	4.5740	4.6230
4	5.2469	5.2648
5	5.7495	5.7533
6	6.1509	6.1472
25	9.1387	9.1268

Moran's model, $N_e/N = \frac{1}{2}$. We may guess, therefore, that, for any model, the mean absorption time (when one A is present initially) will be

$$(N_e/N)2 \log(2N - 1) + \text{a constant}$$

Haldane (1939) proved this result by extension of Fisher's approach. On the other hand, the value of the constant seems quite unrelated to the value of N_e/N (Watterson 1975). However, as Haldane pointed out, the constant will always be fairly small; save perhaps in cases where N is quite small, we conclude that (other things being equal) raising N_e/N substantially will raise the mean absorption time substantially. Hence, as we argued in Chapter 4, the conservationist should try to keep N_e/N as large as is practically possible.

Returning now to the Wright-Fisher model, we consider mean sojourn times. We have

$$\phi(u_n) = \log\left(\frac{dn}{du_n}\right) - \log\left(\frac{dn}{du_0}\right) = \log\left(\frac{dn}{du_n} \bigg/ \frac{dn}{du_0}\right)$$

Expanding dn/du_n in Taylor series gives

$$\frac{dn}{du_n} = \frac{dn}{du_0} + u_n \frac{d^2n}{du_0^2} + \frac{u_n^2}{2!} \frac{d^3n}{du_0^3} + \cdots \qquad (\text{since } u_0 = 0)$$

$$= \frac{dn}{du_0}\left(1 + u_n \frac{d^2n}{du_0^2} \bigg/ \frac{dn}{du_0} + \frac{u_n^2}{2!} \frac{d^3n}{du_0^3} \bigg/ \frac{dn}{du_0} + \cdots\right)$$

Therefore

$$\log\left(\frac{dn}{du_n} \bigg/ \frac{dn}{du_0}\right) = \log\left(1 + u_n \frac{d^2n}{du_0^2} \bigg/ \frac{dn}{du_0} + \frac{u_n^2}{2} \frac{d^3n}{du_0^3} \bigg/ \frac{dn}{du_0} + \cdots\right)$$

Hence, since $x = u_n$

$$\phi(x) = \log\left(1 + x\,\frac{d^2n}{du_0^2}\bigg/\frac{dn}{du_0} + \tfrac{1}{2}x^2\,\frac{d^3n}{du_0^3}\bigg/\frac{dn}{du_0} + \cdots\right)$$

If we expand this in logarithmic series, we obtain the mean sojourn time at frequency $r/(2N)$ as the coefficient of x^r. Fisher gives formulae, analogous to the formulae given earlier for finding a very accurate value of dn/du_0, by which d^2n/du_0^2, $d^3n/du_0^3, \ldots$ may be calculated very accurately, thus giving very accurate values for mean sojourn times (except when r is very close to $2N$) when N is large. These values turn out to be very close to $2/r$ when r exceeds 6. Once again, results prove good even when N is small, as shown in Table 7.2 for $N = 5$.

Table 7.2 Mean sojourn time at frequency $r/(2N)$ when $N = 5$.

r	$2/r$	Fisher value	Exact value
1	2.0000	2.2409	2.2229
2	1.0000	0.9538	0.9561
3	0.6667	0.6719	0.6670
4	0.5000	0.5011	0.4949
5	0.4000	0.3998	0.3933
6	0.3333	0.3332	0.3251
7	0.2857	0.2857	0.2748
8	0.2500	0.2500	0.2335
9	0.2222	0.2222	0.1857

Note that the sum of the entries in the third column for $r = 1$ to 6 (inclusive), less the corresponding sum for the second column, is equal to 0.2007, so that the difference between our two values $\phi(1)$ and $\phi^*(1)$ for the mean absorption time is due almost entirely to differences in mean sojourn time for these values of r.

Clearly, the approximate formula $2/r$, although not very accurate when $r = 1$ or $(2N - 1)$, is not seriously misleading; the same is true for larger values of N. The more general approximate formula

$$2N_e/rN$$

is due to Haldane (1939).

Mean fixation times

Mean sojourn times for the modified process are obtained by multiplying times for the unmodified process by

$$\frac{r}{2N} \div \frac{1}{2N} = r$$

(see the first section of this chapter). Hence when $N_e = N$ the mean sojourn time is approximately $(2/r) \times r = 2$ generations irrespective of the value of r. That is, allele A spends, on average, approximately two generations at every one of the $(2N - 1)$ frequencies $1/(2N), 2/(2N), 3/(2N), \ldots, (2N - 1)/(2N)$. Hence the mean fixation time is approximately $2(2N - 1) = 4N - 2$ generations. In practice, mean fixation times are slightly lower than this (when $N > 1$). However, the discrepancy is not serious in any case of practical interest; we shall, therefore, just indicate briefly how a more accurate statement may be obtained.

Clearly, we shall have to correct for departures of the mean sojourn time from 2 when (a) r is small and (b) r is close to $2N$. Consider first the case r small. If we multiply the mean sojourn times for the unmodified process by r for the cases $r = 1$ to 6 inclusive, we obtain

$$2.2409, \quad 1.9076, \quad 2.0156, \quad 2.0044, \quad 1.9988, \quad 1.9993$$

respectively. These total to 12.1666, rather than the value 12 obtained by taking every mean sojourn time to be 2, so we modify our mean fixation time to

$$4N - 2 + 0.1666 = 4N - 1.8334$$

Now consider r close to $2N$. Put

$$r = 1 - s/(2N)$$

If we take the analogue of Equation 7.3 for the modified process, we find that, when N is very large, this analogue is satisfied if we put the mean sojourn times, for the cases $s = 1$ to 6 inclusive, equal to

$$1.6364, \quad 1.8335, \quad 1.8898, \quad 1.9165, \quad 1.9333, \quad 1.9445$$

respectively, and equal to

$$2 - 1/(3s)$$

for s larger (but not too large), these times being, in fact, just twice the values obtained by Fisher (1930a) for a different problem (not involving sojourn times – see Ch. 3, note 4). Suppose then that we subtract $1/(3s)$ from *every* mean sojourn time, with a small correction to allow for the departure from $2 - 1/(3s)$ when s is very small. That is, we subtract

$$\tfrac{1}{3} \sum_{s=1}^{2N-1} (1/s) + 0.0293 = \tfrac{1}{3} \log(2N - 1) + \tfrac{1}{3}(0.5772) + 0.0293$$

$$= \tfrac{1}{3} \log(2N - 1) + 0.2217$$

from $4N - 1.8334$ to give us

$$4N - \tfrac{1}{3} \log(2N - 1) - 2.0551$$

generations. Clearly, this is too small, since we should not subtract from *every* mean sojourn time, although the error caused by doing so is very small, since $1/(3s)$ is very small when s is large. Our formula, then, provides a lower limit for the mean fixation time, at least when N is large, in contrast to the upper limit, $4N - 2$ generations. In practice, these limits operate for all N. Some numerical values are given in Table 7.3, from which it appears that the exact value, when $N > 1$, is rather closer to our lower limit than to our upper limit, at least when N is small, although the limits do not differ from one another by much in this case. As N becomes large, the limits gradually diverge from one another but the discrepancy is trivial for any case of practical interest. For example, when $N = 10^9$, our limits are 3 999 999 998 and 3 999 999 991. Thus, when N is large, it will be sufficient to take the mean fixation time to be $4N$ generations when $N_e = N$, or, more generally, from Haldane's results, $4N_e$ generations, as first shown by Kimura & Ohta (1969a), who used diffusion methods. The spread of a neutral mutant allele to fixation, then, is a very slow

Table 7.3 Upper and lower limits for, and exact value of, the mean fixation time when one A is present initially; N = population size.

N	Upper limit	Lower limit	Exact value
1	2	1.9449	2.0000
2	6	5.5787	5.7793
3	10	9.4084	9.6560
4	14	13.2963	13.5663
5	18	17.2125	17.4956
25	98	96.6476	96.9712

process when N_e is large. In contrast, the corresponding mean absorption time is small, reflecting the tendency for neutral mutants to be lost rapidly on most occasions.

Continuous approximations

The very accurate methods given so far do not generalize. We consider, therefore, an alternative approach (Fisher 1930a, b, Wright 1931, Feller 1954, Moran 1962, Watterson 1962, Ewens 1963a, 1964a, 1967b, 1969, 1973, 1979, Kimura & Ohta 1969a, b, 1971a, Maruyama 1972a, 1977, Pollak & Arnold 1975), which, while sometimes yielding slightly inaccurate results, can be applied very generally.

We begin with our fundamental equations

$$\bar{s}_{ir} = \sum_{j=0}^{2N} g_{ij}\bar{s}_{jr} + d_{ir}$$

$$\bar{s}_i = \sum_{j=0}^{2N} g_{ij}\bar{s}_j + 1$$

By definition of mean

$$\sum_{j=0}^{2N} g_{ij}\bar{s}_{jr} = \mathop{\mathrm{E}}_{j} \bar{s}_{jr} \quad \text{and} \quad \sum_{j=0}^{2N} g_{ij}\bar{s}_j = \mathop{\mathrm{E}}_{j} \bar{s}_j$$

We rewrite our fundamental equations in terms of allele frequencies, rather than allele numbers. For simplicity, we write the initial frequency $i/(2N)$ as p rather than p_0. We write $j/(2N)$ as $(p + \delta p)$ and $r/(2N)$ as x. Let $\bar{s}(p, x)$ be the mean sojourn time at frequency x, and $\bar{s}(p)$ the mean absorption time, when the initial frequency is p. We have, then,

$$\bar{s}(p, x) = \mathop{\mathrm{E}}_{\delta p} \bar{s}(p + \delta p, x) + d_{px}$$

(where $d_{px} = 1$ when $x = p$ and $d_{px} = 0$ otherwise), with boundary conditions

$$\bar{s}(0, x) = 0 \qquad \bar{s}(1, x) = 0$$

and also

$$\bar{s}(p) = \mathop{\mathrm{E}}_{\delta p} \bar{s}(p + \delta p) + 1$$

with boundary conditions

$$\bar{s}(0) = 0 \qquad \bar{s}(1) = 0$$

In either case, δp is, of course, the change in allele frequency in going from generation 0 to generation 1, so that, as usual, $E_{\delta p}(\delta p)$ will be the change in allele frequency in going from generation 0 to generation 1, calculated deterministically. As in Chapter 5, we shall approximate $E_{\delta p}(\delta p)^2$ by

$$p(1-p)/2N_e$$

and take $E_{\delta p}(\delta p)^r$ as negligible when $r > 2$.

Consider first the mean absorption time. We approximate the allele frequency by a continuous variable and rewrite the mean absorption time as $T(p)$ to emphasize this. We have, then,

$$T(p) = \mathop{E}_{\delta p} T(p + \delta p) + 1$$

$$= \mathop{E}_{\delta p} \left(T(p) + \delta p T'(p) + \frac{(\delta p)^2}{2!} T''(p) + \cdots \right) + 1$$

$$= T(p) + \mathop{E}_{\delta p} (\delta p) T'(p) + \mathop{E}_{\delta p} \frac{(\delta p)^2}{2} T''(p) + 1$$

to a close approximation.

Writing $E_{\delta p}(\delta p) = M$, $E_{\delta p}(\delta p)^2 = V$, we have

$$M \frac{dT(p)}{dp} + \tfrac{1}{2} V \frac{d^2 T(p)}{dp^2} = -1$$

with boundary conditions $T(0) = 0$, $T(1) = 0$.

Naturally, we would like to find a comparable equation for mean sojourn times, but it is not immediately clear how to do this; if we approximate p and x by continuous variables, what do we substitute for d_{px}?

We shall treat this problem informally – see e.g. Karlin & Taylor (1981) for a rigorous discussion. The first step is to represent $\bar{s}(p, x)$ as an area. Let

$$\bar{t}(p, x) = 2N\bar{s}(p, x)$$

so that

$$\bar{s}(p, x) = \bar{t}(p, x)\delta x$$

CONTINUOUS APPROXIMATIONS

where $\delta x = 1/(2N)$. Thus $\bar{s}(p, x)$ may be represented as the area of a rectangle of height $\bar{t}(p, x)$ and width δx; this is shown for values of x from $1/(2N)$ to $5/(2N)$ in Figure 7.1. If the curve $T(p, x)$ is drawn as shown, the mean sojourn time at frequency x will be

$$T(p, x)\delta x = T(p, x)\frac{1}{2N}$$

Our aim is to find a formula for $T(p, x)$. In fact, the $T(p, x)$ supplied by the formula that we shall obtain will not always be quite as good as the ideal $T(p, x)$ shown in the figure; hence some of the mean sojourn times that we shall calculate will be rather approximate.

Now consider the mean absorption time. Obviously, this will be the sum of the areas of all rectangles of the type shown in Figure 7.1. Acting, however, in the spirit of continuous variables, we represent the mean absorption time by an integral rather than a sum, namely the area under the curve $T(p, x)$. From Figure 7.1, a sensible lower limit for this integral is $1/(4N)$, and similarly, we

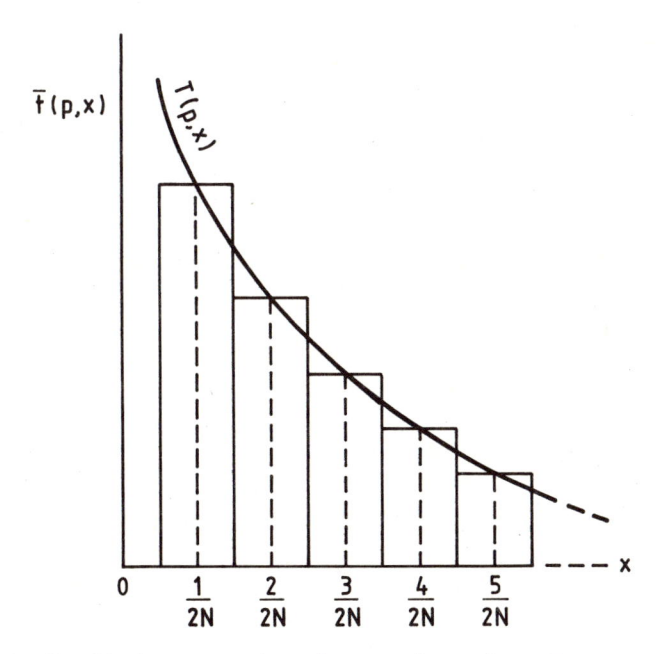

Figure 7.1 Graphical representation of mean sojourn times. Every rectangle has width $\delta x = 1/(2N)$. Mean sojourn time at frequency x is the area of the rectangle of height $\bar{t}(p, x)$. An ideal curve $T(p, x)$ is drawn such that this area equals $T(p, x)\delta x$.

take the upper limit to be $1 - 1/(4N)$. We have then, as a reasonable approximation,

$$T(p) = \int_{1/(4N)}^{1-1/(4N)} T(p, x)\, dx$$

If N is large, $1/(4N)$ will be close to zero and we have the slightly cruder approximation

$$T(p) = \int_0^1 T(p, x)\, dx$$

(where the limits 0, 1 originate from, say, ε, $1 - \varepsilon$, on letting $\varepsilon \to 0$). We shall suppose that N is large and require that $T(p, x)$ obeys the relationship just given.

Consider then the fundamental equation

$$\bar{s}(p, x) = \mathop{E}_{\delta p} \bar{s}(p + \delta p, x) + d_{px}$$

We replace $\bar{s}(p, x)$ by $T(p, x)\delta x$ and $\bar{s}(p + \delta p, x)$ by $T(p + \delta p, x)\delta x$. Further, we replace d_{px} by $h(p, x)\delta x$ where the properties of $h(p, x)$ remain to be determined. This gives us

$$T(p, x)\delta x = \mathop{E}_{\delta p} T(p + \delta p, x)\delta x + h(p, x)\delta x$$

Clearly, we want

$$h(p, x) = 0 \qquad \text{when } x \neq p$$

To identify a second, crucial, property of $h(p, x)$ we continue as follows. Dividing our equation for $T(p, x)\delta x$ by δx gives

$$T(p, x) = \mathop{E}_{\delta p} T(p + \delta p, x) + h(p, x)$$

Now integrate both sides with respect to x from $x = 0$ to $x = 1$ to give

$$\int_0^1 T(p, x)\, dx = \int_0^1 \mathop{E}_{\delta p} T(p + \delta p, x)\, dx + \int_0^1 h(p, x)\, dx$$

In the first term on the right-hand side, the mean E is supposed to be found first and the integration performed next. We shall, however, do the integration first and take the mean next. This implies some restriction on $T(p + \delta p, x)$. For, with δp a continuous variable, finding the mean requires integration as a matter of definition. Thus in the first term on the right-hand side there are two integrations, one with respect to δp and one with respect to x. We are supposing that $T(p + \delta p, x)$ is such that the result is unaffected if we carry out these integrations in the reverse order to that originally written. However, in that the lower and upper limits for either integral are finite, $T(p + \delta p, x)$ would, from standard integration theory, have to be very 'pathological' for our supposition to be wrong. We write, then,

$$\int_0^1 T(p, x)\, dx = \mathop{\mathrm{E}}_{\delta p} \int_0^1 T(p + \delta p, x)\, dx + \int_0^1 h(p, x)\, dx$$

and since

$$\int_0^1 T(p, x)\, dx = T(p) \qquad \int_0^1 T(p + \delta p, x)\, dx = T(p + \delta p)$$

we have

$$T(p) = \mathop{\mathrm{E}}_{\delta p} T(p + \delta p) + \int_0^1 h(p, x)\, dx$$

But we showed earlier that

$$T(p) = \mathop{\mathrm{E}}_{\delta p} T(p + \delta p) + 1$$

Thus

$$\int_0^1 h(p, x)\, dx = 1$$

Now $h(p, x) = 0$ when $x \neq p$, so we can equally say

$$\int_{-\infty}^{\infty} h(p, x)\, dx = 1$$

MEAN SOJOURN, ABSORPTION AND FIXATION TIMES

since the integrand is zero for all impossible values of x. We have shown, then, that

$$h(p, x) = 0 \qquad \text{when } x \neq p$$

$$\int_{-\infty}^{\infty} h(p, x) \, dx = 1$$

Hence $h(p, x)$ is Dirac's delta function $\delta(x - p)$, which we discussed in Chapter 6.

The rest is easy (but see note 2). We have (x being fixed at any desired value)

$$T(p, x) = \underset{\delta p}{\mathrm{E}} T(p + \delta p, x) + \delta(x - p)$$

$$= \underset{\delta p}{\mathrm{E}} \left(T(p, x) + \delta p T'(p, x) + \frac{(\delta p)^2}{2!} T''(p, x) + \cdots \right) + \delta(x - p)$$

differentiation being carried out with respect to p. Thus, to a close approximation

$$M \frac{\partial T(p, x)}{\partial p} + \tfrac{1}{2} V \frac{\partial^2 T(p, x)}{\partial p^2} = -\delta(x - p) \qquad (7.4)$$

with $T(0, x) = 0$, $T(1, x) = 0$ for all x ($0 < x < 1$).

We may compare this with

$$M \frac{dT(p)}{dp} + \tfrac{1}{2} V \frac{d^2 T(p)}{dp^2} = -1 \qquad (7.5)$$

with $T(0) = 0$, $T(1) = 0$.

Although, in the neutral case ($M = 0$), 7.5 can be solved directly by elementary methods (see note 3), in general the most convenient approach is to solve 7.4; $T(p)$, if desired, is then found from

$$T(p) = \int_0^1 T(p, x) \, dx \qquad (7.6)$$

We have thought of $T(p, x)$ in terms of the mean sojourn time and we shall continue to do this. We just note in passing that, faced with the 'boundary value problem'

solve 7.5, with boundary conditions $T(0) = T(1) = 0$

we could, in fact, have reached 7.6 by 'purely mathematical' arguments, in which $T(p, x)$ is *not* given a biological interpretation but is *defined* by 7.4 and its associated boundary conditions. When defined in this way, $T(p, x)$ is known as *Green's function* (after the 19th-century English mathematician) for our particular boundary value problem. Indeed Ewens (1963a), in giving the first general treatment of mean absorption and sojourn times, obtained $T(p)$ by such an argument and only later in his paper interpreted $T(p, x)$ in terms of sojourn times. Our present approach, while of course non-rigorous, does however give the essence of the matter.

Calculating mean sojourn and absorption times

Following Maruyama (1977), we shall find the mean sojourn time, for any chosen value of x, from 7.4. To keep details as simple as possible, we consider first the neutral case. Then $M = 0$, so that

$$\tfrac{1}{2} V \frac{\partial^2 T(p, x)}{\partial p^2} = -\delta(x - p)$$

Putting

$$V = p(1 - p)/(2N_e)$$

and writing T as short for $T(p, x)$, we obtain

$$\frac{\partial^2 T}{\partial p^2} = -\frac{4N_e}{p(1-p)} \delta(x - p) = -\frac{4N_e}{p(1-p)} \delta(p - x)$$

since, as shown in Chapter 6, $\delta(x - p) = \delta(p - x)$.

To find T, we must carry out two successive integrations; this is a little difficult in practice, in view of the Dirac function on the right-hand side. We note first that an indefinite integral can always be written as a definite integral plus a constant. For example

$$\int 2p \, dp = \int_a^p 2\xi \, d\xi + \text{constant}$$

where a is an arbitrary constant and ξ is a dummy variable, since the right-hand side is

$$[\xi^2]_a^p + \text{constant} = p^2 - a^2 + \text{constant}$$
$$= p^2 + (\text{a different constant})$$

Generally, for any function $f(p)$

$$\int f(p)\,dp = \int_a^p f(\xi)\,d\xi + \text{constant}$$

Then, taking $a = 0$, we have

$$\frac{\partial T}{\partial p} = -4N_e \int_0^p \frac{\delta(\xi - x)}{\xi(1 - \xi)}\,d\xi + \text{constant}$$

(by a 'constant' we mean, in the present context, a quantity independent of p, but not necessarily of x). The following approach, suggested to the author by Dr R. C. Jones, while equivalent to Maruyama's, is decidedly easier to follow.

Consider first the case $p < x$. Then ξ, in ranging from 0 to p, never reaches x; hence the definite integral is zero, so that

$$\partial T/\partial p = A$$

where A is a constant.

When $p > x$, the definite integral is, by the 'spotlight' property of the Dirac function (see Ch. 6), $-4N_e/x(1-x)$, and so

$$\frac{\partial T}{\partial p} = -\frac{4N_e}{x(1-x)} + B$$

where B is a constant.

Thus $\partial T/\partial p$ shows a jump, of magnitude

$$-\frac{4N_e}{x(1-x)} + B - A$$

at $p = x$, as shown in Figure 7.2.

It is not immediately obvious that the constants B and A are equal. To see this, we note that the magnitude of the jump equals the value of $\partial T/\partial p$ when p just exceeds x less the value of $\partial T/\partial p$ when p is just less than x; say (with ε positive and tending to zero)

$$\left(\frac{\partial T}{\partial p} \text{ when } p = x + \varepsilon\right) \text{ less } \left(\frac{\partial T}{\partial p} \text{ when } p = x - \varepsilon\right)$$

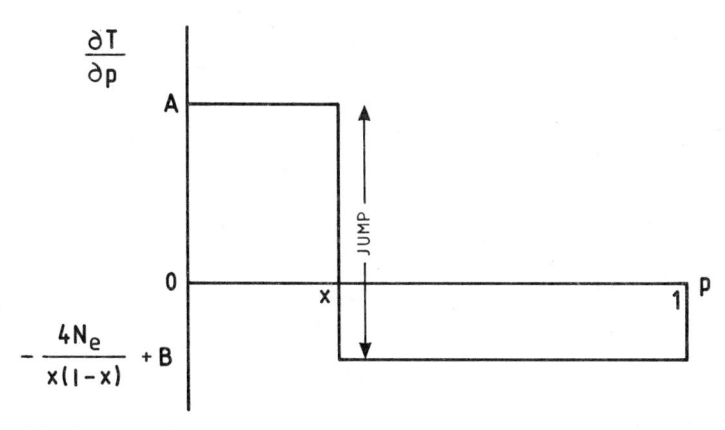

CALCULATING MEAN SOJOURN AND ABSORPTION TIMES

Figure 7.2 Plot of $\partial T/\partial p$ against p (x being fixed at any desired value) showing jump at $p = x$. Note that the area under the curve changes steadily as p increases, even at the jump.

that is

$$\left[\frac{\partial T}{\partial p}\right]_{x-\varepsilon}^{x+\varepsilon} = \int_{x-\varepsilon}^{x+\varepsilon} \frac{\partial^2 T}{\partial p^2}\, dp$$

which (on recalling that $\partial^2 T/\partial p^2 = -4N_e \delta(p-x)/p(1-p)$ equals

$$-4N_e \int_{x-\varepsilon}^{x+\varepsilon} \frac{\delta(p-x)}{p(1-p)}\, dp = -\frac{4N_e}{x(1-x)}$$

from the spotlight property. On comparing our two expressions for the jump, we see at once that $B = A$.

The rest involves only elementary calculus. When $p < x$,

$$\partial T/\partial p = A \qquad T = Ap + C \qquad (C = \text{constant})$$

But when $p = 0$, $T = 0$, and so $C = 0$. Therefore

$$T = Ap$$

When $p > x$

$$\frac{\partial T}{\partial p} = -\frac{4N_e}{x(1-x)} + A \qquad (\text{since } B = A)$$

245

Therefore

$$T = -\frac{4N_e p}{x(1-x)} + Ap + D \qquad (D = \text{constant})$$

But when $p = 1$, $T = 0$, so that

$$D = \frac{4N_e}{x(1-x)} - A$$

Thus

$$T = \left(\frac{4N_e}{x(1-x)} - A\right)(1-p)$$

By elementary calculus, T is just the area under the curve $\partial T/\partial p$. From Figure 7.2, this area changes steadily as p increases, even at $p = x$. In other words, T is a continuous function of p, even at $p = x$; hence our expressions for T when $p < x$ and when $p > x$ must be equal when $p = x$, so that

$$Ax = \left(\frac{4N_e}{x(1-x)} - A\right)(1-x)$$

i.e.

$$A = 4N_e/x$$

and we have, finally (remembering that T is short for $T(p, x)$)

$$T(p, x) = 4N_e p/x \text{ when } p < x$$
$$= 4N_e(1-p)/(1-x) \text{ when } p > x$$

The mean absorption time $T(p)$, which equals

$$\int_0^1 T(p, x) \, dx$$

follows at once. Note that, since $T(p, x)$ changes form when $x = p$, we must split the range of integration for x, 0 to 1, into the ranges 0 to p and p to 1.

Since the integral over a point is zero, no harm is done by including p in both ranges. Hence

$$T(p) = 4N_e\left(\int_0^p \frac{1-p}{1-x}dx + \int_p^1 \frac{p}{x}dx\right)$$
$$= 4N_e\{-(1-p)[\log(1-x)]_0^p + p[\log x]_p^1\}$$
$$= 4N_e[-(1-p)\log(1-p) + 0 + 0 - p\log p]$$
$$= -4N_e[p\log p + (1-p)\log(1-p)]$$

(Watterson 1962, Ewens 1963a).

Mean sojourn times are

$$T(p, x)\delta x = T(p, x)\frac{1}{2N}$$
$$= 2pN_e/xN \qquad \text{when } p < x$$
$$= 2(1-p)N_e/(1-x)N \qquad \text{when } p > x$$
$$= 2N_e/N \qquad \text{when } p = x$$

(Ewens 1963a). We may confirm our formula for $T(x, x)$, the mean sojourn time when $p = x$, by recalling that

$$T(p, x)/T(x, x) = p/x \qquad \text{when } p < x$$
$$= (1-p)/(1-x) \qquad \text{when } p > x$$

given the continuous (or Moran's) model.

If we put $N_e/N = \frac{1}{2}$, we obtain the exact mean sojourn times for Moran's model, calculated earlier by a more elementary method. It is interesting, therefore, to consider the case $N_e = N$, $p = 1/(2N)$ and compare results with Fisher's. We have, in this case, $p \leqslant x$ for all x, so that on writing $2Nx = r$ we have the mean sojourn time at frequency $r/(2N)$ as $2/r$, which, while agreeing with Fisher's result for $r \gtrsim 6$, is less accurate for all smaller r when N is at all large. For the mean absorption time, we have

$$-4N\left[\frac{1}{2N}\log\left(\frac{1}{2N}\right) + \left(1 - \frac{1}{2N}\right)\log\left(1 - \frac{1}{2N}\right)\right]$$
$$= -4N\left[-\frac{\log(2N)}{2N} + \frac{2N-1}{2N}\log\left(1 - \frac{1}{2N}\right)\right]$$
$$= 2\log(2N) - (4N-2)\log\left(1 - \frac{1}{2N}\right)$$

which is less accurate than Fisher's formula. For example, when N is large (greater than about 25) we may approximate $(4N-2)$ by $4N$ and $\log[1-1/(2N)]$ by $[-1/(2N)]$ to give

$$2\log(2N) + 2$$

rather than the more accurate

$$2\log(2N-1) + 1.3551$$

Generally, our formulae for mean sojourn times, while never seriously misleading, will be slightly inaccurate for sojourn times (a) at the initial frequency and (to a much lesser extent) in the neighbourhood of that frequency and (b) at frequencies very close to 0 or 1. However, even in these cases, while the *relative* error can be large, the (biologically much more relevant) absolute error is quite small, never (as far as is known) exceeding one generation. The discrepancy at the initial frequency p is worth taking seriously, since an error here will rather upset the calculation of variances of sojourn times (see below). When $N_e = N$, an improved formula (Pollak & Arnold 1975) for the mean sojourn time at p is

$$\frac{3 - 4[(2N)!/i!(2N-i)!]p^i(1-p)^{2N-i}}{1 - [(2N)!/i!(2N-i)!]p^i(1-p)^{2N-i}}$$

where $i = 2Np$. If N is large and p intermediate, this formula gives approximately three generations, rather than the two generations found earlier. Apart from this slight qualification, we can use our approximate formulae with confidence.

It follows, therefore, that our formula for the mean absorption time should be quite accurate. In fact, the main errors in our approximate mean sojourn times partly cancel when these times are summed. On the other hand, the use of integration between limits 0 and 1 rather than summation will introduce a new small error, as may the accumulation of minor errors. However, from a comparison with exact results in special cases (Ewens 1963b, Burrows & Cockerham 1974), it appears that our formula for the mean absorption time, while slightly inaccurate, is quite good enough for general conclusions. In the case $N_e = N$, we can improve this formula by subtracting

$$\frac{1}{6N}\left[(2N-i)\sum_{j=1}^{i}\left(\frac{1}{j} + \frac{2N}{(2N-j)^2}\right) + i\sum_{j=i+1}^{2N-1}\left(\frac{2N}{j^2} + \frac{1}{2N-j}\right)\right]$$

where $i = 2Np$ and j is a dummy variable. Some numerical values are given in Table 7.4. It will be seen that while this improvement, which is due to Ewens

Table 7.4 Mean absorption time (generations) when two neutral alleles are present initially, in a population of actual and effective size 25.

Initial number of A alleles	Mean absorption time		
	Approximate	Improved	Exact
1	9.80	9.23	9.13
2	16.79	15.99	15.81
3	22.70	21.75	21.56
4	27.88	26.83	26.63
5	32.51	31.37	31.18

(1964a), is helpful, our simple approximate formula is not seriously misleading (see note 4 for further details).

Interpreting the results

In Table 7.4, the mean absorption time rises rapidly as the initial frequency of the rarer allele increases. We shall show that this is true generally when one allele is rare. We have, with $\delta p = 1/(2N)$,

$$T(p + \delta p) - T(p) = \delta p T'(p) + \tfrac{1}{2}(\delta p)^2 T''(p) + \cdots$$

$$= -\frac{N_e}{N}\left[2\log p - 2\log(1-p) + \left(\frac{1}{2Np} + \frac{1}{2N(1-p)}\right) + \cdots\right]$$

$$= \frac{N_e}{N}\left[2\log(2N) - 2\log i + 2\log\left(\frac{2N-i}{2N}\right) - \left(\frac{2N}{i(2N-i)}\right) + \cdots\right]$$

where $i = 2Np$. When i is small compared to $2N$, we may approximate $(2N - i)$ by $2N$, so that the increase in mean absorption time per unit increase in the number i of the rarer allele is approximately

$$\frac{N_e}{N}\left(2\log(2N) - 2\log i - \frac{1}{i}\right)$$

a substantial increase provided N and N_e/N are not very small.

Further (writing T as short for $T(p)$), dT/dp is positive when $p < \tfrac{1}{2}$, zero when $p = \tfrac{1}{2}$ and negative when $p > \tfrac{1}{2}$; also

$$\frac{d^2T}{dp^2} = -4N_e\left(\frac{1}{p} + \frac{1}{1-p}\right) = \text{negative}$$

MEAN SOJOURN, ABSORPTION AND FIXATION TIMES

Hence T increases with increasing p up to a maximum when $p = \frac{1}{2}$ in which case

$$T = -4N_e \log 0.5 = 2.77 N_e$$

and thereafter decreases with increasing p. In contrast the smallest value of T, as given when $p = 1/(2N)$, is

$$\frac{N_e}{N}\left[2 \log(2N) - (4N - 2) \log\left(1 - \frac{1}{2N}\right)\right]$$

The ratio $T_{\text{greatest}}/T_{\text{smallest}}$ equals 7.1 when $N = 25$, 12.4 when $N = 50$ and 22.0 when $N = 100$. Thus polymorphism is lost much more rapidly when one allele is rare than when both alleles are common. We see once again that it is much harder to conserve rare alleles than common alleles. Further, the mean absorption time is directly proportional to N_e. Thus the prospect of conserving rare alleles is particularly poor when N_e is small; see Table 7.5.

In the case of natural populations, we may, as a first approximation, take N_e/N as fixed and investigate how $T(p)$ increases with increasing N. When $p = 1/(2N)$, we have approximately (N being relatively large)

$$T(p) = (N_e/N)[2 \log(2N) + 2]$$

Table 7.5 Approximate mean absorption time (generations) for given population size (N), effective population size (N_e) and initial number (i) of the rarer allele.

		\multicolumn{6}{c}{N_e/N}
N	i	0.2	0.4	0.6	0.8	1.0	2.0
25	1	2.0	3.9	5.9	7.8	9.8	19.6
	2	3.4	6.7	10.1	13.4	16.8	33.6
	3	4.5	9.1	13.6	18.2	22.7	45.4
	4	5.6	11.2	16.7	22.3	27.9	55.8
	5	6.5	13.0	19.5	26.0	32.5	65.0
50	1	2.2	4.5	6.7	9.0	11.2	22.4
	2	3.9	7.8	11.8	15.7	19.6	39.2
	3	5.4	10.8	16.2	21.6	27.0	53.9
	4	6.7	13.4	20.2	26.9	33.6	67.2
	5	7.9	15.9	23.8	31.8	39.7	79.4

which increases very slowly as N increases. On the other hand, when $p = \frac{1}{2}$

$$T(p) = (N_e/N)(2.77N)$$

Now consider the case where three or more neutral alleles are present initially. Littler (1975) has investigated this situation for the case $N_e = N$. For simplicity, we summarize his results for the three-allele case (other cases are essentially similar). If all three alleles are equally common initially, the mean absorption time is $3.24N$ generations, comparable with the $2.77N$ generations for the case of two alleles equally common at the start. In this three-allele case, one allele is lost a considerable time before a second allele is lost; the average time for loss of the first allele being $1.15N$ generations. Particularly interesting is the case where one allele is initially rare and the other two common. The rare allele is usually lost rapidly, while the remaining alleles will usually both be present for a long time. Thus, if initial frequencies are 0.01, 0.49, 0.50, the mean time until one allele is lost (nearly always the rare allele, of course) is $0.17N$ generations, whereas it takes on average $2.80N$ generations for the number of alleles to fall to one.

Mean sojourn times in the modified process

Consider now cases that lead to fixation of A (our modified process) (Maruyama 1972a, Ewens 1973). For initial frequency p, $T^*(p, x)[1/(2N)]$, the mean sojourn time at frequency x, will be

$$T(p, x) \frac{x}{p} \frac{1}{2N} = 2N_e/N \qquad \text{when } p \leqslant x$$

$$= \frac{2(1-p)x}{(1-x)p} \frac{N_e}{N} \qquad \text{when } p \geqslant x$$

to a close approximation, except of course when $x = p$ or x is close to 0 or 1, the error being biologically trivial even in these cases.

The most remarkable result (foreshadowed in our earlier discussion of the special case $p = 1/(2N)$) is that allele A spends, on average, about the *same* time, $2N_e/N$ generations, at *every* frequency exceeding p.

The approximate mean *combined* time at *all* frequencies exceeding p is therefore

$$2\frac{N_e}{N}[2N(1-p)] = 4N_e(1-p)$$

generations.

Since A is ultimately fixed, we might suppose that A spends very little time at frequencies less than p. This is not necessarily correct. For the mean combined time at frequencies less than p is about

$$2N\int_0^p 2\frac{N_e}{N}\frac{1-p}{p}\frac{x}{1-x}dx = 4N_e\frac{1-p}{p}\int_0^p\left(\frac{1}{1-x}-1\right)dx$$

$$= 4N_e\frac{1-p}{p}[-\log(1-x) - x]_0^p$$

$$= -4N_e\left(\frac{(1-p)\log(1-p)}{p} + 1 - p\right)$$

Now, when p is small, $\log(1-p)$ will be close to $(-p - \frac{1}{2}p^2)$ and this mean combined time becomes about $2N_e p$ generations (ignoring a term in p^2) and is thus quite small. On the other hand, if $p = \frac{1}{2}$ we obtain $0.77N_e$ generations, not strikingly less than the $4N_e(\frac{1}{2}) = 2N_e$ generations at frequencies greater than p. In fact, when $p > 0.7968$, A spends, on average, more time overall at frequencies less than p than at frequencies exceeding p.

Mean fixation time

The appropriate mean fixation time, obtained by integrating $T^*(p, x)$ between limits 0 and 1, is

$$-4N_e\left(\frac{(1-p)\log(1-p)}{p}\right).$$

generations, as obtained by Kimura & Ohta (1969a) using an equivalent but slightly different approach.

For small p, we may approximate $\log(1-p)$ by $(-p)$, $(1-p)$ by 1 and the mean fixation time becomes $4N_e$ generations, as noted by Kimura & Ohta. Some numerical values are given in Table 7.6.

Note that our mean time, while of course decreasing with increasing p, is not altogether negligible even when p is very close to unity. Thus when $p = 1 - 1/(2N)$ we obtain

$$\frac{4N_e \log(2N)}{2N - 1}$$

mainly reflecting time spent at frequencies less than p.

EFFECT OF SELECTION

Table 7.6 Mean fixation time for a neutral allele with initial frequency p in a population of effective size N_e.

p	Mean fixation time $\div N_e$
0.001	4.00
0.01	3.98
0.1	3.79
0.3	3.33
0.5	2.77
0.7	2.06
0.9	1.02
0.99	0.19
0.999	0.03

Effect of selection

When selection is acting, we have (Eqn 7.4), writing T as short for $T(p, x)$,

$$M \frac{\partial T}{\partial p} + \tfrac{1}{2} V \frac{\partial^2 T}{\partial p^2} = -\delta(x - p) = -\delta(p - x)$$

where, as usual, M is the mean change in frequency of A in going from generation 0 to generation 1 and $V = p(1 - p)/(2N_e)$. Then

$$\frac{\partial^2 T}{\partial p^2} + \frac{2M}{V} \frac{\partial T}{\partial p} = -\frac{2}{V} \delta(p - x) = -\frac{4N_e}{p(1-p)} \delta(p - x)$$

Multiplying through by

$$H(p) = e^{\int (2M/V)\, dp}$$

and using the usual rule for differentiation of a product, we have

$$\frac{\partial}{\partial p}\left(H(p) \frac{\partial T}{\partial p}\right) = -\frac{4N_e}{p(1-p)} H(p)\delta(p - x)$$

The rest of the argument is very similar to that for the neutral case, so that an outline treatment will suffice. We have

$$H(p) \frac{\partial T}{\partial p} = -\int_0^p \frac{4N_e}{\xi(1-\xi)} H(\xi)\delta(\xi - x)\, d\xi + \text{constant}$$

where ξ is a dummy variable and $H(\xi)$ is $H(p)$ with p replaced by ξ.

When $p < x$

$$H(p) \, \partial T/\partial p = A$$

where A is a constant.

When $p > x$

$$H(p) \frac{\partial T}{\partial p} = -\frac{4N_e}{x(1-x)} H(x) + B$$

where B is a constant.

The jump in $H(p)\partial T/\partial p$ at $p = x$ equals (with ε positive and tending to zero)

$$-\int_{x-\varepsilon}^{x+\varepsilon} \frac{4N_e}{\xi(1-\xi)} H(\xi)\delta(\xi - x) \, d\xi = -\frac{4N_e}{x(1-x)} H(x)$$

whence $B = A$.

Let

$$G(p) = e^{-\int (2M/V) \, dp} = 1/H(p)$$

Then when $p < x$

$$\partial T/\partial p = AG(p)$$

$$T = A \int_0^p G(\theta) \, d\theta + C$$

where θ is a dummy variable and C is a constant. When $p = 0$, the definite integral, having upper limit equal to lower limit, equals 0. Also, of course, when $p = 0$, $T = 0$. Hence $C = 0$, so

$$T = A \int_0^p G(\theta) \, d\theta$$

When $p > x$

$$\frac{\partial T}{\partial p} = -G(p) \frac{4N_e}{x(1-x)} H(x) + AG(p)$$

EFFECT OF SELECTION

(since $B = A$). It will be convenient to express T as a definite integral between limits 1 and p, i.e.

$$T = -\frac{4N_e}{x(1-x)} H(x) \int_1^p G(\theta)\, d\theta + A \int_1^p G(\theta)\, d\theta + D$$

where D is a constant. When $p = 1$, both definite integrals are zero; also, when $p = 1$, $T = 0$, whence $D = 0$.

Putting $H(x) = 1/G(x)$ and noting (from elementary calculus) that if we interchange the limits for integration, only the sign of the definite integral is changed, we have

$$T = \frac{4N_e}{x(1-x)G(x)} \int_p^1 G(\theta)\, d\theta - A \int_p^1 G(\theta)\, d\theta$$

As in the neutral case, T is a continuous function of p for all values of p, so that our expressions for T when $p < x$ and when $p > x$ will be equal when $p = x$. Thus

$$A \int_0^x G(\theta)\, d\theta = \frac{4N_e}{x(1-x)G(x)} \int_x^1 G(\theta)\, d\theta - A \int_x^1 G(\theta)\, d\theta$$

Therefore

$$A\left(\int_0^x G(\theta)\, d\theta + \int_x^1 G(\theta)\, d\theta\right) = A \int_0^1 G(\theta)\, d\theta = \frac{4N_e}{x(1-x)G(x)} \int_x^1 G(\theta)\, d\theta$$

and

$$A = \frac{4N_e}{x(1-x)G(x)} \frac{\int_x^1 G(\theta)\, d\theta}{\int_0^1 G(\theta)\, d\theta}$$

Now, from Chapter 5, $u(p)$, the probability of fixation given initial frequency p, equals

$$\frac{\int_0^p G(\theta)\, d\theta}{\int_0^1 G(\theta)\, d\theta}$$

Similarly, $v(p)$, the probability of loss, equals

$$1 - u(p) = \frac{\int_0^1 G(\theta)\, d\theta - \int_0^p G(\theta)\, d\theta}{\int_0^1 G(\theta)\, d\theta} = \frac{\int_p^1 G(\theta)\, d\theta}{\int_0^1 G(\theta)\, d\theta}$$

Hence, when $p \leqslant x$

$$T(p, x) = \frac{4N_e u(p)}{x(1-x)G(x)} \int_x^1 G(\theta) \, d\theta$$

Noting that

$$1 - \frac{\int_x^1 G(\theta) \, d\theta}{\int_0^1 G(\theta) \, d\theta} = \frac{\int_0^1 G(\theta) \, d\theta - \int_x^1 G(\theta) \, d\theta}{\int_0^1 G(\theta) \, d\theta} = \frac{\int_0^x G(\theta) \, d\theta}{\int_0^1 G(\theta) \, d\theta}$$

we find (by simple algebra) that, when $p \geqslant x$,

$$T(p, x) = \frac{4N_e v(p)}{x(1-x)G(x)} \int_0^x G(\theta) \, d\theta$$

These formulae for $T(p, x)$ are due to Ewens (1963a, 1969); they reduce to the formulae for the neutral case on putting $G(x) = G(\theta) = e^0 = 1$, $u(p) = p$, $v(p) = 1 - p$.

To illustrate the calculation of mean sojourn times under selection, we consider the haploid model. Then

$$M = \alpha p q \qquad V = pq/(2N_e)$$

$$\int (2M/V) \, dp = 4N_e \alpha p \qquad G(p) = e^{-4N_e \alpha p}$$

$$u(p) = \frac{1 - e^{-4N_e \alpha p}}{1 - e^{-4N_e \alpha}} \qquad v(p) = \frac{e^{-4N_e \alpha p} - e^{-4N_e \alpha}}{1 - e^{-4N_e \alpha}}$$

$$\int G(\theta) \, d\theta = -\frac{e^{-4N_e \alpha \theta}}{4N_e \alpha}$$

When $p \leqslant x$

$$T(p, x) = 4N_e \frac{1 - e^{-4N_e \alpha p}}{1 - e^{-4N_e \alpha}} \frac{e^{4N_e \alpha x}}{x(1-x)} \frac{e^{-4N_e \alpha x} - e^{-4N_e \alpha}}{4N_e \alpha}$$

$$= \frac{(1 - e^{-4N_e \alpha (1-x)})(1 - e^{-4N_e \alpha p})}{\alpha x(1-x)(1 - e^{-4N_e \alpha})}$$

and, when $p \geqslant x$

$$T(p, x) = \frac{(e^{4N_e \alpha x} - 1)(e^{-4N_e \alpha p} - e^{-4N_e \alpha})}{\alpha x(1-x)(1 - e^{-4N_e \alpha})}$$

These formulae can sometimes be simplified. Thus in the case $p = 1/(2N)$, $N_e = N$, we have

$$1 - e^{-4N_e\alpha p} = 1 - e^{-2\alpha} \simeq 2\alpha$$

when α is small, whence for $x \geqslant 1/(2N)$

$$T(p, x) = \frac{2(1 - e^{-4N\alpha(1-x)})}{x(1-x)(1 - e^{-4N\alpha})}$$

as given by Fisher (1930b). However, Fisher did not obtain the corresponding formula for the range $x = 0$ to $1/(2N)$ (required in the calculation of mean absorption times) and indeed maintained that a formula valid over this range was unobtainable by diffusion methods! We shall discuss Fisher's views on the reliability of diffusion methods in Chapter 10.

Mean times under genic selection in the modified process

The evolutionist will be mainly interested in mean sojourn times $T^*(p, x)$ and corresponding mean fixation times for those advantageous mutants that are eventually fixed. To obtain $T^*(p, x)$, we must, of course, multiply $T(p, x)$ by (probability of fixation given initial frequency x) ÷ (probability of fixation given initial frequency p) which, for the haploid model, is

$$\frac{1 - e^{-4N_e\alpha x}}{1 - e^{-4N_e\alpha p}}$$

Thus, when $p \leqslant x$

$$T^*(p, x) = \frac{(1 - e^{-4N_e\alpha x})(1 - e^{-4N_e\alpha(1-x)})}{\alpha x(1-x)(1 - e^{-4N_e\alpha})}$$

(Maruyama 1972a, Ewens 1973), which, as expected, is independent of p (in contrast, $T^*(p, x)$ for the case $p > x$ does depend on p and is rather complicated in that case).

For a comprehensive analysis of the process of change of allele frequency under selection and drift, we should consider the effect of dominance, partial or complete. However, the principal features of this process can be found using the relatively simple haploid model. We shall, therefore, give results for this model in some detail.

Effect of selection: some results

We shall concentrate on the biologically most important case, $p = 1/(2N)$. We can, then, introduce a slight simplification; we need consider only the formula for $T^*(p, x)$ when $x \geqslant p$. Obviously, this is true when calculating mean sojourn times, $T^*(p, x)/(2N)$, since we are interested only in $x \geqslant 1/(2N)$.

We shall calculate the *mean* time spent by our advantageous allele in a given range of frequencies, a to b, where, for the moment, $a > 1/(2N)$, $b < 1 - 1/(2N)$. To do this, we integrate $T^*(p, x)$ between appropriate limits. Strictly, if 'the range a to b' includes the frequencies a and b themselves, limits of integration will be $a - 1/(4N)$, $b + 1/(4N)$, whereas if mean time spent *between* a and b is wanted, limits will be $a + 1/(4N)$, $b - 1/(4N)$. We shall ignore this distinction, which has no practical importance, and approximate either mean time by

$$\int_a^b T^*(p, x)\,\mathrm{d}x$$

$T^*(p, x)$ being given by our formula for $x \geqslant p$.

The case $a = 1/(2N)$, $b = 1 - 1/(2N)$ would appear to be an exception. Obviously, if in this case a and b are included in the range, the mean time will be the mean fixation time; following our previous practice, we should then take the limits to be 0, 1 rather than a, b, and it really would be necessary to invoke the (rather complicated) formula for $T^*(p, x)$ when $x \leqslant p$ in order to integrate over the range $x = 0$ to $1/(2N)$. However, this integral will certainly not exceed the corresponding integral in the neutral case, which we showed earlier to be $2N_e p = N_e/N$, thus contributing very little to the mean fixation time. It is easiest, then, and causes very little error, to ignore the range 0 to $1/(2N)$ completely. We take the mean fixation time as approximately

$$\int_{1/(2N)}^1 T^*(p, x)\,\mathrm{d}x$$

Clearly, then, the formula for $T^*(p, x)$ when $x \geqslant p$, namely

$$\frac{(1 - e^{-4N_e \alpha x})(1 - e^{-4N_e \alpha (1-x)})}{\alpha x (1 - x)(1 - e^{-4N_e \alpha})}$$

will be central to our discussion of the case $p = 1/(2N)$; for simplicity, we shall throughout use T^* as short for this formula.

EFFECT OF SELECTION: SOME RESULTS

(1) It may be shown that, for constant N_e, α and N, T^* has minimal value when $x = \frac{1}{2}$. We thus recover the result, familiar from deterministic treatments, that evolution under natural selection is most rapid at intermediate allele frequencies. However, this result is not as helpful as might first seem, since the *magnitude* of the difference in mean time can be quite small; the reader will easily verify this by substituting some values, taking $N_e\alpha$ *small*.

(2) Indeed we might guess, from results on the probability of fixation given in Chapter 5, that when $4N_e\alpha$ is small, mean sojourn times will differ very little from corresponding times for a neutral allele. To verify this, we recall that when $4N_e\alpha$ is small (less than about 0.25) we have the close approximation

$$1 - e^{-4N_e\alpha} = 4N_e\alpha$$

and similarly for the other exponential terms, giving a mean sojourn time close to

$$\frac{4N_e\alpha x 4N_e\alpha(1-x)}{\alpha x(1-x)4N_e\alpha 2N} = 2\frac{N_e}{N}$$

generations, for any x, as in the case of a neutral allele.

(3) Now consider the case $4N_e\alpha$ large; then

$$1 - e^{-4N_e\alpha}$$

will be very close to unity. Further, when x is *very small*

$$1 - e^{-4N_e\alpha(1-x)} \quad \text{and} \quad (1-x)$$

will each be close to unity. However, $4N_e\alpha x$ will be *small*, unless selection is very intense; for example, when $x = 1/(2N)$, $4N_e\alpha x = 2\alpha N_e/N$. We have, then, the close approximation, when $4N_e\alpha x < 1/4$,

$$1 - e^{-4N_e\alpha x} = 4N_e\alpha x$$

and the mean sojourn time at frequency x becomes $2N_e/N$ generations, to a close approximation. We see, then, that *even when $4N_e\alpha$ is large*, the mean sojourn time at frequency x is very close to the corresponding time for a *neutral* allele when x is very close to zero. The same is true when x is very close to unity.

(4) Clearly, then, standard deterministic theory for the change of allele frequencies under selection is quite inappropriate when allele frequencies are

very extreme, since at these frequencies the advantageous allele is behaving as if it were neutral. It is natural, then, to ask: 'Given $4N_e\alpha$ large, is there a range of frequencies for which the deterministic approach is relevant?'

We continue to suppose mild selection and the haploid model. Then, in the deterministic approach, the change in frequency of the advantageous allele in a single generation is

$$\delta x = \alpha x(1-x)$$

to a close approximation. We may write this as

$$\delta x/\delta t = \alpha x(1-x)$$

since $\delta t = 1$ (generation) and approximate $\delta x/\delta t$ by dx/dt in the usual way. Hence the familiar result for this model; the time to go from frequency a to frequency b, given by the deterministic approach, is

$$\int_a^b \frac{1}{\alpha x(1-x)}\,dx$$

We may compare this with the true mean time spent between frequencies a and b (given $4N_e\alpha$ large)

$$\int_a^b \frac{(1-e^{-4N_e\alpha x})(1-e^{-4N_e\alpha(1-x)})}{\alpha x(1-x)}\,dx$$

Note that the exponential terms are necessarily positive, so that the integrand is always *smaller* than the corresponding integrand given by deterministic theory. Hence, if we use the latter to approximate the mean time between a and b, our result will always be *too large*. As we shall see, for some values of a and b, this overestimation is very marked. However, if x lies in the range $2/N_e\alpha$ to $1 - 2/N_e\alpha$ neither exponential term will exceed $e^{-8} = 0.0003$ and the overestimation is very minor (Ewens 1967b, 1969, 1979).

(5) Generally, then, given $N_e\alpha > 4$, so that

$$0 < \frac{2}{N_e\alpha} < 1 - \frac{2}{N_e\alpha} < 1$$

the pattern of mean times is as follows:

(a) for frequencies $1/(2N)$ to $1/(16N_e\alpha)$ and $1 - 1/(16N_e\alpha)$ to 1, much the same as for a neutral allele;

(b) for frequencies $1/(16N_e\alpha)$ to $2/(N_e\alpha)$ and $1 - 2/(N_e\alpha)$ to $1 - 1/(16N_e\alpha)$, affected by both drift and selection;
(c) for frequencies $2/(N_e\alpha)$ to $1 - 2/(N_e\alpha)$, much the same as given by deterministic theory.

We stress that the fraction of the range covered by phase (c) depends on $N_e\alpha$ (rather than on N, as beginners might suppose). Obviously, the use of deterministic theory for phase (a) is quite inappropriate; detailed calculations show that the deterministic approach can give poor results for phase (b) also.

Calculations of mean times under selection: a numerical example

Now consider the actual calculation of mean times. In general, the integration of T^* will have to be done numerically on the computer. If, however, $N_e\alpha$ is large (>20), the following procedure (which supplies a simple formula for the mean fixation time) is convenient. Note, from the symmetry of T^*, that the mean time between a and b is also the mean time between $(1 - b)$ and $(1 - a)$; more formally, this can be shown by substituting $y = 1 - x$ in T^*. Thus it will be sufficient to consider values of a and b in the range $1/(2N)$ to 0.5, inclusive. Since $4N_e\alpha$ is large, the term $e^{-4N_e\alpha(1-x)}$ will be negligible over this range. We now write the integral as a 'deterministic part' minus a 'correction':

$$\int_a^b \frac{1 - e^{-4N_e\alpha x}}{\alpha x(1 - x)} dx = \int_a^b \frac{1}{\alpha x(1 - x)} dx - \int_a^b \frac{e^{-4N_e\alpha x}}{\alpha x(1 - x)} dx$$

For the deterministic part we have the familiar result

$$\int_a^b \frac{1}{\alpha x(1-x)} dx = \int_a^b \frac{1}{\alpha x} dx + \int_a^b \frac{1}{\alpha(1-x)} dx$$
$$= (1/\alpha)[\log x - \log(1-x)]_a^b$$
$$= (1/\alpha)[\log b - \log(1-b) - \log a + \log(1-a)]$$

The correction will be negligible except for x in the range $1/(2N)$ to $2/N_e\alpha$. Over this range ($N_e\alpha$ being large), $1 - x$ will depart little from unity, so that our correction is well approximated by

$$\int_a^b \frac{e^{-4N_e\alpha x}}{\alpha x} dx$$

The relative error in the correction arising from this approximation will always be very small, given $N_e\alpha > 20$; if, in addition, $N_e\alpha^2 > 0.25$, the absolute error will also be very small (see note 6).

On making the substitution $y = 4N_e\alpha x$, our approximate correction becomes

$$\frac{1}{\alpha}\int_{4N_e\alpha a}^{4N_e\alpha b} \frac{e^{-y}}{y}\,dy = \frac{1}{\alpha}\left(\int_{4N_e\alpha a}^{\infty} \frac{e^{-y}}{y}\,dy - \int_{4N_e\alpha b}^{\infty} \frac{e^{-y}}{y}\,dy\right)$$

by elementary calculus.

By a standard mathematical result

$$\int_z^{\infty} \frac{e^{-y}}{y}\,dy = -0.577\,215\,6649 - \log z + z - \frac{z^2}{2\times 2!}$$
$$+ \frac{z^3}{3\times 3!} - \frac{z^4}{4\times 4!} + \cdots (z > 0);$$

this *exponential integral* is extensively tabulated in Abramowitz & Stegun (1965). It will be helpful to note that, when z is small, it is sufficient to take just the first few terms of the series, whereas when z exceeds about 8, the integral is very small in value.

Some mean times are given in Table 7.7. Several features of this table hold generally when $4N_e\alpha$ is large. First, evolution is very much faster at intermediate than at extreme frequencies. Secondly, even mild selection greatly speeds the rate of evolution. Thus, in the table, each range has width 0.02 or so, covering $2N \times 0.02 = 20\,000 \times 0.02 = 400$ frequencies; thus, with $N_e = N$, a neutral allele would spend, on average, about 800 generations in *every* range of frequency. Thirdly, deterministic results for extreme frequencies are decidedly too large; for our range 0.00005 − 0.02, the result would be 1202.3 generations when $\alpha = 0.005$.

Now consider the outcome if we change the value of α, keeping other quantities unchanged. We see at once, from the formula for times calculated deterministically, that *when these calculations are relevant*, multiplying α by a constant divides every mean time by that constant. On the other hand, while changing α will always alter mean times in the *direction* expected, the *magnitude* of the change will not, *for any mean times affected by drift*, be as large as deterministic theory would suggest. For example, changing α from 0.005 to 0.01, but keeping $N_e = N = 10\,000$, reduces the time spent in the range 0.00005–0.02 from 395.5 only to 265.7; however, for all other ranges given in Table 7.7, mean times are, near enough, halved.

Reducing N_e, while keeping N and α constant, also reduces mean times

Table 7.7 Approximate mean times, over various ranges of allele frequency, spent by an advantageous allele under genic selection. Selective advantage $\alpha = 0.005$, actual and effective population size = 10 000.

Range of frequency	Mean time (generations)
0.00005–0.02	395.5
0.02–0.04	142.0
0.04–0.06	85.3
0.06–0.08	61.8
0.08–0.10	49.0
0.10–0.12	41.0
0.12–0.14	35.4
0.14–0.16	31.4
0.16–0.18	28.4
0.18–0.20	26.0
0.20–0.22	24.1
0.22–0.24	22.6
0.24–0.26	21.3
0.26–0.28	20.3
0.28–0.30	19.4
0.30–0.32	18.7
0.32–0.34	18.1
0.34–0.36	17.6
0.36–0.38	17.2
0.38–0.40	16.8
0.40–0.42	16.5
0.42–0.44	16.3
0.44–0.46	16.2
0.46–0.48	16.1
0.48–0.50	16.0

affected by drift. Thus keeping $N = 10\,000$, $\alpha = 0.005$ but reducing N_e from 10 000 to 5000 reduces the mean time in the first two ranges given in Table 7.7 to 266.9 and 133.7, respectively; other mean times are virtually unaffected.

Mean fixation time under selection

The effect of changing α and N_e is particularly apparent when we consider mean fixation times. For the mean time between $1/(2N)$ and 0.5, the deterministic term is

$$\frac{1}{\alpha}\left[\log 0.5 - \log 0.5 - \log\left(\frac{1}{2N}\right) + \log\left(1 - \frac{1}{2N}\right)\right]$$

or about

$$\frac{\log(2N)}{\alpha} - \frac{1}{2N\alpha}$$

The correction term will be negligible except for the range of frequencies $a = 1/(2N)$ to $b = 2/(N_e\alpha)$. Since $4N_e\alpha(1/(2N)) = 2\alpha N_e/N$, $4N_e\alpha(2/N_e\alpha) = 8$, the correction will be

$$\frac{1}{\alpha}\left(\int_{2\alpha N_e/N}^{\infty} \frac{e^{-y}}{y}\,dy - \int_{8}^{\infty} \frac{e^{-y}}{y}\,dy\right)$$

The second integral is very small and will be ignored; for $2\alpha N_e/N$ small (<0.1) the first integral is about

$$-0.5772 - \log\left(2\alpha\frac{N_e}{N}\right) + 2\alpha\frac{N_e}{N} - \alpha^2\left(\frac{N_e}{N}\right)^2$$

Subtracting this from the deterministic term, we obtain the mean time between $1/(2N)$ and 0.5 as

$$(1/\alpha)(\log 2 + \log N + 0.5772 + \log 2 + \log \alpha + \log N_e - \log N)$$

$$-\frac{1}{2N\alpha} - 2\frac{N_e}{N} + \alpha\left(\frac{N_e}{N}\right)^2$$

$$= \frac{1.9635 + \log N_e + \log \alpha}{\alpha} - 2\frac{N_e}{N} - \frac{1}{2N\alpha} + \alpha\left(\frac{N_e}{N}\right)^2$$

The last two terms contribute very little (recall that we are supposing $N_e\alpha > 20$, so that $N\alpha > 20$) and will be ignored. Doubling the remainder will give a close approximation to the mean fixation time. We can add a term for the range $1 - 1/(2N)$ to 1; since our allele is effectively neutral over this range, we may take $T^*(p, x)$ equal to $4N_e$ and our term is

$$\int_{1-1/(2N)}^{1} 4N_e\,dx = 2N_e/N$$

264

Thus, given $N_e \alpha > 20$, the mean fixation time is

$$\frac{3.9270 + 2\log N_e + 2\log \alpha}{\alpha} - 2\frac{N_e}{N}$$

generations, bringing out particularly clearly the contrast between results from the stochastic and the deterministic approaches. The term $2N_e/N$ may clearly be ignored. Some numerical results are given in Table 7.8.

Note once again that the mean time is only very roughly inversely proportional to α. Particularly interesting is the dependence on N_e rather than on N, a result reminiscent of that for the neutral case. However, in that case, the mean fixation time increases linearly with N_e, whereas with selection the mean fixation time increases linearly with the *logarithm* of N_e and hence very slowly with N_e itself. Thus the contrast, noted earlier, between mean times for an advantageous and a neutral allele becomes very marked when N_e is very large.

Finally, to avoid any possible misunderstanding, we note that Haldane, who was responsible for almost all the admirable early work using deterministic methods, was fully aware that the deterministic approach has limitations. Referring to this work, he stated (Haldane 1932): 'we have argued as if the populations dealt with were infinite, and, what is more, as if the numbers of both (or all) the competing types were infinite'. Indeed, we have noted earlier Haldane's contributions to the stochastic theory.

Table 7.8 Approximate mean fixation times (generations) under genic selection; N_e = effective population size, α = selective advantage.

α	N_e		
	10^4	10^6	10^8
0.005	2350	4192	6034
0.01	1314	2235	3156
0.05	327	511	696

Variances of sojourn, absorption and fixation times

To calculate variances of sojourn and other times, we may use the formulae given in Chapter 3. For example, we shall calculate the variance of the fixation time of a neutral allele of initial frequency $1/(2N)$, namely

$$2\sum_x \bar{s}_{px}^* \bar{s}_x^* - \bar{s}_p^* - (\bar{s}_p^*)^2$$

MEAN SOJOURN, ABSORPTION AND FIXATION TIMES

where $\bar{s}_{px}^*$ is the mean sojourn time, in the modified process, at frequency x, given initial frequency p, and $\bar{s}_x^*$, $\bar{s}_p^*$ are the mean fixation times for initial frequencies x, p respectively. For $p = 1/(2N)$ we have approximately

$$\bar{s}_{px}^* = 2N_e/N = 4N_e \delta x \qquad \text{(where } \delta x = 1/(2N)\text{)}$$

Further

$$\bar{s}_x^* = -4N_e \frac{(1-x)\log(1-x)}{x}$$

approximately. Then, approximately

$$2 \sum_x \bar{s}_{px}^* \bar{s}_x^* = -32N_e^2 \sum_{x=1/(2N)}^{1-1/(2N)} \frac{(1-x)\log(1-x)}{x} \delta x$$

We shall approximate the sum by an integral. Ideally limits for the latter should be $1/(4N)$ and $1 - 1/(4N)$ but, for convenience, we shall use limits 0 and 1, it being understood that these arise from $1/(4N)$ and $1 - 1/(4N)$ by letting $1/(4N)$ tend to zero. This seems fairly satisfactory, even when N is fairly small. For example, when $N = 25$, the sum is -0.6348, whereas the integral between limits 0 and 1 is -0.6449. We have, then, the approximation

$$2 \sum_x \bar{s}_{px}^* \bar{s}_x^* = -32N_e^2 \int_0^1 \frac{(1-x)\log(1-x)}{x} dx$$

$$= -32N_e^2 \left(\int_0^1 \frac{\log(1-x)}{x} dx - \int_0^1 \log(1-x) dx \right)$$

By a standard mathematical result, the first integral equals (very surprisingly!) $-\pi^2/6$ and the second integral is

$$[-(1-x)\log(1-x) - x]_0^1 = -1$$

since, by another standard result

$$\lim_{x \to 1}(1-x)\log(1-x) = 0$$

giving in all

$$32N_e^2(\pi^2/6 - 1)$$

Since $p = 1/(2N)$, we may take $\bar{s}_p^*$ equal to $4N_e$, unless N_e is very small. Thus the variance of the fixation time is about

$$32N_e^2(\pi^2/6 - 1) - 4N_e - 16N_e^2$$

Interestingly, this is not quite the same as the formula obtained by diffusion methods (Kimura & Ohta 1969b), in which the term $-4N_e$ does not appear, although the rest of the formula is the same as ours. Generally (Pollak & Arnold 1975, Tavaré 1979) diffusion formulae for variance of sojourn, absorption or fixation times each lack one term, leading to loss of accuracy in some cases. In the present case, however, the discrepancy matters very little if N_e is at all large, since in that case our term in N_e is so small in comparison with terms in N_e^2 that we may as well omit it, obtaining $4.6379N_e^2$. Thus the standard deviation of the fixation time is about $2.15N_e$, comparable in magnitude with the mean fixation time $4N_e$.

Generally, for any frequency, or range of frequencies, for which times are affected (totally or markedly) by drift, standard deviations will be relatively large (sometimes larger than the mean itself). We leave it to the reader to show that for a neutral allele of initial frequency $1/(2N)$ the variance of the absorption time is about

$$8\frac{N_e^2}{N}\frac{\pi^2}{3} - \text{(mean absorption time)} - \text{(mean absorption time)}^2$$

giving a standard deviation substantially larger than the mean when N is large.

Concerted evolution of multigene families

It will be convenient at this stage to discuss, very briefly, the problem of concerted evolution of multigene families, since Ohta (1980) adapted results on mean fixation times when dealing with this problem. In a given species, the same gene (or small group of genes), G say, is often found repeated many times on the same chromosome, tandem fashion

$$G_1 G_2 G_3 G_4 G_5 \ldots$$

where the DNA sequence is much the same for all G's. In cases where the G's have a well defined function, the complete set of G's is called a multigene family.

Now consider a related species. Often, this has a multigene family performing the same function as the G's but with a different unit, F say, the family appearing as

$$F_1 F_2 F_3 F_4 F_5 \ldots$$

where the F's are all much alike. For example, in three species of sea urchin, there is a repeating unit comprising genes H coding, in early embryonic development, for five different histones, together with five distinct non-transcribed spacers NTS

	N		N		N		N		N
H1	T	H4	T	H2B	T	H3	T	H2A	T
	S		S		S		S		S
	1		2		3		4		5

(the H's being transcribed separately). Over 300 tandem repeats of this unit appear in each species. Within a species, the repeats are virtually identical. The H genes are much the same in the different species but each NTS differs markedly between species, both in length and DNA sequence.

Thus, at some stage in the evolution of a given species, the same mutations have spread to all members of the family. Thus the members of the family are evolving in concert rather than independently.

Sometimes, as in the case of genes coding for rRNA in Man, the multigene family is spread over several chromosomes, with a smallish number of tandem repeats on each. In all cases, we have the problem: how can different genes evolve in concert? For a recent review, with extensive references, see Li et al. (1985).

Consider a multigene family made up entirely of tandem repeats. The evolution of the family will be very complicated, involving changes (a) within individual single chromosomes and (b) in the proportions of different types of individual single chromosome in the population. Clearly, however, this evolution must involve some 'homogenizing' process, which tends to make family members on the same individual chromosome alike. Homogeneity might arise through gene conversion, in which two family members exchange single DNA strands, giving a mismatch between complementary strands at places where the two members differed; mismatches are then corrected by appropriate enzymes, probably with a strong bias in the direction of correction. Homogenization in a single generation might, of course, be incomplete and would tend to be broken down by meiotic recombination between chromosomes homogenized for a different repeat. Alternatively, homogeneity could originate from unequal crossing over between sister chromatids at mitosis. This produces duplications and deficiencies of family members. If, then, we start, say, with two different types of family member A, B on the same individual chromosome, the proportion of each in a derived daughter chromosome will sometimes differ from that in the original chromosome. Eventually, by chance, all family members on a given chromosome are the same; note that the number of family members may have changed during this process. Again, the process may well be incomplete in a single generation. Meiosis would presumably be a process of staggering complexity, since pairing chromosomes might differ in number of family members and, even if not, might undergo either equal or unequal crossing over, giving both breakdown of established and creation of new homogeneity (sister chromatid unequal exchange might also occur). Clearly, possibilities for an exact treatment of unequal crossing over seem remote. For gene conversion, however, an exact treatment is possible, although inevitably very complicated; detailed discussion is beyond the scope of this book.

Detailed analysis (involving simplifying assumptions for unequal crossing over) is due mainly to Ohta (see especially Ohta 1980, 1983). She shows that, given plausible assumptions, either gene conversion or unequal crossing over could explain the actual observations. At present, then, we can use only informal arguments to decide which of the two mechanisms is more important. These arguments are not very conclusive. Gene conversion must, presumably, be invoked in cases where the multigene family is spread over several chromosomes. In cases where the multigene family contains a very small number of repeats, a change in this number would presumably be disadvantageous, so that avoidance of unequal crossing over would be favoured; hence homogeneity presumably arises by gene conversion in these cases. On the other hand, large multigene families often differ between species in repeat number, suggesting unequal crossing over. It is interesting that this mechanism has also been invoked to explain the generation of hypervariable DNA (Jeffreys *et al.* 1985a).

Notes and exercises

1 *Fisher's formula for n in terms of u_n.* We have

$$v_n = 1/(1 - u_n)$$

giving

$$v_{n+1} - v_n = \frac{1}{2} + \frac{1}{12}\frac{1}{v_n} - \frac{1}{720}\frac{1}{v_n^3} + \frac{1}{30240}\frac{1}{v_n^5} - \cdots \quad (7.7)$$

We aim to find a formula for v_n in terms of n, and hence a formula for n in terms of v_n, when n is large and then to calculate corrections to this latter formula to accommodate general n. If we write out 7.7 for successive values of n, starting with $n = 0$, sum the resulting equations and recall that $v_0 = 1$, we obtain

$$v_n = 1 + \tfrac{1}{2}n + \tfrac{1}{12}\sum_{i=0}^{n-1} 1/v_i - \tfrac{1}{720}\sum_{i=0}^{n-1} 1/v_i^3 + \tfrac{1}{30240}\sum_{i=0}^{n-1} 1/v_i^5 - \cdots \quad (7.8)$$

We see that $v_n > \tfrac{1}{2}n$. Thus v_n increases fairly rapidly as n increases. Hence $1/v_i$ falls off fairly rapidly and $1/v_i^3, 1/v_i^5, \ldots$ very rapidly, as n increases. In fact, since

$$1/v_i < 2/i$$

it follows, from standard theory of summation of series, that all the sums, apart from the first, in 7.8 approach a finite limit as $n \to \infty$. We find an expression for the first sum by successive approximations.

From 7.8, a first approximation for v_n is $1 + \tfrac{1}{2}n$ giving

$$1/v_n = 2/(n+2)$$

If we insert this in the RHS of 7.7, we obtain in 7.8

$$\tfrac{1}{12}\sum_{i=0}^{n-1} 1/v_i = \tfrac{1}{6}[\tfrac{1}{2} + \tfrac{1}{3} + \cdots + 1/(n+1)]$$

$$= \tfrac{1}{6}[\tfrac{1}{1} + \tfrac{1}{2} + \tfrac{1}{3} + \cdots + 1/(n+1)] - \tfrac{1}{6}$$

Thus an improved approximation for v_n is

$$\tfrac{1}{2}n + \tfrac{1}{6}[\tfrac{1}{1} + \tfrac{1}{2} + \tfrac{1}{3} + \cdots + 1/(n+1)] + \tfrac{5}{6}$$

which, when n becomes large, approaches

$$\tfrac{1}{2}n + \tfrac{1}{6}(\log n + 0.5772) + \tfrac{5}{6} = \tfrac{1}{2}n + \tfrac{1}{6}\log n + 0.9295$$

If we continue this process of successive approximation one stage further, it becomes clear that the discrepancy between our improved approximation above and the true v_n is quite small, sufficiently small in fact to ensure that, as n increases,

$$\tfrac{1}{12}\sum_{i=0}^{n-1} 1/v_i \to \tfrac{1}{6}\log n + \text{(a finite limit)}$$

It follows then that, as $n \to \infty$,

$$\tfrac{1}{2}n - v_n + \tfrac{1}{6}\log n \to \text{a finite limit}$$

Then

$$\log v_n \to \log(\tfrac{1}{2}n + \tfrac{1}{6}\log n + \text{constant})$$

$$= \log\left[(\tfrac{1}{2}n)\left(1 + \frac{1}{3}\frac{\log n}{n} + \frac{2(\text{constant})}{n}\right)\right]$$

$$\to \log(\tfrac{1}{2}n)$$

the terms $(\log n)/n$, constant$/n$ tending to zero. Thus, as $n \to \infty$

$$\tfrac{1}{2}n - v_n + \tfrac{1}{6}\log v_n \to \text{a finite limit}, \; -C \text{ say}$$

Let us write then, for general n,

$$w_n = \tfrac{1}{2}n - v_n + \tfrac{1}{6}\log v_n + C$$

(where $w_n \to 0$ as $n \to \infty$). Our aim is to find w_n in terms of v_n and to evaluate C and thus obtain n in terms of v_n. We have

$$w_{n+1} - w_n = \tfrac{1}{2} - (v_{n+1} - v_n) + \tfrac{1}{6}\log\left(\frac{v_{n+1}}{v_n}\right)$$

$$= \frac{1}{2} - \left(\frac{1}{2} + \frac{1}{12v_n} - \frac{1}{720v_n^3} + \cdots\right)$$

$$+ \frac{1}{6}\log\left(1 + \frac{1}{2v_n} + \frac{1}{12v_n^2} - \frac{1}{720v_n^4} + \cdots\right)$$

from 7.7. This equals (writing v as short for v_n)

$$\frac{-v^{-2}}{144} + \frac{v^{-3}}{720} + \frac{v^{-4}}{24 \times 720} - \frac{v^{-5}}{42 \times 720} - \frac{v^{-6}}{1512 \times 720} + \frac{v^{-7}}{1680 \times 720}$$

$$+ \frac{1473 v^{-8}}{336 \times 720^2} - \frac{v^{-9}}{924 \times 72 \times 720}$$

Now

$$v_{n+1} = v_n \left(1 + \frac{1}{2v_n} + \frac{1}{12v_n^2} - \cdots \right)$$

$$\frac{1}{v_{n+1}} = \frac{1}{v_n}\left(1 + \frac{1}{2v_n} + \frac{1}{12v_n^2} - \cdots \right)^{-1}$$

$$= \frac{1}{v_n}\left(1 - \frac{1}{2v_n} + \frac{1}{6v_n^2} - \cdots \right)$$

Therefore

$$\frac{1}{v_{n+1}} - \frac{1}{v_n} = -\frac{1}{2v_n^2} + \frac{1}{6v_n^3} - \frac{1}{24v_n^4} + \cdots$$

and

$$\frac{1}{v_{n+1}^2} - \frac{1}{v_n^2} = -\frac{1}{v_n^3} + \frac{7}{12v_n^4} - \cdots$$

and so on. Hence

$$w_{n+1} - w_n = -\frac{1}{144v_n^2} + \frac{1}{720v_n^3} + \cdots$$

$$= \frac{1}{72}\left(-\frac{1}{2v_n^2} + \frac{1}{10v_n^3} + \cdots \right)$$

$$= \frac{1}{72}\left(\frac{1}{v_{n+1}} - \frac{1}{v_n} - \frac{1}{6v_n^3} + \cdots + \frac{1}{10v_n^3} + \cdots \right)$$

$$= \frac{1}{72}\left(\frac{1}{v_{n+1}} - \frac{1}{v_n} - \frac{1}{15v_n^3} + \cdots \right)$$

$$= \frac{1}{72}\left[\left(\frac{1}{v_{n+1}} - \frac{1}{v_n}\right) + \frac{1}{15}\left(\frac{1}{v_{n+1}^2} - \frac{1}{v_n^2}\right) + \cdots \right]$$

$$= \frac{1}{72}\left(\frac{1}{v_{n+1}} - \frac{1}{v_n}\right) + \frac{1}{1080}\left(\frac{1}{v_{n+1}^2} - \frac{1}{v_n^2}\right) + \cdots$$

whence

$$w_n = \frac{1}{72v_n} + \frac{1}{1080v_n^2} + \cdots$$

(no additive constant appearing, since we must have $w_n \to 0$ as $n \to \infty$ and with it $v_n \to \infty$). Fisher continued this series for w_n up to terms in $1/v_n^8$; still further terms are negligibly small when $n \geqslant 5$, and make a minute contribution otherwise. Inserting his expression for w_n into

$$n = 2v_n - \tfrac{1}{3}\log v_n - 2C + 2w_n$$

and replacing v_n by $1/(1 - u_n)$, we obtain the expression for n given in the main text, apart from the numerical value of the constant C. If, in this expression, we put $n = 0$, $u_0 = 0$ we obtain

$$C = 1.014\,640\,996$$

This is very slightly inaccurate, owing to the very small error in w_n when $n = 0$. To overcome this difficulty, Fisher put $n = 5$; from

$$u_0 = 0 \qquad u_{n+1} = e^{u_n} - 1$$

$u_5 = 0.731\,923\,1844$, whence

$$C = 1.014\,648\,607$$

Thus the difference between the two values for C is very small; similarly, if we differentiate n successively with respect to u_n and put $u_n = u_0 = 0$ the values obtained for dn/du_0, d^2n/du_0^2, $d^3n/du_0^3,\ldots$ differ little from the exact values given by Fisher.

Finally, our procedure supplies a very accurate value for v_n as $n \to \infty$, namely

$$\tfrac{1}{2}n + \tfrac{1}{6}\log n + 1.014\,648\,607 + \tfrac{1}{6}\log \tfrac{1}{2} = \tfrac{1}{2}n + \tfrac{1}{6}\log n + 0.8991$$

(replacing the slightly approximate expression given earlier) which we quoted, without proof, in Chapter 4.

2 We know from Chapter 6 that $T(p, x)$, considered as a function of p, changes mathematical form at $p = x$. Hence $\partial T(p, x)/\partial p$ in its usual sense is not defined in the special case $p = x$, so that our Taylor expansion does not apply in that case. Nevertheless, Equation 7.4 is valid. To give this

statement some plausibility, argue as follows. The discontinuity in $\partial T(p, x)/\partial p$ arises from the presence of the Dirac function $\delta(x - p)$. Replace the latter throughout with a function, say $h(p, x)$, which tends to $\delta(x - p)$ in the limit, but which permits the Taylor expansion, for all p, in the non-limiting case. This gives us, instead of 7.4,

$$M \frac{\partial T(p, x)}{\partial p} + \tfrac{1}{2}V \frac{\partial^2 T(p, x)}{\partial p^2} = -h(p, x)$$

in the non-limiting case. Plausibly, we can now proceed to the limit and thus replace $h(p, x)$ in this equation by $\delta(x - p)$. Since the latter is not a function in the usual sense, the derivatives cannot be derivatives in the usual sense, but this difficulty can be overcome by appropriate definition (see e.g. Roach 1970, Ch. 7). In practice, however, we can solve 7.4 without worrying about these niceties.

On the other hand, our Taylor expansion is supposed to apply, in the usual sense, for all values of p *other* than $p = x$. This will distort the underlying model in some cases. Thus for the Wright–Fisher model it is possible that

(a) $p < x$ but $(p + \delta p) > x$, or
(b) $p > x$ but $(p + \delta p) < x$.

In either case, the interval p, $p + \delta p$ includes the point $p = x$, at which $T(p, x)$ is not differentiable, and our Taylor expansion is not, in fact, possible. Thus our treatment tacitly supposes that neither (a) nor (b) occurs.

If the allele frequency really were a continuous variable changing over continuous time, this difficulty would not arise. For we could then carry out the whole argument (appropriately modified) over a time unit sufficiently small to make δp arbitrarily small, if necessary.

3 *Mean absorption time: neutral case*. Let $T(p)$ be the mean absorption time in the neutral case. Then

$$\tfrac{1}{2}V \frac{d^2 T(p)}{dp^2} = -1 \qquad \frac{d^2 T(p)}{dp^2} = -\frac{4N_e}{p(1 - p)} = -\frac{4N_e}{p} - \frac{4N_e}{1 - p}$$

Obtain an expression for $T(p)$ by two successive integrations and use the boundary conditions $T(0) = 0$, $T(1) = 0$ to obtain

$$T(p) = -4N_e[p \log p + (1 - p) \log(1 - p)]$$

NOTES AND EXERCISES

Note that, to help to ensure the reliability of our Taylor series for $T(p + \delta p)$ used in deriving our differential equation above, we have tacitly assumed that $T(p)$ is a continuous function of p throughout the range $0 \leqslant p \leqslant 1$, so that our boundary conditions are equivalent to

$$\lim_{p \to 0} T(p) = 0 \qquad \lim_{p \to 1} T(p) = 0$$

4 *Neutral case, $N_e = N$; corrections to $T(p)$* (Ewens 1964a). Use the standard mathematical result

$$\sum_{j=1}^{\infty} 1/j^2 = \pi^2/6$$

to show that, as $N \to \infty$, the correction to be subtracted from the approximate mean absorption time in the cases $i = 1, 2, 3, 4, 5$ becomes approximately 0.55, 0.76, 0.89, 0.99, 1.06, respectively. These are slightly smaller than the corresponding corrections for the case $N = 25$ (see Table 7.4). As these corrections increase with increasing i, it is interesting to consider the case $i = N$ ($p = \tfrac{1}{2}$).

Approximating sums by integrals, show that in this case the correction is about

$$\tfrac{1}{3}(\log N + 1)$$

to be subtracted from $-4N \log \tfrac{1}{2} = 2.77N$. Thus, although the correction in this case *increases* in magnitude with increasing N, its value is quite small for any realistic value of N. For example, if $N = 10^9$ the correction equals 7.24 generations, to be subtracted from 2.77×10^9 generations (note that our $2N$ corresponds to Ewens's N, so that our formula above is slightly different from that in the original paper).

5 *Alternative derivation of $T(p, x)$*. In Chapter 10, we shall show that

$$-\frac{\partial}{\partial x}[MT(p, x)] + \tfrac{1}{2}\frac{\partial^2}{\partial x^2}[VT(p, x)] = -\delta(x - p)$$

Here M and V are our usual M, V but written in terms of x rather than p, so that here

$$V = x(1 - x)/(2N_e)$$

and, as usual, $\delta(x - p)$ is Dirac's function.

The boundary conditions are

$$\lim_{x \to 0}[x(1-x)T(p, x)] = 0 \qquad \lim_{x \to 1}[x(1-x)T(p, x)] = 0$$

Use these results to find $T(p, x)$ in the neutral case (the mathematics is very similar to that used in our main text when finding $T(p, x)$, so this exercise will provide a little practice in using this mathematical approach, which may have seemed daunting at first sight).

6 By elementary algebra

$$\int_a^b \frac{e^{-4N\alpha x}}{\alpha x(1-x)} dx = \int_a^b \frac{e^{-4N\alpha x}}{\alpha x} dx + \int_a^b \frac{e^{-4N\alpha x}}{\alpha(1-x)} dx$$

Now $a > 1/(2N), b \leqslant 2/(N\alpha)$. Thus, given $N\alpha > 20$, x will always be small, so that, on the RHS above, the second integrand and hence the second integral will be small compared to the first and the omission of the second integral will lead only to a small relative error.

Further, since $x \leqslant b$, $1 - x \geqslant 1 - b$,

$$\int_a^b \frac{e^{-4N\alpha x}}{\alpha(1-x)} dx < \frac{1}{1-b} \frac{1}{\alpha} \int_a^b e^{-4N\alpha x} dx = \frac{1}{1-b} \frac{1}{4N\alpha^2}(e^{-4N\alpha a} - e^{-4N\alpha b})$$

$$< \frac{1}{1-b} \frac{1}{4N\alpha^2}(e^{-2\alpha} - e^{-8}) \quad \text{(since } a \geqslant 1/(2N), b \leqslant 2/N\alpha)$$

$$< \frac{1}{1-b} \frac{1}{4N\alpha^2}$$

$$\simeq \frac{1}{4N\alpha^2}$$

(since $b \leqslant 0.1$), which is biologically trivial when $N\alpha^2 > 0.25$ or so.

8

Introduction to probability distributions: probability flux

> For precept must be upon precept, precept upon precept; line upon line, line upon line; here a little, and there a little.
>
> Isaiah 28,10

On the need to calculate probability distributions

We have shown that, for the neutral theory to work, effective population sizes must be very much smaller than was supposed by most workers during the 'pan-selectionist' era of population genetics. For example, the mean fixation time for a neutral mutation, even in the absence of population subdivision, is $4N_e$ generations. While this mean is somewhat inflated by cases in which, by chance, fixation takes an exceptionally long time, it can be shown that fixation before $0.8N_e$ generations is very unlikely to occur (Kimura 1970a, Kimura & Ohta 1971a). Hence, if N_e were very large, very little neutral evolution would occur in the time available. (This does not contradict the notion that the rate of neutral evolution over a long time period equals the neutral mutation rate, since in this context a 'long time period' means a time period that is long compared with the mean fixation time; the latter must be short, on a geological timescale, for the neutral theory to account for the rate of molecular evolution.) The neutralist, therefore, must assume relatively small values of N_e.

The consequences, for any but the most extreme selectionist, are very remarkable. Suppose that one accepts the neutral theory as correct for even a small fraction of the evolution that has occurred. Then one will necessarily accept these comparatively small values of N_e, not merely for the evolution of neutral alleles but for the evolution of *all* alleles. One must, then, consider the possible role of drift in general, *even in cases where natural selection is undoubtedly operating*. As long as it was supposed (very reasonably, on the evidence available at the time!) that values of N_e are typically very large, it could fairly be argued that (except of course for very extreme allele frequencies) effects of drift would usually be small compared to those of selection or even those of recurrent mutation. In spite of the protests of Wright and others, most population geneticists had few qualms about treating, for example, the balance between mutation and selection deterministically. If, however, there is anything in the neutral theory, we shall have to examine the effects of drift, case by case – of course, we shall certainly find some cases where effects of drift can be ignored, as we found for *some* mean sojourn times in Chapter 7.

Now, the most comprehensive way of finding the effect of drift, at any given time, on an allele frequency is to calculate the probability distribution of that allele frequency at the time under consideration. As mentioned earlier, this is often a formidable problem, which we have hitherto managed to avoid. However, it is easily seen that, in many important cases, finding the probability distribution is essential.

Consider, for example, the spread of a fungal pathogen. Suppose a host variety is introduced containing a (major) resistance allele R, which can be overcome by a strain of pathogen containing virulence allele v. It follows from our earlier discussion that the initial rate of spread of pathogen that can grow on our host variety depends strongly on the frequency of allele v immediately before the time when this host variety is introduced. Of course, this frequency is necessarily low (too low to be detected, in fact) since otherwise the host variety would never have been introduced. However, we have shown that the precise value of this low frequency can be critical, given the timescale relevant to the plant breeder. Fairly obviously, the effect of drift on this frequency will depend on the value of N_e. To demonstrate the magnitude of the effect, we must calculate the probability distribution of the frequency of v, just before the resistant host variety was introduced, for different values of N_e. In this way, we decide whether it is important for the fungal epidemiologist to add to his many other problems (such as obtaining an accurate value for the rate of mutation to virulence) the further problem of finding (at least a very rough approximation to) the value of N_e for the pathogen under consideration.

In calculating this distribution, we may, fairly plausibly, suppose that factors affecting the distribution have remained essentially unchanged for a very large number of generations. If this assumption is really justified, it will be sufficient to find the long-term probability distribution – a decided simplification, as we shall see. We should stress, however, that in some cases it is essential to consider short-term results. For example, it is often found (see e.g. many papers in Eriksson *et al.* 1980) that an allele, very harmful when homozygous and very rare in many or most human groups, is unusually frequent in some group or groups. Of course, for some alleles, there is sufficient direct evidence to show that the abnormally high frequency is due to selection or to drift, but in many cases we have to rely on rather tentative indirect arguments. The case of Tay–Sachs disease among Ashkenazi Jews has been much discussed (see e.g. Neel 1979); the frequency of the (recessive) allele causing this disease is about 10–15 times higher in Ashkenazi Jews than in other groups investigated. The increase in frequency appears to have occurred over a relatively small number of generations (perhaps 68, or so). Can this increase be attributed to drift, bearing in mind that we have deliberately picked out a relatively common allele? Clearly, any discussion of this problem requires us to find the probability distribution, generation by generation, of the Tay–Sachs allele.

It is probably unnecessary to discuss further examples. We shall take it that the probability distribution of allele frequencies will turn out to be relevant in a wide range of biological contexts. A comprehensive discussion of these would be well beyond the scope of the present book (indeed, there are many unsolved problems in this general area, but even to discuss the extensive work that has been done would be very demanding). However, while we shall confine ourselves to a restricted range of situations, we hope to convince the reader that the results we shall obtain will be illuminating and will at least suggest factors likely to be important in more general contexts.

Continuous approximation of the allele frequency

We wish to find the probability that a particular allele, A say, has frequency x at time t; call this probability $f(x, t)$. We shall approximate both allele frequency and time by continuous variables. Consider first the allele frequency. For the moment, ignore the cases $x = 0$ or 1. Our procedure then follows that given for mean sojourn times in Chapter 7. We first represent $f(x, t)$ as an area. Let

$$l(x, t) = 2Nf(x, t)$$

so that

$$f(x, t) = l(x, t)\delta x$$

where $\delta x = 1/(2N)$. Thus $f(x, t)$ can be represented as the area of a rectangle of height $l(x, t)$ and width δx. This is shown for values of x from $1/(2N)$ to $6/(2N)$ in Figure 8.1. If the curve $\phi(x, t)$ is drawn as shown, we have

$$f(x, t) = \phi(x, t)\delta x = \phi(x, t)\frac{1}{2N}$$

(Note that in $f(x, t)$ we think of x as a discontinuous variable whereas in $\phi(x, t)$ we think of x as a continuous variable; this will be so whenever we use the symbols $f(x, t)$ and $\phi(x, t)$ in the sequel.)

Alternatively, from Figure 8.1, $f(x, t)$ is well approximated by the area under the curve between limits

$$x - 1/(4N) \qquad x + 1/(4N)$$

To write this as an integral, it is best to replace x in the integrand by a dummy variable, ξ say, thus obviating difficulties that can arise if the same symbol (x in our case) is used both in integrand and in limits for integration. We have, then, the close approximation

$$f(x, t) = \int_{x - 1/(4N)}^{x + 1/(4N)} \phi(\xi, t) \, d\xi$$

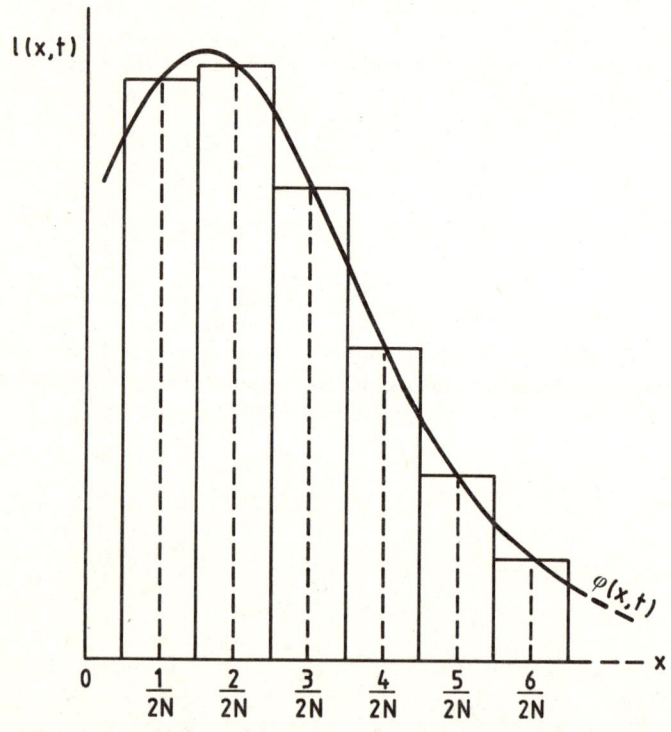

Figure 8.1 Probability distribution of allele frequency x, for any x between 0 and 1. Every rectangle has width $\delta x = 1/(2N)$. The probability $f(x, t)$ that the allele frequency is x at time t is the area of the rectangle of height $l(x, t)$. An ideal curve $\phi(x, t)$ is drawn such that this area equals $\phi(x, t)\delta x$.

In practice, the $\phi(x, t)$ we shall obtain will not be quite as good as the ideal $\phi(x, t)$ shown in Figure 8.1 and the integral just given may, in some cases, give a better approximation to $f(x, t)$ than is given by $\phi(x, t)/(2N)$. However, with N large, the two approximations will usually be very similar numerically; in practice, the simpler approximation $\phi(x, t)/(2N)$ is normally used.

Now consider the probability that the allele frequency lies in some stated interval, say a to b, where $b > a$. From Figure 8.1, we can approximate this probability by integrating $\phi(x, t)$, with respect to x, between appropriate limits. Strictly speaking, we should distinguish the case (i) in which a and b themselves are included in the interval, in which case the limits are

$$a - 1/(4N) \qquad b + 1/(4N)$$

from the case (ii) in which the interval does not include a, b themselves, so that the limits are

$$a + 1/(4N) \qquad b - 1/(4N)$$

In practice, however, the distinction between (i) and (ii) is often ignored and the probability in either case approximated by

$$\int_a^b \phi(x, t) \, dx$$

This rarely leads to serious error.

Approximation of zero and unit frequencies

However, we have a difficulty that we did not encounter in the case of mean sojourn times, where sojourn times at frequencies 0 or 1 had no meaning. We have to find some way of obtaining (approximately, in practice) the probability that the allele frequency is 0 or that this frequency is 1. For the moment, we shall not deal with this problem comprehensively, but will merely make a few remarks on the subject.

Consider first the case of two-way mutation; then allele A, when at frequency 0 or 1, can leave these frequencies by mutation. Then, as we explained in Chapter 6, diffusion theory will represent the allele frequency as lying, at all times, *between* frequencies 0 to 1 (for, according to the diffusion model, even if the allele can reach frequency 0 or 1, it will leave that frequency instantaneously). In standard mathematical fashion, define the *closed interval* 0 to 1 as the set of x for which $0 \leq x \leq 1$ and the *open interval* 0 to 1 as the set of x for which $0 < x < 1$ (i.e. the closed interval minus its end-points). Then, with the probability density function $\phi(x, t)$ as a function of the continuous variable x, x must lies somewhere on the open interval 0 to 1. We have, then, the necessary condition

$$\int_0^1 \phi(x, t) \, dx = 1$$

where the limits of integration 0, 1 are understood as 'just above 0' and 'just below 1'; that is, we regard the 0, 1 as arising as ε, $1 - \varepsilon$ (where ε is positive) and then let $\varepsilon \to 0$.

Now if, in the true distribution of x, the probability that x equals, or is very close to, 0 or 1 is negligibly small, our necessary condition for $\phi(x, t)$ seems quite reasonable. For, in such a case, $\phi(x, t)$ will presumably be negligibly small for extreme values of x. Effectively, then, these extreme values are excluded from the range of integration and our condition is just a continuous version of

$$\sum_x f(x, t) = 1$$

where $f(x, t)$ is negligibly small for extreme values of x. When, however, conditions are such that zero or unit x is quite probable, our necessary condition for $\phi(x, t)$ seems altogether inappropriate. First, it apparently denies the existence of probabilities at $x = 0$ or 1 that are quite large; secondly, the range of integration includes the intervals 0 to $1/(4N)$ and $1 - 1/(4N)$ to 1, for which $\phi(x, t)$ is presumably non-negligible, whereas allele frequencies in these intervals appear to be quite meaningless. In fact, however, the situation is much more promising than the preceding arguments would suggest. Consider the case where allele frequencies are changing by mutation and drift. Then (Moran 1962, Kimura 1968b) it turns out that $\phi(x, t)$ supplies the true mean *and variance* of x in a very wide range of situations, which *include* all cases (likely to occur in practice) where there is a large probability that $x = 0$ or 1 (the authors cited discuss the case where t is very large, but in fact their result holds, near enough, for all t). Clearly (a) our necessary condition, which was introduced by Wright (1931), is very sensible despite all appearances to the contrary and (b) the cases $x = 0$, $x = 1$ must, somehow, have been taken into account by $\phi(x, t)$. Kimura (1968b) investigated in detail the case where t is very large. Developing an approach of Wright (1931), he found the following remarkable approximations

$$f(0, t) = \int_0^{1/(2N)} \phi(x, t) \, dx \qquad f(1, t) = \int_{1-1/(2N)}^1 \phi(x, t) \, dx$$

reliable over a wide range of cases; although these formulae can give values for $f(0, t)$, $f(1, t)$ that have a large *relative* error, this appears to happen only in cases where the true values are very small, so that, even given the large relative error, the approximate values are correctly given as very small also (Watterson 1975, Ewens 1979). We can conclude with some confidence that distortions induced in the two-way mutation case by the use of continuous approximations have little genetical significance, at least for applications where extreme accuracy is not absolutely essential. If very high accuracy is required, it is sometimes possible to alter $\phi(x, t)$ slightly, to give a more complicated but very reliable formula (Ewens 1965).

Now consider the case where mutation is supposed absent. Then $x = 0$, $x = 1$ are exit boundaries; that is, once the allele frequency reaches 0 or 1, it stays at that value permanently. On the other hand, all frequencies between 0 and 1 are mutually accessible, that is the allele can pass from any one such frequency to any other. Further, since absorption must occur eventually, $f(x, t)$ will tend to zero as $t \to \infty$ for all values of x between 0 and 1, whereas when $t \to \infty$, $f(0, t)$, $f(1, t)$ will tend to non-zero values, the probability of ultimate loss and of ultimate fixation, respectively. This suggests that the general mathematical form of $f(x, t)$ will be much the same for all x between 0 and 1, but will take a different form when $x = 0$ or 1; it will be apparent from our discussion in Chapter 3 that this suggestion is indeed correct. We shall have to incorporate all these points into our continuous approximation.

Thinking of x then as a continuous variable, we aim to find a function $\phi(x, t)$, along the lines described earlier in this chapter; the mathematical form of $\phi(x, t)$ will be the same for all mutually accessible x, that is for all x on the open interval (this will have to include x lying in the intervals 0 to $1/(4N)$ and $1 - 1/(4N)$ to 1, for it would be very difficult to carry through a representation in which these intervals were excluded). For x between 0 and 1, $f(x, t)$ will be approximated, as usual, by $\phi(x, t)/(2N)$. The probability that the true allele frequency lies between 0 and 1, in other words the probability that the allele lies in one or other of the mutually accessible frequencies, is approximated by the probability that the continuous variable lies on the open interval, namely

$$\int_0^1 \phi(x, t)\, dx$$

where we understand 0, 1 as 'just above 0' and 'just below 1', respectively. We notice at once a major difference between the two-way mutation case and the present no-mutation case. In the former case, the integral just given had value 1 at all times. In the present case, the value is 1 only at the very start of the process. Thereafter (assuming $\phi(x, t)$ has been calculated correctly!) this value will gradually decline, the 'missing' probability representing (the continuous approximation to) the probability that absorption (= loss + fixation) has occurred. We shall show that this probability of absorption can be partitioned into (a) $f^*(0, t)$, an approximation to the probability of loss $f(0, t)$, and (b) $f^*(1, t)$, an approximation to the probability of fixation $f(1, t)$. At all times, then,

$$f^*(0, t) + \int_0^1 \phi(x, t)\,dx + f^*(1, t) = 1$$

since the expression on the left-hand side covers all possibilities. Here then is a second major difference from the two-way mutation case. There is no possibility in the present case of approximating $f(0, t)$ or $f(1, t)$ by an integral ranging over any portion of the open interval. For that would imply that an allele frequency of 0 or 1 could change to some other value at a later time, which by hypothesis is impossible; a phantom mutation rate would have been subtly introduced in a situation where no mutation exists. We have to go right outside the open interval to obtain approximations for $f(0, t)$ and $f(1, t)$.

Fairly obviously, the inclusion in the no-mutation case of the intervals 0 to $1/(4N)$ and $1-1/(4N)$ to 1 will lead to loss of accuracy. The reader will recall that this inaccuracy is a serious problem for the genetic conservationist in some cases, as we noted in Chapter 4. However, from our discussion there, we may surmise, fairly confidently, that errors arising from our procedure will be rather trivial when we work on an evolutionary timescale; obviously, we shall have to check this surmise in due course.

'We have not yet met our Waterloo, Watson, but this is our Marengo.' In fact, we have now dealt with all the really difficult points and indeed have made much more progress than might appear at first sight. It is to be hoped that the reader will find the rest of our discussion very much easier.

Continuous time approximation

So far, we have discussed the approximation of allele frequency by a continuous variable. As we have noted on several previous occasions, this should work best when N is large (although often yielding good results even when N is fairly small), since with N large, the jump $1/(2N)$ between adjacent frequencies will be very small. Now consider the time. By measuring time on an appropriate scale, we can make the jump in time between two successive generations also equal $1/(2N)$. To do this, we measure time τ in such a way

CONTINUOUS TIME APPROXIMATION

that a time lapse of $2N$ generations corresponds to $\tau = 1$ (Feller 1951). Thus a time lapse of one generation appears, when measured on the τ scale, as $1/(2N)$th of a time unit; we may write this as

$$\delta\tau = 1/(2N)$$

Thus, at least for large N, we can approximate time by a continuous variable.

This approach is essential in some contexts. For example, we must use the τ scale if we wish to fit the theory of changes of allele frequency into standard diffusion theory (which, of course, has applications in many areas of the natural sciences and elsewhere). Hence some authors (e.g. Moran 1962, Ewens 1969, 1979, Karlin & Taylor 1981) prefer to use the τ scale, although usually interpreting their results in terms of generations at appropriate points of the discussion. It will probably be easiest, however, if we do not rescale the time, but rather attempt to approximate time, measured in the usual way in generations, by a continuous variable (see e.g. Fisher 1930b, Wright 1931, Crow & Kimura 1970). This will suffice for the problems we shall discuss (given our informal approach), provided we note the following essential point. If we regard allele frequencies as changing continuously in time, we imply that for very small times we have only very small changes in frequency. Hence we must confine ourselves to *cases where evolution is fairly slow*. From now on, then, we shall suppose conditions are such that allele frequencies change (if at all) only slowly over successive generations.

This being so, we can obtain a very useful approximation. Think first of time t in terms of discrete generations. Consider any given *specified* allele frequency x. Let $\delta\phi$ be the change in $\phi(x, t)$ over time $\delta t = 1$ generation. In general, $\delta\phi$ will itself change with time. If, however, evolution is slow, any factors affecting the value of $\delta\phi$ will hardly change over a period of a few generations. Hence $\delta\phi$ will be much the same for a few successive generations. If, then, we plot $\phi(x, t)$ against t for a few successive values of t, the resulting points will lie, near enough, on a straight line.

Now consider the plot of $\phi(x, t)$ against t for all t. Passing a smooth curve through these points gives the relationship between $\phi(x, t)$ and t, where t is considered now as a continuous variable. Overall, the curve will not be linear, but if evolution is slow, our curve will be *virtually linear in the neighbourhood of any given t*. Hence, near enough, for the curve,

$$\frac{\partial \phi(x, t)}{\partial t} = \frac{\delta\phi}{\delta t}$$

We shall use this result (and analogous results) repeatedly.

INTRODUCTION TO PROBABILITY DISTRIBUTIONS

There is, however, one point about the preceding argument that sometimes worries beginners. If we think in terms of Taylor series, we have

$$\phi(x, t + \delta t) = \phi(x, t) + \delta t \frac{\partial \phi(x, t)}{\partial t} + \frac{(\delta t)^2}{2!} \frac{\partial^2 \phi(x, t)}{\partial t^2} + \cdots$$

so that

$$\delta\phi = \phi(x, t + \delta t) - \phi(x, t) = \delta t \frac{\partial \phi(x, t)}{\partial t} + \frac{(\delta t)^2}{2!} \frac{\partial^2 \phi(x, t)}{\partial t^2} + \cdots$$

Our argument is equivalent to ignoring all but the first term on the right-hand side. Thus here we stop at the *first* derivative whereas in Taylor expansions used earlier we continued to the *second* derivative. It is interesting that essentially the same procedure appears in Einstein's famous theoretical treatment of Brownian motion (first observed on pollen grains of *Clarkia*), from which our approach ultimately derives.

In fact, there is no inconsistency. In our expansion for $\delta\phi$ above, our arguments show that the term

$$\delta t \frac{\partial \phi(x, t)}{\partial t}$$

will very much dominate the right-hand side. On the other hand, let δx be the change in allele frequency in a single generation; in an expression such as

$$(E\delta x) \frac{\partial u}{\partial x} + \frac{E(\delta x)^2}{2!} \frac{\partial^2 u}{\partial x^2} + \cdots$$

the second term may be just as important as the first and may well be the more important of the two. Indeed in the neutral, no-mutation, case $(E\delta x) = 0$. Thus we must go as far as the second derivative. However, terms in $E(\delta x)^3$, $E(\delta x)^4$, ... can nearly always be ignored.

We shall consider the actual calculation of $\phi(x, t)$ for some particular cases. To do this, it is often helpful to calculate the *probability flux*, which we shall now define.

Probability flux

We have to face a minor difficulty. Hitherto, we have defined all concepts in terms of the true allele frequency and regarded the continuous approximation to the allele frequency as a convenient device for solving problems. However, probability flux across a point, which we now discuss, is essentially a notion defined for the continuous variable. A definition in terms of the true allele frequency, although possible and corresponding well to the definition for the continuous frequency, appears rather contrived (see below). For the moment, then, we think of allele frequency as a continuous variable.

It will be convenient to use rather 'physical' imagery to describe changes in the probability distribution over time. The probability that the allele frequency exceeds some *given* value x at time t, which we may call 'the probability to the right of x at time t', is

$$R(x, t) = \int_x^1 \phi(\xi, t) \, d\xi + f^*(1, t)$$

where the last term appears only if allele frequency unity is an exit boundary. The probability to the right of x at a later time $(t + \delta t)$ is

$$R(x, t + \delta t) = \int_x^1 \phi(\xi, t + \delta t) \, d\xi + f^*(1, t + \delta t)$$

(again, the last term appearing only if frequency unity is an exit). We define the net flux of probability, left to right, across x over time δt (generations) as

$$\delta R = R(x, t + \delta t) - R(x, t)$$

that is, the change in R over time δt. Note that, on this definition, a positive net flux means an overall increase in probability on the right of x; a negative net flux means an overall increase in probability on the left of x.

Let

$$P(x, t) = \frac{\partial R(x, t)}{\partial t}$$

Then (evolution being slow) we may approximate $\delta R/\delta t$ by $\partial R(x, t)/\partial t$, provided δt is small, so that, to a close approximation

$$\delta R = P(x, t)\delta t$$

In particular, the net flux, left to right, across x in one generation ($\delta t = 1$) is well approximated by $P(x, t)$.

Now consider the special cases $x \to 1$ and $x \to 0$. In the two-way mutation case, with x confined to lie between 0 and 1, no probability can flow across x when x approaches these limiting values. Thus

$$\lim_{x \to 1} P(x, t) = 0 \qquad \lim_{x \to 0} P(x, t) = 0$$

equivalent to the famous 'flux zero boundary condition' given by Goldberg in his unpublished thesis in 1950.

Now suppose that frequencies 0 and 1 are exit boundaries, meaning that no probability can 'escape' from these boundaries to the open interval. Then, as $x \to 1$, any flux of probability across x is necessarily left to right and having 'nowhere to go but up' will increment the probability that the allele frequency lies to the right of the open interval, that is, the probability that the allele frequency equals 1. In our notation

$$\frac{df^*(1, t)}{dt} = \lim_{x \to 1} \frac{\partial R(x, t)}{\partial t} = \lim_{x \to 1} P(x, t)$$

as pointed out by Watterson (quoted in Moran 1962, p. 78).

To obtain a comparable formula for $f^*(0, t)$, we note that an increase (decrease) of probability on the right of any x over time δt implies a decrease (increase) of equal magnitude on the left of that x. Thus the flux right to left is just $-$ (the flux left to right). As $x \to 0$, any flux across x is necessarily right to left so that, given our definitions, $\partial R(x, t)/\partial t$ will be negative and the flux right to left, being $-\partial R(x, t)/\partial t$, correctly comes out positive; this flux increments the probability that the allele frequency equals 0, whence we have

$$\frac{df^*(0, t)}{dt} = \lim_{x \to 0} \frac{-\partial R(x, t)}{\partial t} = -\lim_{x \to 0} P(x, t)$$

Thus, once we have established a formula for $P(x, t)$, $f^*(1, t)$ and $f^*(0, t)$ can be found by integration with respect to t.

Another very useful formula

$$\frac{\partial P(x, t)}{\partial x} = -\frac{\partial \phi(x, t)}{\partial t}$$

may be derived as follows. We have

$$P(x, t) = \frac{\partial R(x, t)}{\partial t} = \frac{\partial}{\partial t} \left(\int_x^1 \phi(\xi, t) \, d\xi + f^*(1, t) \right)$$

where $f^*(1, t)$ appears only when frequency unity is an exit. Then,

$$P(x, t) = \frac{\partial}{\partial t} \int_x^1 \phi(\xi, t) \, d\xi + \frac{df^*(1, t)}{dt}$$

(our notation being justified since $f^*(1, t)$ is a function of t but not of x). Assuming that we can carry out the differentiation and integration in the reverse order to that written, we have

$$P(x, t) = \int_x^1 \frac{\partial \phi(\xi, t)}{\partial t} \, d\xi + \frac{df^*(1, t)}{dt} = -\int_1^x \frac{\partial \phi(\xi, t)}{\partial t} \, d\xi + \frac{df^*(1, t)}{dt}$$

by elementary calculus. By the fundamental theorem of the integral calculus ('integration is the reverse of differentiation') we have (noting again that $f^*(1, t)$ is not a function of x)

$$\frac{\partial P(x, t)}{\partial x} = -\frac{\partial \phi(x, t)}{\partial t}$$

For a first application of this formula, we check that, for exit boundaries at 0 and 1,

$$\frac{df^*(1, t)}{dt} + \frac{df^*(0, t)}{dt}$$

equals the rate of loss of probability from the open interval, namely

$$-\frac{\partial}{\partial t} \int_0^1 \phi(x, t) \, dx$$

where the 0, 1 are understood in the usual limiting sense. Inverting the order of differentiation and integration gives

$$-\int_0^1 \frac{\partial \phi(x, t)}{\partial t} \, dx = \int_0^1 \frac{\partial P(x, t)}{\partial x} \, dx = [P(x, t)]_0^1$$
$$= \lim_{x \to 1} P(x, t) - \lim_{x \to 0} P(x, t)$$
$$= \frac{df^*(1, t)}{dt} + \frac{df^*(0, t)}{dt}$$

as we established earlier.

Although probability flux appears occasionally in the writings of Fisher and Wright, the notion of probability flux is usually associated with Watterson (1962), who first gave the formula for $P(x, t)$ that we shall derive below; the proof of this formula is due to Kimura (1964). However, formulae equivalent to our formulae for $f^*(1, t)$ and $f^*(0, t)$ were derived earlier by Kimura (1955b), who did not at that time formulate his results in terms of probability flux. Once we have obtained a formula for $P(x, t)$, Kimura's earlier results follow very easily, as indeed he points out (Kimura 1964).

The reader will probably expect some comment on our habit of inverting the order in which we carry out operations involving limits, such as differentiation and integration. This procedure really requires justification, which is beyond the scope of this book. Nevertheless, our procedure is very plausible. In this context, it is a particular pleasure to quote the founder of our subject, G. H. Hardy (Hardy 1952, Appendix III):

> If L and L' are two limit operations then the numbers $LL'Z$ and $L'LZ$ are not *generally* equal, in the strict sense of the word 'general'. We can always, by the exercise of a little ingenuity, find Z so that $LL'Z$ and $L'LZ$ shall differ from one another. But they are equal generally if we use the word in a more 'practical' sense, viz. as meaning 'in the great majority of such cases as are likely to occur naturally'. In practice, a result obtained by assuming that two limit operations are commutative is *probably* true ... But an answer thus obtained must, in default of a further study of the general question, or a special investigation of the particular problem ... be regarded as suggested only and not proved.

In the rest of this book, we shall when necessary invert the order of limiting operations without further comment.

Components of flux

Obviously enough, the overall flux across x is just the balanced cumulative effect of the many individual movements or 'minifluxes' of probability from one side of x to the other. These individual effects will cumulate to give the overall flux, as defined earlier, provided we count minifluxes left to right as positive, right to left as negative. For definiteness, consider movement left to right. We consider a *starting point*, ξ say, where $\xi < x$. As usual for a continuous variable, the probability that, at time t, the allele frequency lies in a very short interval of length $\delta\xi$, centred at ξ, while strictly given by an integral, is well approximated by

$$\phi(\xi, t)\delta\xi$$

the approximation becoming increasingly close, the smaller the value of $\delta\xi$. Equivalently, we say that the probability that, at time t, the allele frequency lies in an interval of length $d\xi$, centred at ξ, equals

$$\phi(\xi, t) \, d\xi$$

This statement may be taken as just a restatement of the previous sentence.

Now consider a *finishing point*, say ρ. There will be a definite probability, f say, that by time $(t + \delta t)$ the allele frequency will have changed from interval $d\xi$, centred at ξ, to interval $d\rho$, centred at ρ. If the latter interval lies above x, the movement from the ξ interval to the ρ interval gives a miniflux across x of value

$$\phi(\xi, t) \, d\xi \, f$$

(think of f as the fraction of $\phi(\xi, t) \, d\xi$ that has moved). The *distance moved* by any point during the process just described will lie between

$$(\rho - \tfrac{1}{2} d\rho) - (\xi + \tfrac{1}{2} d\xi) \quad \text{and} \quad (\rho + \tfrac{1}{2} d\rho) - (\xi - \tfrac{1}{2} d\xi)$$

that is

$$(\rho - \xi) - \tfrac{1}{2}(d\rho + d\xi) \quad \text{and} \quad (\rho - \xi) + \tfrac{1}{2}(d\rho + d\xi)$$

or

$$s - \tfrac{1}{2} ds \quad \text{and} \quad s + \tfrac{1}{2} ds$$

where $s = \rho - \xi$, $ds = (d\rho + d\xi)$. Note that, for a rightward movement, s must be positive. Following Kimura (1964), we shall carry out our discussion in terms of a starting point and distance moved rather than starting and finishing points; in accordance with our standard assumption that evolution is slow, we suppose that the distance moved in a single generation is small. Let $g(s, \xi) \, ds$ be the probability that a frequency in the starting interval $\xi - \tfrac{1}{2} d\xi$, $\xi + \tfrac{1}{2} d\xi$ moves a distance between $s - \tfrac{1}{2} ds$ and $s + \tfrac{1}{2} ds$ in time δt, s being positive. Provided s is sufficiently large to carry our frequency beyond x, our miniflux may be written

$$\phi(\xi, t) \, d\xi \, g(s, \xi) \, ds$$

We shall have to accumulate such expressions for all values of ξ and s that lead to a miniflux across x. At first sight, we might think of accumulation in terms of summation. However, without any further fuss, it will probably be apparent to the reader that our discussion above is tight only in the limit as $\delta\xi \to 0$ and $\delta s \to 0$. Thus our accumulation will involve integration, rather than summation, over appropriate ranges of ξ and s.

Notice that, having 'put the squeeze' on intervals $\delta\xi$, δs (and hence on $\delta\rho$) we can, if necessary, imagine ξ as possibly lying indefinitely close to x on the left-hand side of x and ρ as possibly lying indefinitely close to x on the right-hand side of x. Further, since the area over a point is zero, no probability attaches to point x itself. Thus all rightward movements of probability have been taken into account.

Leftward movements may be analysed in a very similar fashion. We continue to take $\xi - \frac{1}{2}d\xi$, $\xi + \frac{1}{2}d\xi$ as the starting interval and $\rho - \frac{1}{2}d\rho$, $\rho + \frac{1}{2}d\rho$ as the finishing interval; as before s is defined as $\rho - \xi$. Now, however, $\xi > x$, $\rho < x$ so that s is negative. As before, we write $g(s, \xi)\,ds$ for the probability that a frequency in starting interval $\xi - \frac{1}{2}d\xi$, $\xi + \frac{1}{2}d\xi$ moves a distance between $s - \frac{1}{2}ds$ and $s + \frac{1}{2}ds$ in time δt (but with s now negative). Provided s is sufficiently large in magnitude to carry our frequency beyond x, we have a right to left miniflux

$$\phi(\xi, t)\,d\xi\, g(s, \xi)\,ds$$

identical in mathematical form with our earlier left to right miniflux, but for a different range of ξ and of s.

We are now ready to find a formula for $P(x, t)$. To do this we find the combinations of ξ and s that lead to a miniflux across x.

Calculating the probability flux

To make a start, we consider minifluxes left to right, so that $\xi < x$ and s is positive. Of course, there will still be many combinations of ξ and s giving a miniflux. To list these systematically, we first take any positive value of s and find the appropriate values of ξ.

Consult Figure 8.2. Remembering that, for the moment, s is fixed, we see that, if ξ is too far to the left of x there will be no miniflux across x. In fact, from the figure, there will be a miniflux, left to right, across x only if

$$x - s < \xi < x$$

Formally, we confirm this as follows. By assumption $\xi < x$. For a miniflux across x, we must have $\rho > x$; by definition $s = \rho - \xi$, so that $\rho = s + \xi$, $s + \xi > x$, $\xi > x - s$.

CALCULATING THE PROBABILITY FLUX

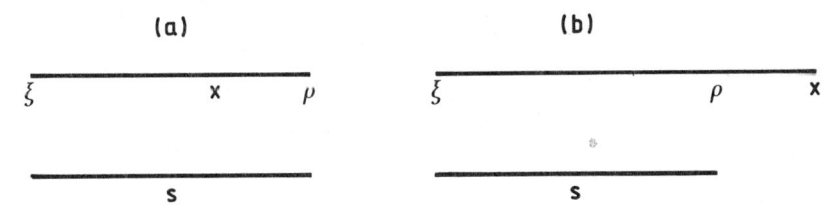

Figure 8.2 Possibilities for a miniflux, left to right, across x. Starting interval $\xi - \frac{1}{2} d\xi$, $\xi + \frac{1}{2} d\xi$ and finishing interval $\rho - \frac{1}{2} d\rho$, $\rho + \frac{1}{2} d\rho$; $s = \rho - \xi$, with s fixed at a given value. (a) Miniflux across x. (b) No miniflux across x, ξ being too far to the left of x.

Still keeping s fixed, but now taking all values of ξ within the range $(x - s)$ to x into account, we see that the flux across x will be obtained by integration with respect to ξ, over limits $\xi = x - s$ to $\xi = x$. We rearrange the integrand slightly to emphasize that ξ is the variable of integration and obtain

$$\int_{\xi = x - s}^{\xi = x} [\phi(\xi, t) g(s, \xi)\, ds]\, d\xi$$

Let $\varepsilon = x - \xi$, $\xi = x - \varepsilon$, $d\xi/d\varepsilon = -1$. From the usual rules for integration by substitution, we obtain

$$\int_{\varepsilon = 0}^{\varepsilon = s} \phi(x - \varepsilon, t) g(s, x - \varepsilon)\, ds]\, d\varepsilon$$

Expanding in Taylor series gives

$$\int_{\varepsilon = 0}^{\varepsilon = s} \left(\phi(x, t) g(s, x)\, ds - \varepsilon \frac{\partial}{\partial x} [\phi(x, t) g(s, x)]\, ds + \cdots \right) d\varepsilon$$

(It will be clear almost at once that, in this particular case, it will be sufficient to stop our Taylor expansion at the first derivative.) This Taylor expansion appears sensible only if we assume that $s < (1 - x)$. This will usually be covered by our supposition that s is small but, for x very close to unity, rather distorts what happens to the true allele frequency when we think of time in generations. On the other hand, in a continuous time model, we can make s as small as we wish by choosing a sufficiently small value of δt (or $\delta \tau$). We see once again that the diffusion model is not completely satisfactory when true allele frequencies are very extreme. We shall take it that s obeys the condition above.

The actual integration is very elementary, since terms such as $\phi(x, t)g(s, x) \, ds$, with s for the moment fixed, are constants as far as ε is concerned. It yields

$$s\phi(x, t)g(s, x) \, ds - \tfrac{1}{2}s^2 \frac{\partial}{\partial x}[\phi(x, t)g(s, x)] \, ds + \cdots$$

this being the contribution to the overall net flux across x from cases where s takes any specified positive value (obeying our condition). If we integrate this contribution, with respect to s, over all such values of s, we have the complete contribution to the net flux, from cases where s is positive.

Flux right to left is treated in a manner very similar to that just given. We first suppose that s takes a specified negative value. The reader will readily construct the analogue of Figure 8.2, with $\xi > x$, $\rho < x$, $s = \rho - \xi$. We have $\rho < x$, $s + \xi < x$, $\xi < x - s$. Thus the contribution to the right to left flux from cases where s takes a specified negative value is

$$\int_{\xi=x}^{\xi=x-s} [\phi(\xi, t)g(s, \xi) \, ds] \, d\xi$$

Let $\varepsilon = \xi - x$, $\xi = x + \varepsilon$, $d\xi/d\varepsilon = 1$. This gives

$$\int_{\varepsilon=0}^{\varepsilon=-s} [\phi(x + \varepsilon, t)g(s, x + \varepsilon) \, ds] \, d\varepsilon$$

We shall take it that $-s < x$.

Expanding in Taylor series gives

$$\int_{\varepsilon=0}^{\varepsilon=-s} \left(\phi(x, t)g(s, x) \, ds + \varepsilon \frac{\partial}{\partial x}[\phi(x, t)g(s, x)] \, ds + \cdots \right) d\varepsilon$$

$$= -s\phi(x, t)g(s, x) \, ds + \tfrac{1}{2}s^2 \frac{\partial}{\partial x}[\phi(x, t)g(s, x)] \, ds + \cdots$$

to be integrated over all negative values of s (obeying the condition $-s < x$) to give the overall right to left flux, from cases where s is negative. We notice that this integrand has essentially the same mathematical form as when s was positive, but with the sign reversed. But this time we have a right to left flux, so that our integral will have to be *subtracted* from the previous integral to obtain, finally, the net flux, left to right across x. Thus we reverse the sign of our present integral (conveniently done by reversing the sign of its integrand)

and add the result to our previous integral. Then, irrespective of whether s is positive or negative, the integrand will be

$$s\phi(x, t)g(s, x) \, ds - \tfrac{1}{2}s^2 \frac{\partial}{\partial x}[\phi(x, t)g(s, x)] \, ds + \cdots$$

In fact, since the integral over a point is zero, we may as well include $s = 0$ in the range of integration, which is therefore all s (subject to the conditions for s noted earlier). Reversing, when necessary, the order of differentiation and integration, we have

$$\int_s \phi(x, t)sg(s, x) \, ds - \tfrac{1}{2} \frac{\partial}{\partial x}\left(\int_s \phi(x, t)s^2 g(s, x) \, ds\right) + \cdots$$

We note that $\phi(x, t)$ is not a function of s and that, given our earlier definitions, $g(s, x)$ is the probability density function for the distance moved when the starting interval is centred at x. By definition of Es, Es^2

$$\int_s sg(s, x) \, ds = Es \qquad \int_s s^2 g(s, x) \, ds = Es^2$$

Es being the mean change, Es^2 the mean square change, in allele frequency, over time δt, when the starting interval is centred at x. Since s is small, we may ignore terms in Es^3, Es^4, ..., which are just the terms appearing as $+ \ldots$ in the expression above. Indeed, according to Maruyama (1977), the omission of these terms, when δt is sufficiently small, is justified merely by our assumption of a continuous frequency changing in continuous time. Then, writing as before δR for the net flux, left to right across x, over time δt, we have the approximation

$$\delta R = (Es)\phi(x, t) - \tfrac{1}{2} \frac{\partial}{\partial x}[(Es^2)\phi(x, t)]$$

To obtain an exact expression, we proceed as follows. Let $V(s)$ be the variance of the change in allele frequency, over time δt, when the starting interval is centred at x. By the usual result for variances

$$V(s) = Es^2 - (Es)^2$$

Now, from our discussion, our formula for δR becomes increasingly accurate as $\delta t \to 0$. With this in mind, we define

$$M = \lim_{\delta t \to 0} \frac{Es}{\delta t} \qquad V = \lim_{\delta t \to 0} \frac{V(s)}{\delta t}$$

so that, with increasing accuracy as $\delta t \to 0$,

$$\text{E}s = M\delta t \qquad \text{V}(s) = V\delta t \qquad \text{E}s^2 = V\delta t + (M\delta t)^2$$

Inserting these expressions for $\text{E}s$ and $\text{E}s^2$ in our formula for δR, dividing through by δt and letting $\delta t \to 0$, we have the exact result

$$P(x, t) = \frac{\partial R(x, t)}{\partial t} = M\phi(x, t) - \tfrac{1}{2}\frac{\partial}{\partial x}[V\phi(x, t)]$$

Notice that the term $(M\delta t)^2 = M^2(\delta t)^2$ has disappeared on dividing by δt and taking the limit as $\delta t \to 0$. Hence we would have obtained the same ultimate result if we had defined V as the limit, when $\delta t \to 0$, of $\text{E}s^2/\delta t$ rather than $\text{V}(s)/\delta t$, a point we shall take up shortly.

Probability flux for discontinuous frequency

We now discuss how our formula for $P(x, t)$ may be applied to a discontinuous allele frequency, changing in generations. We first define flux over a point x for this situation. Consider allele frequencies ξ, x, ρ with $\xi < x < \rho$. Obviously, on any sensible definition, movements of probability from ξ to ρ contribute to the flux, left to right across x; however, it is not immediately clear how to deal with movements of probability to or from frequency x itself. Figure 8.1 provides a guide to a suitable procedure. With probability represented in histogram fashion, half the probability that the allele *frequency* is x lies to the left, and half lies to the right of *point x*, shown on the diagram. Thus if frequency x is the starting point for a movement of probability up to frequency ρ, only half the moving probability crosses point x, since the remaining half was to the right of point x before movement began. Similar considerations apply in other cases involving movements to or from frequency x. This suggests that we define the net flux, left to right, across point x as follows. To the overall increase, over time δt, in probability that the allele frequency exceeds x we add one-half of the increase in probability that the allele frequency equals x. Decreases count as negative increases. Thus minifluxes from frequencies on one side of x to frequencies on the other side of x count fully, but minifluxes from frequency x as starting point or to frequency x as finishing point are halved. This definition is not really as arbitrary as first appears. For example, in the important case where the probability distribution does not change over time, the net flux, as just defined, will correctly be zero.

Our definitions of flux for the continuous and true allele frequency correspond closely. We shall take it, therefore, that (apart from some distortion for very extreme frequencies) our formulae give a reliable representation of events for the true frequency. In particular, the true flux over a single generation will usually be well approximated by $P(x, t)$, as given by our formula above. We must, however, decide upon appropriate expressions for M and V.

Expressions for M and V

We recall that, with increasing accuracy as $\delta t \to 0$, $\mathrm{E}s = M\delta t$. Since evolution is slow, we take this as near-enough correct when $\delta t = 1$. Then $M = \mathrm{E}s$ over one generation, which is just the continuous analogue of the mean change, M(x) say, in the true allele frequency over a single generation, when the starting frequency is x. Hence we put M equal to M(x). For example, given genic selection, $M = \alpha x(1 - x)$.

V, however, is more difficult. We recall that our formula for $P(x, t)$ would have been the same whether we had defined V as the limit, when $\delta t \to 0$, of either $\mathrm{E}s^2/\delta t$ or V(s)$/\delta t$. Should we, then, equate V to the mean square change MS(x), or to the variance of the change V(x), in the true allele frequency over a single generation (for starting frequency x)?

Now, with the evolution slow, M(x) and V(x) are both small. Suppose M(x) is not too large, relative to V(x). Then $[\mathrm{M}(x)]^2 \ll \mathrm{V}(x)$, $[\mathrm{M}(x)]^2 \ll \mathrm{MS}(x) - [\mathrm{M}(x)]^2$, $2[\mathrm{M}(x)]^2 \ll \mathrm{MS}(x)$, so that

$$[\mathrm{M}(x)]^2 \ll \mathrm{MS}(x)$$

and MS(x), V(x) are much the same. Hence it scarcely matters which we equate to V and we are free to choose V(x), which proves more convenient. However, if M(x), while still small, is large relative to V(x), the equating of V(x) to V, which is customary, is more difficult to justify. A strong argument in favour of this common practice has, however, been advocated by Kimura (1955b). Adapting his approach to our present context, we argue as follows.

We should like $P(x, t)$ to give the flux in as wide a range of situations as possible. In particular, we should like our formula to be correct for a purely deterministic situation, in which case V(x), but not MS(x), is zero for all x. Suppose, in addition to $M = \mathrm{M}(x)$, we put $V = \mathrm{V}(x)$. Then, when $V = 0$ for all x, we have

$$P(x, t) = \mathrm{M}(x)\phi(x, t)$$

If we can show, by an *independent* argument, that this formula is correct for a deterministic process, our substitution of V(x), rather than MS(x), for V will be justified.

INTRODUCTION TO PROBABILITY DISTRIBUTIONS

Let x_t be the allele frequency, calculated deterministically, at time t. In the usual way, we have

$$dx_t/dt = M(x_t)$$

where $M(x_t)$ is our $M(x)$, with x_t instead of x. Thus, for genic selection, $M(x_t) = \alpha x_t(1 - x_t)$.

For a deterministic process

$$\phi(x, t) = \delta(x - x_t)$$

where δ is Dirac's delta function. From the 'function of a function' rule, we have (a prime denoting differentiation with respect to $x - x_t$)

$$\frac{\partial \phi(x, t)}{\partial t} = \delta'(x - x_t)\left(-\frac{dx_t}{dt}\right) = -M(x_t)\delta'(x - x_t)$$

Thus

$$\frac{\partial P(x, t)}{\partial x} = -\frac{\partial \phi(x, t)}{\partial t} = M(x_t)\delta'(x - x_t)$$

$$P(x, t) = \int M(x_t)\delta'(x - x_t)\, dx$$

$$= M(x_t)\delta(x - x_t) + C$$

(easily verified using the 'function of a function' rule, noting that $M(x_t)$ is not a function of x), where C is independent of x. Now $\delta(x - x_t) = 0$ when $x \neq x_t$. Hence non-zero C would imply a flux across x even when x is remote from x_t, which, for a deterministic process, is impossible. Thus $C = 0$. Further, since $\delta(x - x_t) = 0$ when $x \neq x_t$

$$M(x)\delta(x - x_t) = M(x_t)\delta(x - x_t)$$

as we proved formally in Chapter 6, note 6.

We have then

$$P(x, t) = M(x)\phi(x, t)$$

as we wished to prove.

EXPRESSIONS FOR M AND V

This result sometimes puzzles beginners, since when written

$$P(x, t) = \frac{\partial R(x, t)}{\partial t} = M(x_t)\delta(x - x_t)$$

it seems to imply that δR, over time δt, is infinite. The difficulty here is that, with a deterministic process, $R(x, t)$ can suddenly jump, over time, from zero to unity; hence the approximation $\delta R = \partial R/\partial t$ is not valid. Rather we have, with ξ as dummy variable,

$$\delta R = R(x, t + \delta t) - R(x, t) = \int_t^{t+\delta t} M(x_\xi)\delta(x - x_\xi) \, d\xi$$

To evaluate this integral, we introduce *Heaviside's unit function*, $H(z)$ say, which takes value

$$H(z) = 0 \quad \text{when } z < 0$$
$$= 1 \quad \text{when } z > 0$$

It rarely matters mathematically what value we give to $H(z)$ when $z = 0$; we shall find it convenient to define $H(0) = \frac{1}{2}$.

It may be shown that $H'(z) = \delta(z)$; see note 1. Thus, since

$$dx_\xi/dt = M(x_\xi)$$

$$\delta R = [H(x - x_\xi)]_{t+\delta t}^{t}$$
$$= H(x - x_t) - H(x - x_{t+\delta t})$$

(use the 'function of a function' rule).

It is easily verified, by listing all possibilities, that this formula for δR represents a deterministic process. For example, if $x_t < x < x_{t+\delta t}$, the whole distribution passes across x, left to right, in time δt; thus the flux is unity. Our formula gives

$H(\text{positive number}) - H(\text{negative number}) = 1 - 0 = 1$ and is thus correct. In case $x < x_t < x_{t+\delta t}$, we obtain $0 - 0 = 0$, and so on.

We have shown that the substitution of $V(x)$ for V is valid in the two extreme cases (a) $M(x)$ not large relative to $V(x)$ and (b) $V(x) = 0$. This suggests that our substitution will yield satisfactory results in most cases of slow evolution and this does seem to be so. We shall, therefore, write

$$P(x, t) = M(x)\phi(x, t) - \tfrac{1}{2}\frac{\partial}{\partial x}[V(x)\phi(x, t)]$$

We shall follow the standard practice of putting

$$V(x) = x(1 - x)/(2N_e)$$

with N_e calculated for the neutral, no-mutation, case, although for other cases this formula for $V(x)$ will be slightly approximate and will occasionally lead to difficulties, as we have discussed. We have, then

$$dV(x)/dx = (1 - 2x)/(2N_e)$$

Values of $P(x, t)$ when x takes limiting values

It is now easy to obtain an expression for $P(x, t)$ when $x \to 0$ or 1, in cases where 0, 1 are exit boundaries. From the usual rules for differentiation of a product, $P(x, t)$ may be written

$$M(x)\phi(x, t) - \tfrac{1}{2}V(x)\frac{\partial \phi(x, t)}{\partial x} - \tfrac{1}{2}\phi(x, t)\frac{dV(x)}{dx}$$

Given the flux into the exit boundaries, probabilities should not accumulate very much, or change very drastically, in the neighbourhood of an exit boundary. Hence both $\phi(x, t)$ and $\partial\phi(x, t)/\partial x$ will remain finite as $x \to 0$ or 1. Further, mutation being absent, both $M(x)$ and $V(x)$ tend to zero when $x \to 0$ or 1. Thus

$$\lim_{x \to 0} P(x, t) = -\frac{1}{4N_e} \lim_{x \to 0} \phi(x, t) \qquad \lim_{x \to 1} P(x, t) = \frac{1}{4N_e} \lim_{x \to 1} \phi(x, t)$$

Writing, as before, $f(x, t)$ for the probability that the allele frequency is x at time t, approximated by $\phi(x, t)/(2N)$ when $0 < x < 1$ and by $f^*(0, t), f^*(1, t)$ respectively when $x = 0, 1$, and recalling that

$$\frac{df^*(0, t)}{dt} = -\lim_{x \to 0} P(x, t) \qquad \frac{df^*(1, t)}{dt} = \lim_{x \to 1} P(x, t)$$

we have

$$\frac{df^*(0, t)}{dt} = \frac{1}{4N_e} \lim_{x \to 0} \phi(x, t) \qquad \frac{df^*(1, t)}{dt} = \frac{1}{4N_e} \lim_{x \to 1} \phi(x, t)$$

(Kimura 1964).

If we *assume* that $\partial \phi(x, t)/\partial x$ is not merely finite but *also small in magnitude* in the neighbourhood of an exit boundary, we can obtain an alternative approximation due to Wright (1931); see also Kimura (1964, 1968b). We note, with Wright, that this approximation is not universally valid. Given, then, that $\phi(x, t)$ does not change much with x near the exit boundaries, $\lim_{x\to 0} \phi(x, t)$ will be close to $\phi(1/(2N), t)$ and $\lim_{x\to 1} \phi(x, t)$ will be close to $\phi(1 - 1/(2N), t)$. Thus we may replace

$$\frac{1}{4N_e} \lim_{x\to 0} \phi(x, t) \quad \text{by} \quad \tfrac{1}{2}f\left(\frac{1}{2N}, t\right)\frac{N}{N_e}$$

$$\frac{1}{4N_e} \lim_{x\to 1} \phi(x, t) \quad \text{by} \quad \tfrac{1}{2}f\left(1 - \frac{1}{2N}, t\right)\frac{N}{N_e}$$

without serious loss of accuracy. Thus, when $N_e = N$, we have Wright's famous result: the flux into an exit boundary over a single generation equals, approximately, half the probability for the corresponding 'subterminal class', i.e. for frequency $1/(2N)$ or $1 - 1/(2N)$. When $N_e \neq N$, we multiply this probability by N/N_e. We stress that this result, although sometimes surprisingly accurate, should not be used uncritically, as Wright (e.g. Wright 1939a) was careful to point out.

Notes and exercises

1   To verify that $H'(z) = \delta(z)$, multiply both sides by test function $\theta(z)$ and integrate from $-\infty$ to ∞. The right-hand side gives

$$\int_{-\infty}^{\infty} \delta(z)\theta(z)\,dz = \theta(0)$$

The left-hand side gives

$$\int_{-\infty}^{\infty} H'(z)\theta(z)\,dz$$

Integrating by parts and remembering that $\theta(z)$ is defined to equal zero outside a finite range of z, we have

$$-\int_{-\infty}^{\infty} H(z)\theta'(z)\,dz = -\int_{0}^{\infty} H(z)\theta'(z)\,dz$$

since $H(z)$ is zero when z is negative. Putting $H(z) = 1$ for $z > 0$, we have

$$-\int_0^\infty \theta'(z) \, dz = [\theta(z)]_\infty^0 = \theta(0)$$

since $\theta(z)$ is zero outside a finite range.

2 Similarly, show that $H'(z - a) = \delta(z - a)$, where a is a constant.

9
Stationary distributions: Frequency spectra

> A pie may be produced any number of times.
> Stephen Leacock, *Boarding House Geometry*

Long-term distributions

We can now determine long-term probability distributions of allele frequencies. Of course, when both boundaries are exits, ultimate absorption is certain and our problem reduces to finding the probability of ultimate loss and of ultimate fixation, which we discussed in detail earlier. In this chapter, we consider long-term distributions when the effects of mutation are taken into account. As usual, our distribution represents the probability distribution of allele frequencies over an indefinitely large set of biologically similar populations.

Generally, we imagine, for a given locus, a set of K *possible* alleles, A_1, $A_2, \ldots, A_K$, which can mutate *inter se*. This representation will also be appropriate for a given nucleotide site (rather than a locus proper), in which case K, of course, equals 4. For simplicity, however, we shall often suppose $K = 2$ and write $A_1 = A$, $A_2 = a$, with A mutating to a at rate u, a mutating to A at rate v. This 'classical model' is not nearly as restrictive as it might seem at first. For example, with K neutral alleles, we may concentrate on the probability distribution for a single allele, A say, lumping the remaining $(K - 1)$ alleles together as a single 'allele' a; mutation rates will then be accommodated by the two-allele scheme described above. Similar considerations apply to a nucleotide site. For example, A could represent the presence of, and a the absence of, a specific base forming part of a recognition sequence for a given restriction endonuclease. Ewens *et al.* (1981) used this classical representation when finding a correction for an ascertainment bias arising when one attempts to estimate the degree of polymorphism at the DNA level using restriction endonucleases; see Ewens (1983) for further discussion of this problem.

In practice, the number of possible alleles at a locus is very large; this suggests an alternative approach. Since K is very large, the chance that a mutation will produce an allele present in the population, currently or in the past, is very small indeed. In the extreme case $K \to \infty$, every allele appearing by mutation will be 'new'. Thus, in this 'infinite allele model' of Kimura & Crow (1964), while a given allele may mutate, back-mutation is impossible, so that we may speak of mutation as 'irreversible'. Hence allele frequency 0 is an exit, but allele frequency 1 (in the diffusion model) is an entrance or regular. This infinite allele model is often considered in its own right, as a close approximation to the true situation in which K is very large. It is natural to suppose that results from the infinite allele model will be close to those from a model in which K is finite but large. Fortunately, this assumption is correct (even though frequency 0 is an exit in one case but not in the other, so that the two cases are not mathematically equivalent). The two different approaches are, in fact, complementary.

For the moment, we shall take K as finite. To introduce basic notions, we consider a two-allele case and suppose selection to be absent; A has forward mutation rate u, back-mutation rate v. To facilitate reference to Kimura's work, we write x for the frequency of A. We aim to find the long-term probability distribution of x.

Intuitively, it seems unlikely that the initial frequency of A will matter in the long run. We may guess that the ultimate probability distribution will depend solely on our mutation rates and on N, N_e. But, by hypothesis, none of these changes over time. Hence, while, *within any one population*, the allele frequency varies over time, even in the long term, the *probability distribution* of the frequency over populations eventually reaches a shape that does not change thereafter. Such a distribution, unchanging with time and independent of the initial frequency, is known as a 'stationary distribution'.

To verify this intuitive conclusion, we appeal to standard results in the theory of stochastic processes (see e.g. Cox & Miller 1965). The true allele frequency is a Markovian variable (see Ch. 2), changing in discrete time (generations); in *any* generation, the allele frequency may take *any* one of a *finite* number of discrete values $(0, 1/(2N), 2/(2N), \ldots, 1)$. Now, given two-way mutation, *all frequencies are mutually accessible*; that is, the allele frequency can pass over time from any one value to any other. Then, by standard theory, we conclude that a stationary distribution is ultimately attained.

For a distribution that does not change over time, the probability flux, $P(x, t)$, across *any* frequency x is necessarily zero. Hence to obtain the continuous approximation to the stationary distribution, we put

$$P(x, t) = 0 \qquad \text{for all } x$$

(see below for details).

Once we introduce selection, as well as two-way mutation, the allele frequency is not always Markovian in the strict sense. Fortunately, this does not affect, in any practical way, the conclusions that we have drawn (Watterson 1962, Norman 1975a). It is important to appreciate that a stationary distribution will ultimately be obtained even in cases where a *deterministic* treatment yields more than one stable equilibrium, implying, wrongly in fact, that the final outcome would depend upon the initial allele frequency. Generally, the circumstances under which deterministically obtained results are appropriate are not obvious and have to be considered case by case.

There is another way of looking at stationary distributions that is very illuminating. Suppose we consider just one population, at a sufficiently large number of generations from the start for the stationary distribution to have become (near-enough) attained. We can regard this one population as drawn at random from our indefinitely large set of populations. At any sufficiently late time, our population is equally likely to be any individual population in the set. In the words of the immortal Stephen Leacock (*Boarding House Geometry*):

If there be two boarders on the same flat, and the amount of side of the one be equal to the amount of side of the other, each to each, and the wrangle between one boarder and the landlady be equal to the wrangle between the landlady and the other, then shall the weekly bills of the two boarders be equal also, each to each.

For if not, let one bill be the greater.

Then the other is less than it might have been – which is absurd.

Suppose, for definiteness, any specified allele frequency, say 0.5, with probability in the stationary distribution 0.1, say. Thus, at any late time, one-tenth of the populations have allele frequency 0.5. Hence, on any occasion on which we draw our population from the set, there is a probability one-tenth that we draw a population with allele frequency 0.5. If we draw in this way, generation after generation, we shall obtain a frequency 0.5 in one-tenth of our drawings. Thus, provided we consider a sufficiently long time period at a sufficiently late stage, our population will have allele frequency 0.5 for *one-tenth of the time*.

Generally, a probability π for frequency x in the stationary distribution implies that any specified single population, existing for an indefinitely large number t of late generations, will have an allele frequency x in πt of these generations; see Cox & Miller (1965) for a rigorous proof of this result.

Yet another way to regard the stationary distribution amounts to saying that, if our π is very small, we expect a long time to elapse between successive appearances of frequency x in the same population (with an obvious extension to other values of π). It can be shown that the *mean recurrence time for frequency x*, in a given population at a late stage, equals $1/\pi$.

So far, we have taken N and N_e as constants. We now consider what happens when N varies cyclically, N_e/N remaining constant. Suppose that the cycle is k generations long; we number the generations in a cycle $1, 2, 3, \ldots, k$. Clearly, there will be allele frequencies that can appear at some stages of the cycle but not at others. Thus there cannot be a stationary distribution in the strict sense. Suppose, however, that we consider allele frequencies at times k generations apart. For example, we consider only generation 1 in every cycle. Fairly obviously, we find that, in the long term, the probability distribution reaches a form that is the same for generation 1 in every subsequent cycle; similarly for every other generation in the cycle. Generally, these long-term distributions will differ with generation in the cycle. If, however, evolution is slow and the cycle fairly short, the probability distribution will not change very markedly during the cycle (at stages in which a particular frequency is necessarily absent, the probability associated with that frequency, at other stages, moves to frequencies fairly close by). It can be shown rigorously (Dr A. J. Girling, personal communication) that the diffusion approximation for the long-term distribution is the *same* for every generation in the cycle; this approximate distribution is identical with that for a population of fixed size, provided we replace N_e by N_e^*, the harmonic mean of the N_e's at different stages of the cycle. Presumably, this distribution will give a reasonable idea of the general features of the true distribution at any stage of the cycle, provided cycles are fairly short and fluctuations of N not too drastic.

Wright's formula for the stationary distribution

Since we are dealing with a distribution unchanging over time, we can write $\phi(x)$ rather than $\phi(x, t)$ for the probability density function. Then, given zero flux

$$M\phi(x) - \tfrac{1}{2}\frac{d}{dx}[V\phi(x)] = 0$$

where M, V are the mean and variance of the change in allele frequency in a single generation.

Dividing through by $V\phi(x)$ we have

$$\frac{1}{V\phi(x)}\frac{d}{dx}[V\phi(x)] = \frac{2M}{V}$$

or

$$\frac{d}{dx} \log[V\phi(x)] = \frac{2M}{V}$$

$$\log[V\phi(x)] = \int (2M/V) \, dx + \log C \qquad (C = \text{constant})$$

Therefore

$$\phi(x) = (C/V) \, e^{\int (2M/V) \, dx}$$

C being chosen to make

$$\int_0^1 \phi(x) \, dx = 1$$

Wright (1931), by the exercise of astonishing ingenuity and intuition, obtained several special cases of this distribution. Later (Wright 1937, 1938b) he derived the general formula for $\phi(x)$, in a very pleasing manner, from the fact that the mean and variance of the stationary distribution are unchanged in successive generations. Finally (Wright 1952) he rederived the formula for $\phi(x)$ from the consideration that every moment of the distribution remains the same in sucessive generations. In the present author's opinion, however, our proof from the probability flux, which is due to Kimura (1964), reveals more of the underlying factors in operation and, being more direct than Wright's proof, is probably easier to understand. Yet another proof, which requires rather more than the others in previous mathematical knowledge, is given by Maruyama (1977).

Stationary distributions (case of no selection): mean and variance

We now consider, in detail, the case of two-way mutation, when selection is absent. It will be very helpful first to calculate the *exact* mean and variance of the stationary distribution.

We follow Kimura (1970b). Suppose that the frequency of a given allele A in generation t equals x_t. Conditional on this, the allele frequency in generation $(t + 1)$ will be

$$x_{t+1} | x_t = x_t(1 - u) + (1 - x_t)v + e$$

STATIONARY DISTRIBUTIONS: FREQUENCY SPECTRA

where the symbol | stands for 'conditional on', u and v are the forward and back-mutation rates and e is a random variable, representing the effect of drift. Then $Ee = 0$ and we have the mean of x_{t+1}, conditional on x_t,

$$E(x_{t+1}|x_t) = x_t(1-u) + (1-x_t)v = x_t(1-u-v) + v$$

We obtain the unconditional mean of x_{t+1}, in the usual way, by averaging over all values of x_t, giving

$$Ex_{t+1} = Ex_t(1-u-v) + v$$

Here Ex_{t+1}, Ex_t are unconditional means. Now suppose that the stationary distribution has been attained. Let μ be the mean of the stationary distribution; μ, of course, does not change with time. Then the unconditional means of x_t, x_{t+1} will be the same, each equalling μ, so that

$$\mu = \mu(1-u-v) + v$$

whence

$$\mu = v/(u+v)$$

Thus the (exact) mean of the stationary distribution equals the well known deterministic equilibrium frequency. This was to be expected from our discussion in Chapter 2, since $E(x_{t+1}|x_t)$ is a linear function of x_t in the present case. (We stress that this equality will not necessarily hold for other cases.)

We can also find the exact variance of the stationary distribution, although this is slightly more difficult. Writing X_t as short for

$$x_t(1-u-v) + v$$

so that

$$x_{t+1}|x_t = X_t + e$$

we have

$$E(x_{t+1}^2|x_t) = E[(X_t^2 + 2X_t e + e^2)|x_t]$$

Now, conditional on x_t, X_t^2 is a constant, $Ee = 0$ and

$$Ee^2 = Ee^2 - (Ee)^2 = \text{(variance of } e) = \frac{X_t(1-X_t)}{2N_e}$$

STATIONARY DISTRIBUTIONS: MEAN AND VARIANCE

Notice that we have *not* made the usual approximations when calculating the drift variance; our expression is exact. We have, then

$$E(x_{t+1}^2 | x_t) = X_t^2 + \frac{X_t(1-X_t)}{2N_e} = \frac{X_t}{2N_e} + X_t^2\left(1 - \frac{1}{2N_e}\right)$$

We shall average both sides over all values of x_t.

In the stationary distribution, (the unconditional mean of x_{t+1}^2) = (the unconditional mean of x_t^2) = μ_2', say. Note that μ_2' is *the mean frequency of AA homozygotes*. Also, the unconditional mean of X_t will be

$$\mu(1 - u - v) + v$$

and the unconditional mean of X_t^2, that is, of

$$x_t^2(1-u-v)^2 + 2x_t v(1-u-v) + v^2$$

will be

$$\mu_2'(1-u-v)^2 + 2\mu v(1-u-v) + v^2$$

Putting these points together, we have

$$\mu_2' = \frac{\mu(1-u-v) + v}{2N_e} + \left(1 - \frac{1}{2N_e}\right)[\mu_2'(1-u-v)^2 + 2\mu v(1-u-v) + v^2]$$

so

$$2N_e \mu_2' = \mu(1-u-v) + v + (2N_e - 1)\mu_2'(1-u-v)^2$$
$$+ (2N_e - 1)[2\mu v(1-u-v) + v^2]$$

Now we showed above that $\mu = v/(u+v)$; simple algebra now yields

$$\mu_2' = \frac{v}{u+v} \frac{1 + (2N_e - 1)v(2 - u - v)}{2N_e - (2N_e - 1)(1 - u - v)^2}$$

On subtracting μ^2 we obtain, after further simple algebra, the exact variance of the stationary distribution:

$$\sigma^2 = \frac{uv}{(u+v)^2} \frac{1}{2N_e - (2N_e - 1)(1-u-v)^2}$$

$$= \frac{uv}{(u+v)^2} \frac{1}{1 + 2(2N_e - 1)(u+v) - (2N_e - 1)(u+v)^2}$$

Given the values of N_e found in practice, we can, without appreciable loss of accuracy, replace $(2N_e - 1)$ by $2N_e$. We thus obtain the formula for σ^2 given by Wright (1937), who used an approximate approach. In practice, the term $2N_e(u + v)^2$ will (usually at least) be negligibly small, so that, to a very close approximation

$$\sigma^2 = \frac{uv}{(u+v)^2} \frac{1}{1 + 4N_e u + 4N_e v}$$

as noted by Wright. Similarly, a very close approximation for μ'_2 is

$$\frac{v}{u+v} \frac{1 + 2N_e v(2)}{2N_e - (2N_e - 1)(1) - 2N_e(-2u - 2v)} = \frac{v}{u+v} \frac{1 + 4N_e v}{1 + 4N_e u + 4N_e v}$$

As a first application, we consider variability at nucleotide sites.

Variation at nucleotide sites: effective number of bases

We suppose that substitutions at our site are neutral. We assume that every base mutates at the same overall rate u and, on mutation, is equally likely to mutate to any of the other three bases. Thus any specified base has forward mutation rate u, backward mutation rate $u/3$; these assumptions, while not strictly correct, will do as a first approximation.

How much variability should we expect at our site in a given generation? At first thought, we might imagine that, in view of the symmetry of our model, an equilibrium would be reached in which, in any population, each base had frequency $\frac{1}{4}$. In fact, such an outcome, or anything resembling it, is most unlikely unless N_e is extremely large. This statement may seem surprising, since, from the previous section (with 'base' replacing 'allele'), the mean frequency of any base at equilibrium is

$$\frac{v}{u+v} = \frac{u/3}{u + u/3} = \tfrac{1}{4}$$

However, it is easily seen that this mean frequency is not very helpful, since it would take value $\tfrac{1}{4}$ in either of the extreme situations: (a) every population in our set has all four bases at equal frequency; (b) every population in our set is monomorphic, with $\tfrac{1}{4}$ of the populations monomorphic for any specific base. The reader will easily imagine many other situations giving a mean of $\tfrac{1}{4}$.

VARIATION AT NUCLEOTIDE SITES

Now, the probability that an individual is homozygous for a given base (= the mean frequency of individuals homozygous for that base) is

$$\mu'_2 = \frac{v}{u+v} \frac{1+4N_e v}{1+4N_e u+4N_e v} = \frac{1}{4} \frac{3+4N_e u}{3+16N_e u}$$

on using our very close approximation for μ'_2 and putting $v = u/3$. Since homozygosity for one base or another are mutually exclusive possibilities, and since our formula for μ'_2 holds for any of the four bases, the overall mean frequency of homozygotes (= the probability that an individual is homozygous) at our site equals

$$\sum_{\text{bases}} \mu'_2 = 4\mu'_2 = (U+3)/(4U+3)$$

where $U = 4N_e u$. If all four bases were equally frequent in all populations, the mean frequency of homozygotes would be

$$\sum_{\text{bases}} (\tfrac{1}{4})^2 = 4(\tfrac{1}{4})^2 = \tfrac{1}{4}$$

which is substantially less than the value calculated above unless U is fairly large, implying (since u is very small) a very large value of N_e.

The point is often put as follows. Imagine a hypothetical situation in which n equally frequent bases were present. Each base then has frequency $1/n$, giving rise to a homozygote with expected frequency $(1/n)^2$, so that in all the expected frequency of homozygotes would be

$$\sum_{\text{bases}} (1/n)^2 = n(1/n)^2 = 1/n$$

The 'effective number of bases' n_e is defined as the number of bases that, *if present at equal frequency*, would give rise to the actual mean frequency of homozygotes; that is

$$1/n_e = 4\mu'_2 \quad\text{giving}\quad n_e = (4U+3)/(U+3)$$

Hence, n_e will be close to 4 only when U is quite large. For U small (< 0.1) we may neglect terms in $U^2, U^3, \ldots$ to give the close approximation

$$n_e = \tfrac{1}{3}(4U+3)\left(1+\frac{U}{3}\right)^{-1} = \tfrac{1}{3}(4U+3)\left(1-\frac{U}{3}\right) = 1+U$$

311

STATIONARY DISTRIBUTIONS: FREQUENCY SPECTRA

Table 9.1 Effective number n_e of bases present, and mean heterozygosity, in populations of effective size N_e, at a nucleotide site having forward mutation rate $u = 5 \times 10^{-9}$; $U = 4N_e u$.

N_e	U	n_e	Mean heterozygosity
5×10^4	0.001	1.0010	0.0010
5×10^5	0.010	1.0100	0.0099
1.25×10^6	0.025	1.0248	0.0242
2.5×10^6	0.05	1.0492	0.0469
5×10^6	0.1	1.0968	0.0882
2.5×10^7	0.5	1.4286	0.3000
5×10^7	1	1.7500	0.4286
2.5×10^8	5	2.8750	0.6522
5×10^8	10	3.3077	0.6977
10^9	20	3.6087	0.7229

implying that, when U is small, one base will normally be present at high frequency. Some numerical values are given in Table 9.1, in which, following Kimura (1983), we take $u = 5 \times 10^{-9}$ as a representative value. The mean frequency of heterozygotes

$$1 - \frac{U+3}{4U+3} = \frac{3U}{4U+3}$$

is also tabulated.

From the work of Maruyama (especially Maruyama 1970b), these results are not seriously affected by population subdivision, so that N_e is the effective size of the whole species under consideration.

It is particularly interesting that the mean heterozygosity depends on N_e but not on N. This is unexpected, since increasing N, while keeping N_e and u fixed, must increase the mean number of new mutants appearing per generation. Note that N_e must be very large for the mean heterozygosity to approach its greatest possible value 0.75, which would obtain if all four bases were present at equal frequency. In practice, estimated mean heterozygosities (in Man and *Drosophila*) lie in the range 0.001 to 0.019 (Jeffreys 1979, Ewens et al. 1981, Langley et al. 1982, Kreitman 1983, Birley 1984, Cross & Birley 1986, Aquadro et al. 1988). Of course, our formula for mean heterozygosity was derived on the assumption that all mutants at a nucleotide site are neutral. However, even if we make a generous allowance for reduction of heterozygosity by negative selection (say by reducing our u by a factor of 10), the estimated mean heterozygosities do suggest a relatively small value of N_e.

THE STATIONARY DISTRIBUTION FOR NO SELECTION

Despite many uncertainties in interpreting these results (for example, we do not know if the populations sampled were in a stationary state), we are probably justified in concluding that, given the neutral theory in its strict form, really large values of N_e are not compatible with the observed variability, a conclusion first reached from a study of protein polymorphism (see Kimura & Ohta 1971b). Of course, Ohta's (1974, 1976) theory of very slightly disadvantageous mutants was designed to overcome this difficulty; however, this theory, proposed originally for coding sequences, would not help the neutralist much in the present context, if it really is the case that a fair portion of the genome is 'junk'.

It is remarkable that we have been able to extract so much from the moment Ex^2 of the stationary distribution. On the other hand, mean values do have limitations. Suppose, for example, that the mean heterozygosity is low. This, in principle, would be consistent with, say, (a) a low frequency of heterozygotes at a site at *all* times within a population or (b) monomorphism, or near-monomorphism, at our site for most of the time but a high frequency of heterozygotes occasionally. Clearly, (b) is required; we need monomorphism, or near-monomorphism, most of the time to account for species-specific, or near-specific, DNA sequences but a high frequency of heterozygotes at times when one base is replacing another. To identify the circumstances under which (b) would be attained on the neutral theory, we must first calculate the probability of monomorphism (= fraction of times for which our site is monomorphic in a given population). We turn, therefore, to the stationary distribution.

Finding the stationary distribution in the case of no selection

We return to the classical model and consider the stationary distribution for allele A. If the allele frequency in a given generation is x, the frequency one generation later, calculated deterministically, is

$$x(1 - u) + (1 - x)v = x(1 - u - v) + v$$

Hence M, the mean change in frequency, being equal to the deterministic change, equals

$$x(1 - u - v) + v - x = -x(u + v) + v$$

The variance of the change in allele frequency will be, as usual

$$x(1 - x)/(2N_e)$$

For the rest of this chapter, we shall write this variance as V(x), reserving V for a different quantity defined below. Hence, noting that

$$\frac{1}{x(1-x)} = \frac{1}{x} + \frac{1}{1-x}$$

we have

$$\int \frac{2M}{V(x)} dx = \int \frac{4N_e[-x(u+v)+v]}{x(1-x)} dx$$

$$= 4N_e \int \left(\frac{-u}{1-x} + \frac{v}{x}\right) dx$$

$$= 4N_e[u \log(1-x) + v \log x]$$

$$e^{\int [2M/V(x)] dx} = (1-x)^{4N_e u} x^{4N_e v}$$

Therefore

$$\phi(x) = \frac{C}{V(x)} e^{\int (2M/V) dx} = \frac{2N_e C}{x(1-x)} (1-x)^{4N_e u} x^{4N_e v}$$

Writing $2N_e C = D$, $U = 4N_e u$, $V = 4N_e v$ we have

$$\phi(x) = D(1-x)^{U-1} x^{V-1}$$

To find D, we use

$$\int_0^1 \phi(x) \, dx = 1$$

Now, by a standard mathematical result, $B(m, n)$, the *beta function* of m and n ($m > 0$, $n > 0$), defined as

$$B(m, n) = \int_0^1 x^{m-1}(1-x)^{n-1} dx$$

equals

$$\Gamma(m)\Gamma(n)/\Gamma(m+n)$$

where $\Gamma(m)$ is the *gamma function* of m ($m > 0$), defined as

$$\Gamma(m) = \int_0^\infty x^{m-1} e^{-x}\, dx$$

Note that

$$\Gamma(1) = \int_0^\infty e^{-x}\, dx = 1$$

A very useful aid to calculations with $\Gamma(m)$ appears on integrating $\Gamma(m)$ by parts, namely

$$\Gamma(m) = (m-1)\Gamma(m-1)$$

Repeated integration by parts gives

$$\Gamma(m) = (m-1)\Gamma(m-1) = (m-1)(m-2)\Gamma(m-2)$$
$$= (m-1)(m-2)(m-3)\Gamma(m-3)$$

and so on. Hence if m is a positive integer

$$\Gamma(m) = (m-1)(m-2)\cdots(3)(2)(1)\Gamma(1)$$
$$= (m-1)!$$

since $\Gamma(1) = 1$. We may write this result as

$$\Gamma(m+1) = m!$$

(provided $m > -1$); it is customary to *define* $m!$ as $\Gamma(m+1)$, for *any* m exceeding -1, this reducing to the usual definition in the special case where m is a positive integer. The 'mysterious' result $0! = 1$ follows immediately.

We have then

$$1 = \int_0^1 \phi(x)\, dx = D\,\frac{\Gamma(U)\Gamma(V)}{\Gamma(U+V)}$$

and so

$$\phi(x) = \frac{\Gamma(U+V)}{\Gamma(U)\Gamma(V)}(1-x)^{U-1} x^{V-1}$$

This *beta distribution* has mean

$$E(x) = \int_0^1 x\phi(x)\,dx = \frac{\Gamma(U+V)}{\Gamma(U)\Gamma(V)} \int_0^1 (1-x)^{U-1} x^{(V+1)-1}\,dx$$

$$= \frac{\Gamma(U+V)}{\Gamma(U)\Gamma(V)} \frac{\Gamma(U)\Gamma(V+1)}{\Gamma(U+V+1)}$$

$$= \frac{\Gamma(U+V)}{\Gamma(V)} \frac{V\Gamma(V)}{(U+V)\Gamma(U+V)}$$

$$= V/(U+V)$$

$$= v/(u+v) \qquad \text{(since } U = 4N_e u, V = 4N_e v\text{)}$$

the *exact* mean of the stationary distribution, which we calculated earlier. Also

$$E(x^2) = \int_0^1 x^2 \phi(x)\,dx = \frac{\Gamma(U+V)}{\Gamma(U)\Gamma(V)} \frac{\Gamma(U)\Gamma(V+2)}{\Gamma(U+V+2)}$$

$$= \frac{(V+1)V}{(U+V+1)(U+V)}$$

On subtracting $[E(x)]^2$ we find the variance

$$\frac{UV}{(U+V)^2(U+V+1)} = \frac{uv}{(u+v)^2} \frac{1}{1 + 4N_e u + 4N_e v}$$

which, as we saw earlier, is very close to the true variance of the stationary distribution, provided $2N_e \gg 1$, $2N_e u^2$ is very small. This agreement of mean and variance with the true values strongly suggests that our formula for $\phi(x)$ will be reliable. Further, and very remarkably, the agreement in variance, which clearly holds even in cases where there is a large probability that $x = 0$ or 1 (e.g. very low mutation rates, N_e smallish, so that the outcome is dominated by the effects of drift) means that frequencies 0, 1 have been taken into account in a reasonably accurate way. It seems, then, that the probabilities that $x = 0, 1$, which we deliberately ignored when deriving our formula for $\phi(x)$, are represented in $\phi(x)$ and should be extractable from it.

Stationary distribution (continued): allele frequencies zero and unity

Consider first an argument due to Ewens (1965). Suppose N is not too small and that $N_e = N$. If i A alleles are present in generation t, the mean number of

STATIONARY DISTRIBUTION: ALLELE FREQUENCIES 0, 1

A alleles present in generation $(t + 1)$ is

$$i(1 - u) + (2N - i)v$$

We aim to calculate $f(0)$, the probability that the frequency of A is zero; we shall do this, for our stationary distribution, by equating $f(0)$ in $(t + 1)$ with $f(0)$ in t. Now, with evolution slow, $f(0)$ in $(t + 1)$ will depend solely on probabilities in t for frequencies close to, or equal to, zero. Thus we need consider only the case i small and so can neglect iu, iv, so that the mean number of A alleles in generation $(t + 1)$ is, near enough,

$$i + 2Nv$$

With $N_e = N$ not too small and $(i + 2Nv)$ small, the number of A in $(t + 1)$ will follow the Poisson distribution. Hence, if i A alleles are present in generation t, the probability that 0 A alleles are present in generation $(t + 1)$ is, to a close approximation,

$$e^{-(i+2Nv)} = e^{-i}e^{-V/2}$$

where, as usual, $V = 4Nv$. Let $f(i/(2N))$ be the probability that i A alleles are present in generation t. Then $f(0)$ in $(t + 1)$, which, given a stationary distribution, is the same as $f(0)$ in t, is given by

$$f(0) = f(0)\, e^{-V/2} + f\!\left(\frac{1}{2N}\right) e^{-1} e^{-V/2} + f\!\left(\frac{2}{2N}\right) e^{-2} e^{-V/2} + \cdots$$

Therefore

$$e^{V/2} f(0) = f(0) + f\!\left(\frac{1}{2N}\right) e^{-1} + f\!\left(\frac{2}{2N}\right) e^{-2} + \cdots$$

$$f(0) = \frac{f(1/(2N))\, e^{-1} + f(2/(2N))\, e^{-2} + \cdots}{e^{V/2} - 1}$$

This, then, is $f(0)$ 'as determined by the Poisson law'. Similarly

$$f(1) = \frac{f(1 - 1/(2N))\, e^{-1} + f(1 - 2/(2N))\, e^{-2} + \cdots}{e^{U/2} - 1}$$

where $U = 4Nu$. The procedure, then, is to approximate $f(x)$ by $\phi(x)/(2N)$ for all x except 0 and 1; substitution of these approximations in our formulae gives $f(0)$ and $f(1)$. Finally, Ewens proposes to multiply all calculated f's by a constant chosen to make the f's add to unity. With appropriate modifications when the Poisson distribution is not applicable, this method is general for stationary distributions and seems to be very accurate, apart from some distortion (of no great biological significance) for very extreme frequencies.

A generally less accurate, but in some ways more convenient, approach is due to Wright (1931) and Kimura (1968b). These authors do not multiply by a constant at the end of the procedure and, further, calculate $f(0), f(1)$ by a slightly different method from that above. We consider here just our case

$$\phi(x) = \frac{\Gamma(U+V)}{\Gamma(U)\Gamma(V)} (1-x)^{U-1} x^{V-1}$$

and suppose first that $N_e = N$ not too small. For x small (and realistic u), we may take $(1-x)^{U-1}$ equal to 1. Writing ST for the probability for the subterminal frequency $1/(2N)$

$$\text{ST} = f\left(\frac{1}{2N}\right) = \frac{\phi(1/(2N))}{2N} = \frac{\Gamma(U+V)}{\Gamma(U)\Gamma(V)} \left(\frac{1}{2N}\right)^V$$

and substituting in our formula for $f(0)$, we find

$$f(0) = \frac{\text{ST}(1^{V-1}e^{-1} + 2^{V-1}e^{-2} + 3^{V-1}e^{-3} + \cdots)}{e^{V/2} - 1}$$

It is easily seen that when V is small, the term in parentheses is approximately 0.5 (since this term increases with V, it is sufficient to note that, as $V \to 0$, the term comes out as 0.46, as noted by Wright, whereas when $V = 0.5$, we get 0.51). Further, for V small ($\leqslant 0.5$, say), the denominator

$$e^{V/2} - 1 = [1 + \tfrac{1}{2}V + \tfrac{1}{2}(\tfrac{1}{2}V)^2 + \cdots] - 1$$

is close to $\tfrac{1}{2}V$. Thus when V is small, an approximation for $f(0)$ is

$$f^*(0) = \frac{\tfrac{1}{2}\text{ST}}{\tfrac{1}{2}V} = \frac{\tfrac{1}{2}\text{ST}}{2Nv}$$

This is Wright's formula. Although we supposed that $V \leqslant 0.5$, the formula will work fairly well for V up to about 2, the errors in numerator and denominator tending to compensate. For $V > 2$, however, Wright's formula rapidly loses accuracy as V increases. But, given a realistic value of N (> 500, say) and realistic mutation rates, it turns out that both $f(0)$ and $f^*(0)$ are very small when $V > 2$, so that the large *relative* error in $f^*(0)$ need not cause concern. Thus Wright's formula, although not strictly accurate, should be good enough for practical conclusions. The same applies to his approximation for $f(1)$

$$f^*(1) = \frac{\tfrac{1}{2}\text{ST}'}{2Nu} \qquad \text{where ST}' = f\left(1 - \frac{1}{2N}\right)$$

An intuitive interpretation of Wright's formulae is as follows. With low mutation rates, the flux of probability into frequency zero depends almost entirely on drift only and is given by

$$f\left(\frac{1}{2N}\right)e^{-1} + f\left(\frac{2}{2N}\right)e^{-2} + \cdots$$

or, approximately, by $\tfrac{1}{2}\text{ST}$. For stationarity, the flux into 0 equals the flux out of 0. The latter will be $f(0) \times$ the probability of leaving frequency 0. The latter term will be about $2Nv$. For, given frequency 0, there are $2N$ alleles that *could* mutate to A but, with low mutation rates, it is unlikely that more than one will actually do so, so that the chance that a mutation to A occurs is (approximately) the probability that one or other of $2N$ mutually exclusive events, each having probability v, will occur; that is, $2Nv$. Thus approximately

$$\tfrac{1}{2}\text{ST} = f(0) \times 2Nv$$

and similarly for $f(1)$. This argument makes it clear that, when $N_e \neq N$, we must multiply $\tfrac{1}{2}\text{ST}$ (and $\tfrac{1}{2}\text{ST}'$) by N/N_e to obtain the correct flux into 0 (or 1), as discussed earlier, but that the flux out of 0 (or 1) is still written in terms of $2Nv$ (or $2Nu$). Thus, according to Kimura

$$f^*(0) = \frac{\tfrac{1}{2}\text{ST}(N/N_e)}{2Nv} \qquad f^*(1) = \frac{\tfrac{1}{2}\text{ST}'(N/N_e)}{2Nu}$$

It is plausible that these formulae are about as accurate as in the case $N_e = N$; we assume for the moment that this is so. Consider now, with $U = 4N_e u$, $V = 4N_e v$,

$$\int_0^{1/(2N)} \phi(x)\,dx = \int_0^{1/(2N)} \frac{\Gamma(U+V)}{\Gamma(U)\Gamma(V)}(1-x)^{U-1}x^{V-1}\,dx$$

Since x, in ranging over 0 to $1/(2N)$, will (given realistic N) be small, we approximate $(1-x)^{U-1}$ by 1, as is justified by the fact that u is small (e.g.

STATIONARY DISTRIBUTIONS: FREQUENCY SPECTRA

when $x = 1/(2N)$, we get approximately $e^{-(U-1)/(2N)} = e^{-2uN_e/N + 1/(2N)}$, which is very close to $e^0 = 1$). By elementary calculus, our integral then becomes

$$\frac{\Gamma(U+V)}{\Gamma(U)\Gamma(V)}\left[\frac{x^V}{V}\right]_0^{1/(2N)} = \frac{1}{V}\frac{\Gamma(U+V)}{\Gamma(U)\Gamma(V)}\left(\frac{1}{2N}\right)^V$$

$$= \frac{1}{V}\mathrm{ST}$$

$$= \frac{\frac{1}{2}\mathrm{ST}}{2Nv}\frac{N}{N_e}$$

$$= f^*(0)$$

Using $\Gamma(V+1) = V\Gamma(V)$ this may be written as

$$f^*(0) = \frac{\Gamma(U+V)}{\Gamma(U)\Gamma(V+1)}\left(\frac{1}{2N}\right)^V$$

Similarly,

$$f^*(1) = \int_{1-1/(2N)}^{1} \phi(x)\,\mathrm{d}x = \frac{\Gamma(U+V)}{\Gamma(U+1)\Gamma(V)}\left(\frac{1}{2N}\right)^U$$

as given by Kimura. We have thus fulfilled our promise to demonstrate that the probabilities that $x = 0, 1$ are given by the area under $\phi(x)$ over an appropriate range of x on the open interval (the dual role of the interval $1/(2N)$–$1/(4N)$, which contributes to our approximations for both $f(0)$ and $f(1/(2N))$, causes only minor difficulties in interpretation; similarly for the interval $[1 - 1/(2N)]$–$[1 - 1/(4N)]$).

For tables and formulae for calculating numerical values of gamma functions, see Abramowitz & Stegun (1965). The reader, however, will probably prefer to use a standard computer program for this function. We just note the useful approximations, valid when Z is small (<0.01, say)

$$\Gamma(1+Z) = 1 \qquad \Gamma(Z) = 1/Z$$

Accuracy of the results

In Table 9.2, we give the stationary distribution, in cases $N = 5$ and $N = 10$, for a classical model with $u = v = 0.001$, $N_e = 5$. For calculation of the exact

ACCURACY OF THE RESULTS

Table 9.2 Stationary distributions. Forward and back-mutation rates each equal 0.001, effective population size $N_e = 5$. N = actual population size, x = allele frequency, f = probability. Approximate f calculated by Kimura's method.

| \multicolumn{3}{c}{$N = 5$} | \multicolumn{3}{c}{$N = 10$} |

x	Exact f	Approximate f	x	Exact f	Approximate f
0	0.4724	0.4778	0	0.4679	0.4712
			0.05	0.0093	0.0099
0.1	0.0114	0.0106	0.10	0.0050	0.0053
			0.15	0.0036	0.0038
0.2	0.0057	0.0060	0.20	0.0029	0.0030
			0.25	0.0024	0.0026
0.3	0.0046	0.0046	0.30	0.0022	0.0023
			0.35	0.0020	0.0021
0.4	0.0040	0.0040	0.40	0.0019	0.0020
			0.45	0.0019	0.0020
0.5	0.0038	0.0039	0.50	0.0018	0.0019
			0.55	0.0019	0.0020
0.6	0.0040	0.0040	0.60	0.0019	0.0020
			0.65	0.0020	0.0021
0.7	0.0046	0.0046	0.70	0.0022	0.0023
			0.75	0.0024	0.0026
0.8	0.0057	0.0060	0.80	0.0029	0.0030
			0.85	0.0036	0.0038
0.9	0.0114	0.0106	0.90	0.0050	0.0053
			0.95	0.0093	0.0099
1.0	0.4724	0.4778	1.00	0.4679	0.4712

distribution, see note 1; the approximate distribution was found by the Wright–Kimura method, using the last of the formulae derived above for $f^*(0)$, $f^*(1)$. Clearly, the Wright–Kimura formulae, although not strictly accurate, yield very reasonable approximations in these cases. As expected, with low values of U and V (0.02), the flux into frequencies 0, 1 is slightly overestimated, giving approximations for $f(0)$ and $f(1)$ that are slightly too large (note, however, from our previous discussion, that when U and V are large, the corresponding approximations would be too small). The approximate probabilities sum, in either case, to about 1.01 rather than 1, but again this discrepancy is quite trivial. The absolute error in any approximate probability is quite small. These results suggest that we may use the approximate formulae with confidence in cases where computation of exact

STATIONARY DISTRIBUTIONS: FREQUENCY SPECTRA

results would be impracticable; this conclusion is strengthened by simulations carried out by Kimura, to be mentioned later.

For some properties of stationary distributions in more general cases, see notes 3 to 7. Our immediate need is to find the probability of monomorphism at a nucleotide site.

Nucleotide sites (resumed): probability of monomorphism

Since, on our model, u, v and hence $f^*(1)$ are the same for every base and since frequencies of unity, at a given time, for different bases are mutually exclusive, the probability of monomorphism is

$$4f^*(1) = \frac{4\Gamma(U+V)}{\Gamma(U+1)\Gamma(V)}\left(\frac{1}{2N}\right)^U$$

where $V = U/3$. Note particularly that this probability depends on N as well as N_e and u. It is easily seen that a high probability of monomorphism is very likely when U is small. For in that case, we may use our approximations for $\Gamma(Z)$ given above and our probability reduces to

$$\frac{4V}{U+V}\left(\frac{1}{2N}\right)^U = \left(\frac{1}{2N}\right)^U$$

Some numerical values are given in Table 9.3. Notice that the probability of monomorphism declines with increasing N when other factors are held constant. However, the effect on within-population variability is rather

Table 9.3 Probability of monomorphism at a nucleotide site when $U = 4 \times$ effective population size $\times$ forward mutation rate is small. $N =$ actual population size.

		U	
$2N$	0.001	0.005	0.01
10^4	0.99	0.95	0.91
10^5	0.99	0.94	0.89
10^6	0.99	0.93	0.87
10^7	0.98	0.92	0.85
10^8	0.98	0.91	0.83

minor. For, whereas the probability of monomorphism in the strict sense depends on N, the probability that one or other allele has frequency exceeding $(1 - q)$, where q is a *fixed small* quantity, which probability we may call the probability of quasi-monomorphism, is about

$$4 \int_{1-q}^{1} \phi(x) \, dx$$

and is thus *independent of N*. In fact, it will be about q^U when U is small. Thus, for our set of populations, the main effect of increasing N, while keeping other factors constant, is to 'convert' a fraction of the monomorphic populations into populations with one very common wild type base and the occasional mutant(s). Note again that the mean frequency of heterozygotes is unaffected. We shall see that similar considerations apply when we consider alleles rather than bases. The customary practice of working in terms of quasimonomorphism (usually taking $q = 0.01$), rather than monomorphism proper, is much to be recommended, if our principal interest lies in within-population variability. Some numerical values are given in Table 9.4.

Clearly, high probabilities of quasi-monomorphism, or even of monomorphism proper, at a given site will obtain, provided U is small. Thus, given small U, a population will be monomorphic at a given site for most of the time, thus ensuring species-specific, or near-specific, sequences, as pointed out by Kimura (1983), who gives a very interesting discussion of the case $u = 5 \times 10^{-9}$, $N_e = N = 10^5$, $U = 0.002$ (see his pp. 194-202).

Table 9.4 Probability that one base has a frequency exceeding 0.99 in populations of effective size N_e. $U = 4 N_e \times$ forward mutation rate.

U	Probability
0.001	0.9954
0.005	0.9772
0.010	0.9550
0.025	0.8915
0.05	0.7954
0.1	0.6341
0.5	0.1098
1	0.0133
5	—
10	—
20	—

Further, in that case, the mean heterozygosity is 0.001 995, and the probability of monomorphism is 0.975 883. Thus the heterozygosity, *averaged over times when the population is polymorphic*, equals

$$0.001\,995/(1 - 0.975\,883) = 0.0827$$

Also, the probability of quasi-monomorphism is 0.990 832, so that the probability that one base is present with frequency between 0.99 and 1 equals $0.990\,832 - 0.975\,883 = 0.014\,949$. In that case, the mean heterozygosity is at most $2 \times 0.99 \times 0.01 = 0.0198$, giving a contribution to the overall mean heterozygosity of at most $0.014\,949 \times 0.0198 = 0.000\,296$. Thus times at which no base has frequency exceeding 0.99 contribute at least $0.001\,995 - 0.000\,296 = 0.001\,699$. The heterozygosity averaged over these times is thus at least

$$0.001\,699/(1 - 0.990\,832) = 0.1853$$

Thus, as required, high levels of heterozygosity will appear on the occasions when the population is polymorphic. In fact, from the work of Stewart (1976), we can conclude with confidence that a high frequency of heterozygotes is particularly likely to be found on such occasions.

Needless to say, this scenario, with monomorphism most of the time but high polymorphism occasionally, would not arise, on neutral theory, if N_e were really large. Note again that, under most circumstances, N_e here is the world population size of the species.

Multiple alleles: infinite alleles model

Consider now a complete locus, with K possible alleles, neutral *inter se*. For simplicity, we suppose that every allele has forward mutation rate u, backmutation rate v. As we have shown, the mean frequency of any such allele eventually reaches $v/(u + v)$, the equilibrium allele frequency calculated deterministically, which, by symmetry, will be $1/K$ in the present case. Thus, in this model, $v = u/(K - 1)$. The mean frequency of individuals homozygous for a given allele equals, in our previous notation, μ'_2 ($= \mathrm{E}x^2$). Arguing as we did for bases, we see that the overall mean frequency of homozygotes is $K\mu'_2$ and we define n_e, the effective number of alleles, as $1/(K\mu'_2)$. Using our very close approximation for μ'_2, we find that the mean frequency of homozygotes eventually reaches

$$\frac{1 + 4N_e v}{1 + 4N_e u + 4N_e v}$$

Now, K is very large, so it is natural to let $K \to \infty$, thus achieving simpler formulae. However, since $v = u/(K - 1)$, there will be no back-mutation in this limiting case and our stationary distribution $f(x)$ no longer exists. Nevertheless, in the long run a balance will be struck between mutational input of new alleles and loss of old alleles by drift, so that our set of populations will still exhibit long-term stationary properties. For example, the mean frequency of homozygotes will eventually settle down to the value $1/(1 + 4N_e u)$ obtained by letting $K \to \infty$ (and hence $v \to 0$) in our formula above, as shown by Malécot (1948) and Kimura & Crow (1964), whose derivations assume an infinite number of alleles *ab initio*; Malécot gives a formal proof that an equilibrium will be attained. At equilibrium, then

$$n_e = 1 + 4N_e u \qquad \text{mean frequency of heterozygotes} = \frac{4N_e u}{1 + 4N_e u}$$

The essence of the matter is as follows. Any allele will remain in the population for a finite time only; once lost, it never reappears. Thus, if we inspect a population at two widely separated times, we are most unlikely to see the same alleles on both occasions. Nevertheless, certain statistical properties of the population, such as the frequency of heterozygotes present, averaged over a large number of generations, eventually reach an equilibrium form. As far as these properties are concerned, alleles present at different epochs stand in for one another.

Mean number of alleles present

Another feature that eventually reaches equilibrium form under the infinite alleles model is n_a, the mean number of different alleles present.

We argue as follows. If an allele survives in the population for g generations on average, then at any time after equilibrium has been established, the number of alleles present, on average, will be those that arose by mutation during the last g generations. But, on average, $2Nu$ new mutations (all different, *ex hypothesi*) appear in the population per generation. Thus

$$n_a = 2Nug$$

(Ewens 1964b, 1969). Here g is the mean survival time, given initial frequency $p = 1/(2N)$, which follows easily once we have calculated $T/(2N)$, the mean sojourn time, at frequency x, for initial frequency p. The situation is rather different from that in Chapter 7, since $x = 1$ is no longer an exit, but much of the argument is the same. Writing, as before,

$$G(p) = e^{-\int [2M/V(p)] dp}$$

STATIONARY DISTRIBUTIONS: FREQUENCY SPECTRA

we find (as before):

$$\text{when } p < x \qquad T = A \int_0^p G(\theta) \, d\theta$$

$$\text{when } p > x \qquad \frac{1}{G(p)} \frac{\partial T}{\partial p} = -\frac{4N_e}{x(1-x)G(x)} + A$$

where θ is a dummy variable, A is an expression independent of p, and $G(\theta)$, $G(x)$ are $G(p)$ with p replaced by θ, x respectively.

Now loss of any allele is inevitable. We showed, in Chapter 6, that if fixation is inevitable, T is independent of p for all x greater than p. Similarly, we can show that, with loss inevitable, T is independent of p for all x less than p; in this case, T will be the same as when the initial frequency is x.

Thus, when $p > x$

$$\frac{\partial T}{\partial p} = 0 \qquad \text{whence} \qquad A = \frac{4N_e}{x(1-x)G(x)}$$

Hence, when $p < x$

$$T = \frac{4N_e}{x(1-x)G(x)} \int_0^p G(\theta) \, d\theta$$

and since the expressions for T when $p < x$ and when $p > x$ must agree when $p = x$, we see that, when $p \geqslant x$

$$T = \frac{4N_e}{x(1-x)G(x)} \int_0^x G(\theta) \, d\theta$$

as given by Ewens (1964a). With forward mutation rate u and no back-mutation, M and $V(p)$, rewritten in terms of θ, are respectively $-u\theta$, $\theta(1-\theta)/(2N_e)$, whence

$$G(\theta) = e^{\int [4N_e u/(1-\theta)] \, d\theta} = e^{-4N_e u \log(1-\theta)} = (1-\theta)^{-4N_e u}$$

$$\int_0^p G(\theta) \, d\theta = \left[-\frac{(1-\theta)^{1-4N_e u}}{1-4N_e u} \right]_0^p = \frac{1-(1-p)^{1-4N_e u}}{1-4N_e u}$$

$$= \frac{1 - [1-(1-4N_e u)p + \text{terms in } p^2, p^3 \cdots]}{1-4N_e u}$$

on expanding in binomial series. Since p is small, we may neglect terms in p^2, $p^3,\ldots$ Hence, to a close approximation

$$\int_0^p G(\theta)\,d\theta = p$$

Then, since $G(x) = (1-x)^{-4N_e u}$

$$T = \frac{4N_e(1-x)^{4N_e u - 1}}{2Nx} \quad \text{when } \frac{1}{2N} \leq x < 1$$

and

$$T = \frac{4N_e(1-x)^{4N_e u - 1}}{x} \cdot \frac{1-(1-x)^{1-4N_e u}}{1-4N_e u}$$

$$= \frac{4N_e}{x(1-4N_e u)}[(1-x)^{4N_e u - 1} - 1] \quad \text{when } 0 < x \leq \frac{1}{2N}$$

In the latter case, x will be small, so that, near enough

$$T = \frac{4N_e}{x(1-4N_e u)}(-x)(4N_e u - 1) = 4N_e$$

Thus

$$g = \int_0^1 T\,dx = \int_0^{1/(2N)} 4N_e\,dx + \int_{1/(2N)}^1 \frac{4N_e(1-x)^{4N_e u - 1} x^{-1}\,dx}{2N}$$

$$= \frac{4N_e}{2N}\left(1 + \int_{1/(2N)}^1 (1-x)^{4N_e u - 1} x^{-1}\,dx\right)$$

Therefore

$$n_a = 2Nug = \theta\left(1 + \int_{1/(2N)}^1 (1-x)^{\theta - 1} x^{-1}\,dx\right)$$

where $\theta = 4N_e u$ (Ewens 1969). As expected from our earlier experience with mean absorption times, this formula slightly overestimates the true value of n_a. The slightly simpler formula

$$n_a = \theta \int_{1/(2N)}^1 (1-x)^{\theta - 1} x^{-1}\,dx$$

given by Ewens (1964b) as an approximation to our first formula above and derived by Kimura (1968b) by a different method (see note 5) slightly underestimates the true n_a, as Kimura shows from simulations. The difference between the two formulae has little practical significance and we shall, following normal practice, use the simpler formula, unless stated otherwise.

Yet another approximation (Ewens 1972) is

$$n_a = \frac{\theta}{\theta} + \frac{\theta}{\theta+1} + \frac{\theta}{\theta+2} + \cdots + \frac{\theta}{\theta+2N-1}$$

which, for practical values of N, equals our simpler $n_a + \gamma\theta$, where $\gamma = 0.5772\ldots$. For computation of n_a, the use of this form will usually be more convenient than the numerical integration used by Kimura.

Since

$$\frac{1}{\theta+1} + \frac{1}{\theta+2} + \cdots + \frac{1}{\theta+2N-1} < \frac{1}{1} + \frac{1}{2} + \cdots + \frac{1}{2N-1}$$
$$= \gamma + \log(2N-1)$$

then

$$n_a < 1 + \theta[\gamma + \log(2N-1)]$$

so that, for practical values of N, n_a will not exceed 1 by much when θ is small (<0.004, say).

On the other hand, if, in our simpler formula, we expand $(1-x)^{\theta-1} = (1-x)^{4N_e u - 1}$ by the binomial theorem, integrate term by term and ignore terms in $[1/(2N)]^2$, $[1/(2N)]^3$,... we obtain

$$n_a = \theta\left(\log(2N) + \frac{\theta-1}{2N} - (\theta-1) + \frac{(\theta-1)(\theta-2)}{2\times 2!}\right.$$
$$\left. - \frac{(\theta-1)(\theta-2)(\theta-3)}{3\times 3!} + \cdots\right)$$

which, as θ becomes large, approaches

$$\theta\left(\log(2N) + \frac{\theta-1}{2N} - (\theta-1) + \frac{(\theta-1)^2}{2\times 2!} - \frac{(\theta-1)^3}{3\times 3!} + \cdots\right)$$

Using the standard result mentioned earlier, i.e.

$$\int_x^\infty \frac{e^{-u}}{u} \, du = -\gamma - \log x + x - \frac{x^2}{2 \times 2!} + \frac{x^3}{3 \times 3!} - \cdots$$

and noting that this integral tends to 0 as x becomes large, we have the approximation

$$n_a = \theta\left(\log(2N) + \frac{\theta - 1}{2N} - \log(\theta - 1) - \gamma\right)$$

This gives results close to Kimura's when $N \geqslant 500$, $\theta \geqslant 4$. The reader will easily verify that n_a becomes very large when θ is large. Thus, when $\theta = 40$, $N = 500$, our formula gives $n_a = 108.2$ (Kimura's result is $n_a = 107.7$). Also n_a increases with N, although not as dramatically, in view of the logarithmic term for N. Thus, when $\theta = 40$, $N = 10^6$, we obtain $n_a = 410.7$ (Kimura's result is 410.2). Now $n_e = 1 + \theta$. Numerical calculations show that, in all cases

$$n_a > n_e$$

(the inequality being very marked when θ is large). It is interesting that the case n_a close to n_e, with $n_e \geqslant 2$, does not occur. A little caution is required in interpreting this, since, for a single population, n_a is the *mean*, over time, of the number of alleles present and n_e the reciprocal of the *mean* frequency of homozygotes, over time. However, simulations (Ewens & Ewens 1966, Kimura 1968b) indicate that radical departures of actual values from mean values will be infrequent. We can conclude with confidence that, for most of the time equal, or approximately equal, frequencies of all alleles present will not occur; this conclusion is strengthened by the detailed results of simulations, presented in graphical form by Kimura.

Mean number of alleles in a given frequency range

We can take the matter further (Ewens 1964b, 1969). By a simple extension of the preceding arguments, the mean number of alleles present at any time whose frequency lies between a and b equals

$$4N_e u \int_a^b (1 - x)^{4N_e u - 1} x^{-1} \, dx$$

We shall (following Ewens 1969, 1972) interpret this formula to obtain a 'typical' representation of the population structure, once equilibrium has been attained; this representation will not be valid for *every* possible occasion.

Note that the integrand depends on N_e but not on N. If, then, we divide the overall range of frequencies, $1/(2N)$ to 1, into $(n + 1)$ intervals

$$1/(2N) \text{ to } k_1, \quad k_1 \text{ to } k_2, \quad k_2 \text{ to } k_3, \quad \ldots, \quad k_n \text{ to } 1$$

where the k's are fixed quantities, a change in N can affect only the mean number of alleles in the first interval. An increase in N obviously increases the number of new mutants appearing in the population every generation, so that if N is increased, while other factors are held constant, the mean number of alleles present rises. However, only the mean number of alleles in the interval $1/(2N)$ to k_1 is affected; thus increasing N by itself just increases the mean number of different types of rare allele. It will be sufficient, then, to consider just one value of N, say 10^6; see Table 9.5 (which is largely self-explanatory, so that a few comments will suffice).

If θ is very small, the probability of quasi-monomorphism, q^θ, will be close to unity (see note 4). Thus, typically, there will be one very common allele, together with a small number of very low-frequency alleles. On the other hand, when θ is at all large (> 4, say), it is most unlikely that any allele will be

Table 9.5 Mean number of alleles present, at frequencies indicated, in a population of actual size $N = 10^6$. $\theta = 4 \times$ effective population size $\times$ forward mutation rate; a (–)indicates a number less than 0.01.

Range of frequencies	\multicolumn{9}{c}{θ}
	0.004	0.1	0.5	1.0	2.0	4.0	8.0	16.0
$1/(2N)$ to 0.01	0.04	0.99	4.95	9.90	19.79	39.49	78.68	156.14
0.01 to 0.02	–	0.07	0.35	0.69	1.37	2.65	5.01	8.93
0.02 to 0.05	–	0.09	0.47	0.92	1.77	3.32	5.82	8.97
0.05 to 0.1	–	0.07	0.36	0.69	1.29	2.22	3.30	3.70
0.1 to 0.2	–	0.08	0.37	0.69	1.19	1.74	1.91	1.20
0.2 to 0.3	–	0.05	0.23	0.41	0.61	0.70	0.46	0.11
0.3 to 0.4	–	0.04	0.18	0.29	0.38	0.32	0.12	0.01
0.4 to 0.5	–	0.04	0.15	0.22	0.25	0.15	0.03	–
0.5 to 0.6	–	0.04	0.14	0.18	0.16	0.07	0.01	–
0.6 to 0.7	–	0.04	0.13	0.15	0.11	0.03	–	–
0.7 to 0.8	–	0.05	0.13	0.13	0.07	0.01	–	–
0.8 to 0.9	–	0.07	0.15	0.12	0.04	–	–	–
0.9 to 1.0	0.98	0.80	0.33	0.11	0.01	–	–	–

present at high frequency. With very large θ (not shown in the Table), a large number of alleles will be present, all rare (although frequencies will be unequal enough to give a very large discrepancy between n_a and n_e). Thus, when $\theta \ll 1$ or when $\theta \gg 1$, the population is most unlikely to contain any allele at intermediate frequency; such alleles appear only when θ lies in a fairly restricted range and even then there may well be only one intermediate-frequency allele present.

The frequency spectrum

Clearly, we may neglect the possibility that two or more different alleles of exactly equal frequency are present; this permits an interesting development of the preceding theory. To avoid any possible confusion, we shall speak of 'number of allelic types' in place of the looser term 'number of alleles' used previously. The mean number of allelic types present at frequency x is

$$4N_e u \int_{x-1/(4N)}^{x+1/(4N)} (1-x)^{4N_e u - 1} x^{-1} \, dx$$

or (replacing the area under the curve by the area of a corresponding rectangle) about

$$4N_e u (1-x)^{4N_e u - 1} x^{-1} \delta x$$

where $\delta x = 1/(2N)$. This mean arises by averaging over times when one allele is present at frequency x and times when no such allele is present. Thus if P is the probability (= proportion of times) that an allele is present at frequency x, the mean number of allelic types present at that frequency equals

$$1 \times P + 0 \times (1-P) = P$$

Thus, with $\theta = 4N_e u$

$$P = \theta (1-x)^{\theta - 1} x^{-1} \delta x$$

Note, however, that

$$\theta (1-x)^{\theta - 1} x^{-1}$$

known as the 'frequency spectrum' (Ewens 1972), is *not* a probability density function; its properties, at first sight rather surprising, are easily understood from a special case.

STATIONARY DISTRIBUTIONS: FREQUENCY SPECTRA

Suppose we have a population of size 3. The genetic composition of this population is given in terms of the number of copies of each allelic type present. Thus (6) will represent any case in which six copies of a single allelic type (not necessarily the same type in different cases) are present. Similarly, (51) represents five copies of any one allelic type and one copy of any different type; and (321) represents three copies of any one allelic type, two copies of a different type and one copy of yet another type. We suppose (admittedly artificially in this special case) that two or more allelic types will not occur at the same frequency at the same time. Then there are just four possible compositions for our population:

Number of allelic types present	Composition	Probability
1	(6)	P_1
2	(51)	P_2
2	(42)	P_3
3	(321)	P_4

The mean number of allelic types present is thus

$$n_a = P_1 + 2P_2 + 2P_3 + 3P_4$$

Also we have the following 'frequency table':

Frequency	Probability that any allelic type is present at that frequency
1	P_1
5/6	P_2
4/6	P_3
3/6	P_4
2/6	$P_3 + P_4$
1/6	$P_2 + P_4$
	$P_1 + 2P_2 + 2P_3 + 3P_4 = n_a$

Thus, if we write down, frequency by frequency, the probability that an allelic type is present at that frequency, these probabilities sum to n_a, the mean number of allelic types present. But these probabilities are approximated by

$$\theta(1-x)^{\theta-1}x^{-1}\delta x$$

so that the continuous analogue of the above is

$$\int_{1/(2N)}^{1} \theta(1-x)^{\theta-1} x^{-1} \, dx = n_a$$

Consider now, composition by composition, the sum of the squared allelic frequencies, $\Sigma_i x_i^2$ say, namely

$$1^2, \quad (\tfrac{5}{6})^2 + (\tfrac{1}{6})^2, \quad (\tfrac{4}{6})^2 + (\tfrac{2}{6})^2, \quad (\tfrac{3}{6})^2 + (\tfrac{2}{6})^2 + (\tfrac{1}{6})^2$$

which are $\tfrac{36}{36}, \tfrac{26}{36}, \tfrac{20}{36}$ and $\tfrac{14}{36}$, respectively. The mean value of $\Sigma_i x_i^2$ is thus

$$\tfrac{1}{36}(36P_1 + 26P_2 + 20P_3 + 14P_4)$$

Precisely the same result would be obtained by squaring every frequency, multiplying by the probability that an allele is present at the frequency, and adding, to give

$$\tfrac{1}{36}(36P_1 + 25P_2 + 16P_3 + 9P_4 + 4P_3 + 4P_4 + P_2 + P_4)$$
$$= \tfrac{1}{36}(36P_1 + 26P_2 + 20P_3 + 14P_4)$$

The continuous analogue here is

$$E \sum_i x_i^2 = \int_0^1 x^2 \theta(1-x)^{\theta-1} x^{-1} \, dx = \int_0^1 x\theta(1-x)^{\theta-1} \, dx$$

which, on integration by parts, yields the correct result $1/(\theta + 1)$.

Generally, given frequency spectrum

$$\theta(1-x)^{\theta-1} x^{-1}$$

and $\Sigma_i \phi(x_i)$, where $\phi(x_i)$ is an arbitrary function of x_i (e.g. x_i^2 in the case above),

$$E \sum_i \phi(x_i) = \int_0^1 \phi(x)\theta(1-x)^{\theta-1} x^{-1} \, dx$$

(Ewens 1972). The use of zero for the lower limit of integration is justified provided $\phi(x)$ is at most comparable with x as $x \to 0$. The reader is referred to Ewens (1979) for further discussion of the frequency spectrum and its use in developing tests of the neutral theory; note that tests of this kind suppose that the population has reached stationarity. We give an application of the frequency spectrum in the next section.

Infinite sites model

We now ask: 'For a given locus, how many nucleotide sites, on average, will be polymorphic (in the strict sense)?' And what is the mean number of heterozygous sites, within the locus, per individual? (Kimura 1969, Kimura & Ohta 1971a, Watterson 1975, Ewens 1979).

The number of sites is so large ('effectively infinite') that we may assume that every new mutant appearing in the population arises at a previously (strictly) monomorphic site. Thus, at any given site, *at most two* different base pairs will be present in the population.

We consider just the neutral case; the argument is much the same as in the previous section. Let x be the frequency of the mutant (initial frequency $p = 1/(2N)$ at any site; the site ceases to be polymorphic once $x = 0$ or $x = 1$ is reached. Thus the average time that any site remains polymorphic is just the mean absorption time, derived in Chapter 7, for the case where $x = 0, 1$ are exit boundaries, namely, with $p = 1/(2N)$

$$-4N_e \left[\frac{1}{2N} \log\left(\frac{1}{2N}\right) + \left(1 - \frac{1}{2N}\right) \log\left(1 - \frac{1}{2N}\right) \right]$$

or, N being large, about

$$(N_e/N)[2 \log(2N) + 2] = g \qquad \text{say}$$

If we look at the population at any time after stationarity has been established, the mean number of sites polymorphic will be the number that became polymorphic during the last g generations. If u is the mutation rate for the *complete locus*, $2Nu$ sites, on average, become polymorphic every generation (on our assumption that all mutants appearing in the population arise at previously monomorphic sites). Thus the mean number of polymorphic sites in the population is

$$2Nug = 2N_e u[2 \log(2N) + 2]$$

In the case $N_e = N$, a more accurate formula is

$$2Nu[2 \log(2N - 1) + 1.355\,076\,40]$$

INFINITE SITES MODELS

It is interesting that Fisher's formula for the mean absorption time, to which we devoted much effort in Chapter 7, should appear in such a 'modern' context.

While these formulae can give very large values for the mean number of polymorphic sites (e.g. 61 in the case $N_e = N = 10^6$, $u = 10^{-6}$, it must be remembered that polymorphism is defined in the strict sense here and includes cases where one base pair is very infrequent.

Now consider the mean number of heterozygous sites per individual. Note first that the mean number of sites at which the mutant nucleotide has frequency x in the population $(1/(2N) \leqslant x < 1)$ equals

$$2Nu \times \text{mean sojourn time at } x, \text{ given initial frequency } p = 1/(2N)$$

which, from Chapter 7, with $p = 1/(2N)$, is

$$2Nu \frac{2p}{x} \frac{N_e}{N} = \frac{4N_e u}{x} \delta x$$

where $\delta x = 1/(2N)$. Thus the frequency spectrum in this case is $4N_e u/x$. The number of heterozygous sites in an individual is ΣX_i, where $X_i = 1$ if the individual is heterozygous at the ith polymorphic site, $X_i = 0$ otherwise, and the sum is taken over all polymorphic sites. Then, for a population in which the mutant has frequency x_i at the ith polymorphic site,

$$EX_i = 1[2x_i(1-x_i)] + 0[1 - 2x_i(1-x_i)] = 2x_i(1-x_i)$$

and the mean number of heterozygous sites per individual in a *specific* population with mutant frequencies $x_1, x_2, x_3, \ldots$ at individual polymorphic sites is

$$2x_1(1-x_1) + 2x_2(1-x_2) + 2x_3(1-x_3) + \ldots$$

The mean number of heterozygous sites per individual is therefore

$$E \Sigma 2x(1-x)$$

which, on adapting the frequency spectrum argument of the previous section, becomes

$$\int_0^1 2x(1-x)(4N_e u/x) \, dx = 8N_e u[x - \tfrac{1}{2}x^2]_0^1 = 4N_e u$$

335

Our formulae for the *mean* number of polymorphic sites and the *mean* number of heterozygous sites per individual are independent of the amount of recombination between sites.

Frequency spectrum for transposable elements

Langley *et al.* (1983) proposed an interesting application of the frequency spectrum in the population genetics of transposable elements. Consider a given chromosome site at which a specific transposable element may appear; let x_i be the frequency of chromosomes in a population carrying the element at the ith site. In *Drosophila melanogaster*, the x_i are always small, at least for the elements studied so far (Montgomery & Langley 1983); in humans, however, the *Alu* element showed sample frequency unity at all sites studied. It is natural then to ask: 'What prediction can we make for the mean number, per haploid genome, of sites for which our element has a stated population frequency, x say?'

Langley *et al.* assumed a model (which has some empirical support) in which, in a given genome, an element transposes, by replication, at a rate per generation that declines with increasing number of copies of the element already present in that genome; elements may also be deleted. It is assumed that every copy of the element has the same probability μ of being deleted in any generation, where μ is independent of the copy number.

Suppose that fitness is unaffected by the presence or absence of our element at any site. The mean number of sites with element frequency x was shown to be

$$\Lambda \theta x^{-1}(1-x)^{\theta-1} \delta x$$

where Λ is the mean number of copies per haploid genome, $\theta = 4N_e\mu$, $\delta x = 1/(2N)$. The same result was obtained, using a different approach, by Charlesworth & Charlesworth (1983) as a special case; their model allows for the possibility that the presence of a transposable element at a given site reduces fitness.

This brief account does not, of course, do justice to the extensive literature on the population genetics of transposable elements; see especially Charlesworth (1985) and Brookfield (1986).

Stationary distributions under selection

We return to two-allele models, but now suppose that selection is acting; as before, we concentrate on the frequency of one allele, A say, having forward mutation rate u, backward mutation rate v. Let the allele frequency among the zygotes in some generation t be x; we write x_{sel} for the allele frequency that would obtain one generation later if selection, but no other factors, were operating. Then, with selection and mutation acting, the mean change in allele frequency from zygotes in t to zygotes in $(t + 1)$ is

$$M = x_{sel}(1 - u) + (1 - x_{sel})v - x = (x_{sel} - x)(1 - u - v) - ux + v(1 - x)$$

Since, in practice, $u \ll 1$, $v \ll 1$, we have, to a very close approximation,

$$M = x_{sel} - x - ux + v(1 - x)$$

or, using Wright's (1937) well known formula for $x_{sel} - x$,

$$M = \frac{x(1-x)}{2\bar{W}} \frac{d\bar{W}}{dx} - ux + v(1-x)$$
$$= \tfrac{1}{2}x(1-x) \frac{d\log \bar{W}}{dx} - ux + v(1-x)$$

where $\bar{W}$ is the mean fitness of our three genotypes. For example, if we write the fitnesses of AA, Aa, aa, as $1 + s$, $1 + sh$, 1 respectively, then

$$\bar{W} = x^2(1 + s) + 2x(1 - x)(1 + sh) + (1 - x)^2$$
$$= 1 + sx^2 + 2shx(1 - x)$$

Provided that M is small when stationarity has been attained, we may write the variance of the change in allele frequency as

$$x(1 - x)/(2N_e)$$

It follows easily, using our general formula for $\phi(x)$ established earlier, that in this case

$$\phi(x) = C\bar{W}^{2N_e}(1 - x)^{U-1}x^{V-1}$$

where U, V are $4N_e u$, $4N_e v$ respectively and C is a constant (Wright 1937). Here $\phi(x)$ is the distribution among newly formed zygotes.

It is natural to suppose that, when N_e is very large, a deterministic treatment will be appropriate. To investigate this possibility, we first show that, when N_e is sufficiently large, a maximum of $\phi(x)$ corresponds to a stable

STATIONARY DISTRIBUTIONS: FREQUENCY SPECTRA

deterministic equilibrium value of x, whereas a minimum of $\phi(x)$ corresponds to an unstable equilibrium value.

Let Δx be the change in x in a single generation, calculated deterministically; then

$$\Delta x = \frac{x(1-x)}{2\bar{W}} \frac{d\bar{W}}{dx} - ux + v(1-x)$$

Let $\Delta(x_e)$ and $\Delta'(x_e)$ be the values of Δx and its derivative with respect to x when x takes a deterministic equilibrium value x_e, so that $\Delta(x_e) = 0$. For x close to an equilibrium value, we may ignore terms in $(x - x_e)^2$, $(x - x_e)^3, \ldots$ and write

$$\Delta x = \Delta(x_e) + (x - x_e)\Delta'(x_e)$$
$$= (x - x_e)\Delta'(x_e)$$

since $\Delta(x_e) = 0$. For a stable deterministic equilibrium, Δx must be negative when $x > x_e$, positive when $x < x_e$, which will happen if $\Delta'(x_e)$ is negative; positive $\Delta'(x_e)$ implies an unstable equilibrium.

Now suppose N_e is sufficiently large to make U, V very large; then, in $\phi(x)$ we may replace $U - 1$, $V - 1$ by U, V without appreciable loss in accuracy. Writing ϕ as short for $\phi(x)$ and ignoring the cases $x = 0, 1$, which will give $\phi(x) = 0$ (since $U > 1$, $V > 1$), we have

$$\phi = C\bar{W}^{2N_e}(1-x)^U x^V$$

$$\frac{d\phi}{dx} = C\left(2N_e \bar{W}^{2N_e - 1} \frac{d\bar{W}}{dx}(1-x)^U x^V - \bar{W}^{2N_e}U(1-x)^{U-1}x^V\right.$$

$$\left. + \bar{W}^{2N_e}(1-x)^U V x^{V-1}\right)$$

$$= \phi\left(\frac{2N_e}{\bar{W}} \frac{d\bar{W}}{dx} - \frac{U}{1-x} + \frac{V}{x}\right)$$

$$= \frac{4N_e \phi}{x(1-x)} \Delta x$$

$$\frac{d^2\phi}{dx^2} = \Delta x \frac{d}{dx} \frac{4N_e \phi}{x(1-x)} + \frac{4N_e \phi}{x(1-x)} \Delta'(x)$$

STATIONARY DISTRIBUTIONS UNDER SELECTION

Thus, when $d\phi/dx = 0$ we necessarily have $\Delta x = 0$, $x = x_e$ and also

$$\frac{d^2\phi}{dx^2} = \frac{4N_e\phi}{x(1-x)}\Delta'(x)$$

so that the second derivative and $\Delta'(x)$ have the same sign. Thus a maximum of ϕ occurs at $x = x_e$, with $\Delta'(x_e)$ negative, that is, at a stable equilibrium value of x on deterministic theory. Similarly, a minimum of ϕ occurs at an unstable deterministic equilibrium value of x.

If, on deterministic theory, there is just *one* stable equilibrium point, the outcome is quite straightforward. As N_e becomes very large, $\phi(x)$ becomes very 'peaky' at $x = x_e$ and very small when x is not close to x_e. Thus the allele spends virtually all its time, once equilibrium has been established, at, or very close to, x_e and the use of deterministic theory in this case is justified.

On the other hand, deterministic theory can be very misleading, in some cases at least, when two or more stable equilibria occur. Consider the case of a dominant disadvantageous allele A:

	AA	Aa	AA
viability	$1-s$	$1-s$	1

For illustration, we take the case $s = 10^{-3}$, $u = 10^{-5}$. It is easily shown (for example, by plotting Δx against x) that, as long as v is not too large, there are *two* stable equilibrium points. Thus, when $v = 10^{-8}$

$$x = 0.010\,102 \qquad x = 0.998\,874$$

are such points. The equilibrium frequency attained will, on deterministic theory, depend on the 'initial' frequency, that is, the frequency at the time when the current pattern of fitness was established. It would appear then that the fate of allele A is strongly dependent on the past history of selection at the locus. If A has always been disadvantageous and thus has always been rare, it will attain the low equilibrium frequency 0.010 102; if, however, A has previously been advantageous and thus present at very high frequency it will attain frequency 0.998 874 and thus remain very common, in spite of current selection against A, an extreme example of Lewontin's 'problem of historicity in evolution'.

In fact, this will not happen. Although $\phi(x)$ has a maximum at each of these values of x, the peak at $x = 0.010\,102$ is very large, whereas the peak at $x = 0.998\,874$ is very small. For example, if $N_e = 2.5 \times 10^8$, the ratio of the two values of $\phi(x)$ is about $e^{444\,075}$. In fact, $\phi(x)$ is exceedingly small for x outside the range 0.0095 to 0.0107, so that, at almost all times, A will be rare once stationarity has been established.

Harmful recessives

Suppose now (Wright 1937, Nei 1968, Crow & Kimura 1970) that a is a harmful recessive:

	AA	Aa	aa
frequency in zygotes	$(1-x)^2$	$2x(1-x)$	x^2
viability	1	1	$1-s$

so that

$$\overline{W} = 1 - sx^2$$

Mutation rates are

$$a \underset{v}{\overset{u}{\rightleftarrows}} A$$

We suppose s sufficiently large to keep x small once stationarity has been attained (this will certainly be so if $s \gg v$ and $N_e s$ is large). Then we may ignore mutation of a to A, i.e. we put $u = 0$; also (in our usual notation) M will be small at stationarity, even if s is as large as unity (recessive lethal). Then, from Wright's formula for $\phi(x)$,

$$\phi(x) = C(1 - sx^2)^{2N_e}(1-x)^{-1}x^{4N_e v - 1}$$

Since x is small, we may use the approximations

$$(1 - sx^2)^{2N_e} = e^{-2N_e sx^2} \qquad (1-x)^{-1} = 1$$

so that

$$\phi(x) = Ce^{-2N_e sx^2} x^{4N_e v - 1}$$

Given $N_e s$ large, the form of $\phi(x)$ just given will ensure that $\phi(x)$ will be close to zero when x is moderate or large, as required. For practical purposes, then, this form may be taken as valid for all x, thus enabling us to calculate the constant C from the condition

$$\int_0^1 \phi(x)\, dx = 1$$

In fact, since $\phi(x)$ is negligibly small for x at all large, no harm is done by replacing our condition by

$$\int_0^\infty \phi(x)\,dx = 1$$

which enables us to write C in a simple form.

Make the substitution $y = 2N_e sx^2$, giving

$$\frac{dy}{dx} = 4N_e sx \qquad \frac{dx}{dy} = \frac{1}{4N_e sx} = \frac{1}{2\sqrt{(2N_e sy)}}$$

to give

$$\int_0^\infty Ce^{-y}\left(\frac{y}{2N_e s}\right)^{1/2(4N_e v - 1)} \frac{1}{2\sqrt{(2N_e sy)}}\,dy = \frac{C}{2(2N_e s)^{2N_e v}} \int_0^\infty y^{2N_e v - 1} e^{-y}\,dy = 1$$

The integral is, by definition of the gamma function, $\Gamma(2N_e v)$. Hence

$$\phi(x) = \frac{2(2N_e s)^{2N_e v}}{\Gamma(2N_e v)} e^{-2N_e sx^2} x^{4N_e v - 1}$$

As usual, $f(x)$, the probability that the allele frequency is x, will be approximated by $\phi(x)/(2N)$ when x lies between 0 and 1. To obtain an approximation to $f(0)$, we note that when x is very small, $e^{-2N_e sx^2}$ will be very close to unity, so that $\phi(x)$ takes the form

$$\text{constant} \times x^{4N_e v - 1}$$

just as in the no-selection case discussed earlier. Hence the argument used in that case to find an approximation to $f(0)$ applies, unmodified, to the present case, so that our approximation to $f(0)$ is

$$\frac{\frac{1}{2}f(1/(2N))N/N_e}{2Nv} = \frac{f(1/(2N))}{4N_e v}$$

Note that, when $4N_e v < 1, f(0) > f(1/(2N))$.

The theory just given assumes a two-allele model, but this theory may be adapted to give an indication of what will happen in more realistic cases. We consider recessive lethals ($s = 1$). Although many biochemically distinct recessive lethal mutants may arise at a given locus, we can ignore the odd case where two such distinct mutants, a_1 and a_2 say, complement to give a non-lethal phenotype for the a_1/a_2 heterozygote. Thus we can treat all recessive lethal mutants at our locus as if they were the same recessive lethal. As a first approximation, we treat all other alleles at our locus as a single wild type allele. Then the stationary distribution of 'the lethal' has probability density function

$$\phi(x) = \frac{2(2N_e)^{2N_e v}}{\Gamma(2N_e v)} e^{-2N_e x^2} x^{4N_e v - 1}$$

where v is the rate of mutation to recessive lethality. For x very small, the term $e^{-2N_e x^2}$ will be very close to unity. The probability that $x = 0$ (population free of lethals) is thus about

$$\frac{2(2N_e)^{2N_e v}}{\Gamma(2N_e v)} \left(\frac{1}{2N}\right)^{4N_e v - 1} \left(\frac{1}{2N}\right) \frac{1}{4N_e v} = \frac{(2N_e)^{2N_e v}}{\Gamma(2N_e v + 1)} \left(\frac{1}{2N}\right)^{4N_e v}$$

We shall be particularly interested in the reliability of the deterministic approach for this model. Now, by elementary calculus

$$\frac{d\phi(x)}{dx} = \phi(x)\left(-4N_e x + \frac{4N_e v - 1}{x}\right)$$

which will be negative, for all x, if $4N_e v < 1$.

Thus, if $4N_e v < 1$, $\phi(x)$ decreases steadily with increasing x; further, $f(0) > f(1/(2N))$, as shown above, so that the most probable value of x is zero. In detail, consider the case $v = 10^{-5}$, $N_e = N$. For $N = 10^4, 10^3, 10^2$, $f(0) = 0.15, 0.87, 0.99$ approximately, as given by Wright. Even when 'the lethal' is present, its frequency will usually be very much lower than the deterministic equilibrium frequency $\sqrt{v}$. Clearly, the deterministic approach is altogether misleading when $4N_e v < 1$.

Further, and in contrast with the no-selection case, the *mean* of the stationary distribution will *not*, in general, equal the deterministic equilibrium frequency. We show, following Wright, that this discrepancy becomes very marked when $4N_e \ll 1$. The mean $\bar{x}$ is

$$\int_0^1 x \frac{2(2N_e)^{2N_e v}}{\Gamma(2N_e v)} e^{-2N_e x^2} x^{4N_e v - 1} dx$$

If, as before, we extend the range of integration to ∞ and substitute $y = 2N_e x^2$, we obtain

$$\bar{x} = \frac{1}{\sqrt{(2N_e)}\Gamma(2N_e v)} \int_0^\infty y^{(2N_e v + 1/2) - 1} e^{-y} dy = \frac{\Gamma(2N_e v + \frac{1}{2})}{\sqrt{(2N_e)}\Gamma(2N_e v)}$$

by definition of the gamma function. When $2N_e v$ is very small, we have the approximations

$$\Gamma(2N_e v) = 1/(2N_e v) \qquad \Gamma(2N_e v + \tfrac{1}{2}) = \Gamma(\tfrac{1}{2})$$

and since, by a standard mathematical result, $\Gamma(\tfrac{1}{2}) = \sqrt{\pi}$, we find that

$$\bar{x} = v\sqrt{(2\pi N_e)}$$

in cases where $2N_e v$ is very small. Obviously, $\bar{x}$ will be very different from $\sqrt{v}$ in these cases.

Generally, the ratio, R say, of the mean to the deterministic equilibrium frequency $\sqrt{v}$,

$$R = \frac{\Gamma(2N_e v + \tfrac{1}{2})}{\sqrt{(2N_e v)}\Gamma(2N_e v)}$$

will be substantially less than unity when $N_e v$ is small, as noted by Wright. Thus when $N_e v = 0.1$, $R = 0.63$. However, when $N_e v = 0.25$, $R = 0.80$ and when $N_e v = 0.5$, $R = 0.89$; thus, *provided N_e is comparable with or larger than the reciprocal of the mutation rate*, deterministic theory will give the mean essentially correctly. In fact for $N_e v$ 'large' (>1, for our purposes) we may use the approximation (Abramowitz & Stegun 1965, p. 258)

$$\frac{\Gamma(n + \tfrac{1}{2})}{\sqrt{(n)}\Gamma(n)} = 1 - \frac{1}{8n} + \frac{1}{128n^2} - \cdots$$

with $n = 2N_e v$, which makes it clear that $R \to 1$ as $N_e v \to \infty$; in fact R will be close to 1 when $N_e v$ exceeds 4 or so and very close when $N_e v > 10$, as again pointed out by Wright.

When $N_e v > 0.25$, $\phi(x)$ will be zero when $x = 0$; further $\phi(x)$ will have a mode at

$$x = \sqrt{[(4N_e v - 1)/(4N_e)]}$$

so that, when $4N_e v \gg 1$, mean and mode will, near enough, coincide at value $\sqrt{v}$; in such cases the variance of the distribution will also be fairly small. However, we stress that, for $N_e v > 0.25$ but still fairly small (1, say), the variance will be large and deterministic theory will thus be unsatisfactory.

To sum up: if the values of N_e suggested by the neutral theory are correct, the deterministic treatment of the balance between mutation and selection could well give a misleading result. Similar considerations apply in the case of a haploid, as the reader may care to demonstrate.

STATIONARY DISTRIBUTIONS: FREQUENCY SPECTRA

Notes and exercises

1 *Exact stationary distributions.* Let **f** be a single row matrix of probabilities for frequencies of allele A in the stationary distribution. Let g_{ij} be the probability that A goes from frequency $x = i/(2N)$ in one generation to frequency $j/(2N)$ in the next, where i, j take values from 0 to $2N$, inclusive. Finally, let **Q** be a matrix with element g_{ij} in its ith row, jth column, with rows and columns numbered 0 to $2N$, inclusive. Then, since we are dealing with a stationary distribution

$$\mathbf{f} = \mathbf{fQ} \quad\quad \text{whence} \quad\quad \mathbf{f(I-Q)} = \mathbf{0}$$

and the probabilities making up **f** are easily found by standard computer programs, on noting that these probabilities must sum to unity.

We find the g_{ij} as follows. Suppose A has forward mutation rate u, back-mutation rate v. Let

$$x^* = x(1 - u) + (1 - x)v$$

When $N_e = N$

$$g_{ij} = \frac{(2N)!}{j!(2N-j)!}(1 - x^*)^{2N-j} x^{*j}$$

When $N_e < N$, we use the probability distribution given by Moran (1961), which we mentioned in Chapter 5.

Let

$$a = \frac{2N_e - 1}{2N - 1} \quad\quad b = 1 - a$$

$$A = \frac{a}{b} 2Nx^* \quad\quad B = \frac{a}{b} 2N(1 - x^*)$$

Then

$$g_{ij} = \frac{\Gamma(A+j)\Gamma(B+2N-j)\Gamma(2N+1)\Gamma(A+B)}{\Gamma(j+1)\Gamma(A)\Gamma(2N-j+1)\Gamma(B)\Gamma(A+B+2N)}$$

2 Consider the exact stationary distributions in Table 9.2. Note that the mean allele frequency, 0.5, is the *least* probable of all frequencies, confirming that, in some cases, the mean gives a very misleading

impression of the long-term allele frequency. Verify that the mean value of $2x(1 - x)$ is independent of the value of N.

3 Note the high probability of monomorphism for the distributions in Table 9.2. It is easily shown that, for the K-allele model described in the main text, for which $v = u/(K - 1)$, a high probability of monomorphism is a very likely outcome when our U (and, *a fortiori*, our V) is small. Verify that, in that case, the probability of monomorphism, which is approximately $Kf^*(1)$, will be close to $[1/(2N)]^U$ and calculate this for a few values of $2N$ (say, 10^4, 10^5, 10^6, 10^7, 10^8) and smallish U (say 0.001, 0.005, 0.01).

4 Consider now our K-allele model, with K very large, so that V is very small. Show that, in this case also, the probability of monomorphism, for general U, as given by $Kf^*(1)$, is $[1/(2N)]^U$ (Kimura 1971). A very accurate formula for this probability, derived by Watterson (1975), is

$$e^{-0.1003U}\Gamma(1 + U)[1/(2N)]^U$$

Kimura's formula agrees closely with this for U up to about 1, but for larger U underestimates the probability of monomorphism. Verify that, when $U > 1$, the probability of monomorphism will be very small, for all values of N of practical interest.

In practice, then, Kimura's formula, though very inaccurate in the *relative* sense when U is large, will not mislead in that case. This formula can also be deduced from the infinite allele model (Ewens 1979). The probability of quasi-monomorphism (that is, the probability that one or other allele will have frequency exceeding $1 - q$, where q is a fixed small quantity) is approximately q^U and is thus independent of N.

5 With K finite, the probability that any given allele is present equals $1 - f(0)$, where $f(0)$ is the probability that the allele frequency is zero. Thus n_a, the mean number of alleles present, is approximately

$$K[1 - f^*(0)] = \frac{K\Gamma(U + V)}{\Gamma(U)\Gamma(V)} \int_{1/(2N)}^{1} (1 - x)^{U-1} x^{V-1} \, dx$$

(Kimura 1968b). Let $K \to \infty$ and thus obtain Kimura's formula

$$n_a = 4N_e u \int_{1/(2N)}^{1} (1 - x)^{4N_e u - 1} x^{-1} \, dx$$

6 Show that, in all cases where $U \gg V$ and $V < 1$, the most probable frequency for any *given* allele is zero. Note that the mean frequency, while small, will often still be an unrepresentative value in these cases, especially

when V is very small. For example, when $u = 10^{-6}$, $v = 10^{-9}$, $N = N_e = 125\,000$, the mean allele frequency is about $250/(2N)$, whereas, with 99.5 per cent probability, the allele frequency lies in the range 0 to $25/(2N)$, inclusive.

Generally, in cases $U \gg V$, $V < 1$, the form of the stationary distribution will depend on the value of U. If $U > 1$, the probability declines steadily with increasing allele frequency. When $U < 1$, the probability will at first decline with increasing allele frequency and then rise to give a probability for allele frequency 1 very much less than the probability for frequency 0.

7   Finally, consider the case where both U and V exceed unity. Elementary calculus shows that, in this case, our stationary distribution is unimodal, with mode at allele frequency $(V - 1)/(U + V - 2)$. Strictly, this equals the mean allele frequency only in the highly artificial case $U = V$. However, in other cases, as N_e becomes very large, giving large values of U, V, the mean and mode approach coincidence. In fact, from the properties of the beta distribution (Wright 1931) or, more rigorously, from the large-population theory developed by Norman (1975b), the allele frequency is normally distributed when N_e is sufficiently large (with diminishing standard deviation σ as N_e increases). We stress the word *sufficiently*. For example, when $u = 10^{-6}$, $v = 10^{-9}$, $N_e = 2.5 \times 10^9$, $U = 10\,000$, $V = 10$ we find

$$\text{mean} = 9.9900 \times 10^{-4} \quad \text{mode} = 8.9928 \times 10^{-4} \quad \sigma = 3.1574 \times 10^{-4}$$

so that the mean differs from the mode by about 0.32σ. The (approximate) median, calculated from Pearson's empirical rule

$$\text{mean} - \text{mode} = 3(\text{mean} - \text{median})$$

equals 9.6576×10^{-4}. Note that σ, relative to the mean, is quite large. Thus, even with this large value of N_e, the deterministic approach is not very satisfactory.

10
Diffusion methods

> What is the difference between method and device?
> A method is a device which you use twice.
> Apocryphal saying of the traditional mathematics
> professor, given in G. Polya, *How To Solve It*

Aims: notation

In the last chapter, we considered long-term probability distributions of allele frequencies. We now take up the more difficult task of finding distributions for any given time t. A comprehensive treatment is beyond the scope of this book. Rather, our aim is to give enough to enable the reader to follow the extensive literature in this area, in particular the work of Kimura (see especially Kimura 1964), whose notation we shall usually follow.

In general, the probability that the allele frequency takes value x at time t will depend on the initial allele frequency p; to emphasize this, we could write this probability as $f(p, x, t)$. The continuous approximation to this, for x on the open interval (that is $0 < x < 1$), would then be written $\phi(p, x, t)\delta x$, where $\delta x = 1/(2N)$. Clearly, however, we should try to avoid this cumbersome notation wherever possible. When no ambiguity arises, we shall often drop the p or the x, or even just write ϕ, with the understanding that in all such cases $\phi(p, x, t)$ is intended.

Similarly for $x = 1$ or 0; we shall write our continuous approximations to $f(p, 1, t), f(p, 0, t)$ as $u(p, t), v(p, t)$ respectively, rather than use the more cumbersome $f^*(p, 1, t), f^*(p, 0, t)$ suggested by our notation of Chapter 8.

Forward equation

In Chapter 8, we developed the formula for the flux across x ($0 < x < 1$)

$$P(x, t) = M(x)\phi(x, t) - \tfrac{1}{2}\frac{\partial}{\partial x}[V(x)\phi(x, t)]$$

where M(x) and V(x) are the mean change in allele frequency and variance of the change in allele frequency, over a single generation, written in terms of x (to emphasize the latter, we write M(x) and V(x) rather than our earlier M and V). Further, we showed that

$$\frac{\partial \phi(x, t)}{\partial t} = -\frac{\partial P(x, t)}{\partial x}$$

Hence

$$\frac{\partial \phi(x, t)}{\partial t} = -\frac{\partial}{\partial x}[M(x)\phi(x, t)] + \tfrac{1}{2}\frac{\partial^2}{\partial x^2}[V(x)\phi(x, t)]$$

This equation, which is often found by a different method (see note 1), is known as the *diffusion equation, forward Kolmogorov equation* or *Fokker-Planck equation*; the last name is used mainly by physicists, but appears occasionally in the genetical literature. Planck, no doubt, will be a familiar name to the reader; according to Pais (1982), Fokker's principal interests, outside physics, were 31-tone music and the purity of the Dutch language.

For any given set-up, once expressions for M(x) and V(x) have been substituted on the right-hand side, the equation may be 'solved', that is, an expression for $\phi(x, t)$ may be found, although in practice obtaining such a solution can be a very formidable problem, especially in cases where natural selection is acting.

Note that the equation applies only for x on the open interval. However, for $x = 0, 1$ we may (when mutation is absent) use the relationships established in Chapter 8, i.e.

$$\frac{du(p, t)}{dt} = \frac{1}{4N_e}\lim_{x \to 1} \phi(x, t) \qquad \frac{dv(p, t)}{dt} = \frac{1}{4N_e}\lim_{x \to 0} \phi(x, t)$$

and these equations are solved by integration with respect to t.

Backward equation

We can, however, obtain a different equation for $\phi(p, x, t)$ involving partial derivatives with respect to p rather than with respect to x. Let $h(p, p + \delta p)$ be the probability that the allele frequency, p in generation 0, changes to $p + \delta p$ in generation 1. Then (with $\delta t = 1$)

$$f(p, x, t + \delta t) = \sum_{\delta p} h(p, p + \delta p)f(p + \delta p, x, t)$$

since reaching frequency x in $(t + \delta t)$ generations, starting at generation 0, is the same as reaching frequency x in t generations, starting at generation 1. Then, by definition of mean,

$$f(p, x, t + \delta t) = \mathop{\mathrm{E}}_{\delta p} f(p + \delta p, x, t)$$

The continuous approximation of this is (on dropping the x)

$$\phi(p, t + \delta t) = \mathop{\mathrm{E}}_{\delta p} \phi(p + \delta p, t)$$

The argument now follows a familiar path. We expand the right-hand side in Taylor series and write

$$M(p) \text{ for } \mathop{\mathrm{E}}_{\delta p} \delta p \qquad V(p) \text{ for } \mathop{\mathrm{E}}_{\delta p} (\delta p)^2$$

Note that here M and V are to be written in terms of p, rather than x; for example, for genic selection

$$M(p) = \alpha p(1 - p)$$

Further,

$$V(p) = p(1 - p)/(2N_e)$$

will be used in practice; this use of the variance of the change in allele frequency, rather than the mean square change, can be justified along the lines given in Chapter 8. Then, ignoring, as usual, terms in $E(\delta p)^r$ when $r > 2$, we obtain

$$\phi(p, t + \delta t) = \phi(p, t) + M(p) \frac{\partial \phi(p, t)}{\partial p} + \tfrac{1}{2} V(p) \frac{\partial^2 \phi(p, t)}{\partial p^2}$$

Now, with $\delta t = 1$, we may write

$$\phi(p, t + \delta t) - \phi(p, t) = \frac{\delta \phi(p, t)}{\delta t}$$

DIFFUSION METHODS

and, if evolution is slow, we may, as in Chapter 8, treat t as a continuous variable and replace $\delta\phi(p, t)/\delta t$ by $\partial\phi(p, t)/\partial t$, giving us

$$\frac{\partial \phi(p, t)}{\partial t} = M(p) \frac{\partial \phi(p, t)}{\partial p} + \tfrac{1}{2} V(p) \frac{\partial^2 \phi(p, t)}{\partial p^2}$$

This is known as the *backward Kolmogorov equation*. Like the forward equation, the backward equation may be used to obtain an expression for $\phi(p, t)$.

Since $\phi(x, t)$ and $\phi(p, t)$ are really the same, both being short for $\phi(p, x, t)$, it may puzzle the reader that we have bothered to obtain two equations, each of which will yield $\phi(p, x, t)$. In practice, for a given situation, solutions from the forward and backward equations are comparable in difficulty and indeed involve essentially the same mathematics. It is, then, mainly a matter of personal taste which equation is used; we have mentioned both in order to facilitate reference to the literature. However, according to Karlin & Taylor (1981), a rigorous derivation of the backward equation is decidedly easier than a rigorous derivation of the forward equation (a point not obvious in our derivations above, in which we have ignored some subtleties) and, for this reason, the use of the backward equation is preferable if we wish to be very scrupulous.

Generality of the backward equation

A more practical point is as follows. When dealing with the forward equation, we had to derive special equations to deal with the cases $x = 0$ and $x = 1$. However, our equation

$$f(p, x, t + \delta t) = \underset{\delta p}{E} f(p + \delta p, x, t)$$

is true even if $x = 1$ or 0; further, since x has the same value on either side of the equation and is thus a constant as far as this equation is concerned, no special problems arise in cases $x = 1, 0$. Consider the case $x = 1$. Clearly, we have the approximation

$$u(p, t + \delta t) = \underset{\delta p}{E} u(p + \delta p, t)$$

Expanding the right-hand side in Taylor series and putting, as usual,

$$\underset{\delta p}{E} \delta p = M(p) \qquad \underset{\delta p}{E} (\delta p)^2 = V(p) \qquad \underset{\delta p}{E} (\delta p)^r = 0 \quad \text{when } r > 2$$

we find

$$\frac{\partial u(p, t)}{\partial t} = M(p) \frac{\partial u(p, t)}{\partial p} + \tfrac{1}{2}V(p) \frac{\partial^2 u(p, t)}{\partial p^2}$$

which is our backward equation with $u(p, t)$ instead of $\phi(p, t)$; similarly for $v(p, t)$. Thus $u(p, t)$ and $v(p, t)$ 'obey the backward equation'. So indeed do many other quantities of interest. Let $\theta(x)$ be an arbitrary function of x; some particularly relevant special cases are $\theta(x) = x$, $\theta(x) = 2x(1 - x)$. Let $\{E\theta(x), p, t\}$ be the mean value of $\theta(x)$ at time t, given initial frequency p; then, by definition of mean,

$$\{E\theta(x), p, t\} = \sum_x \theta(x) f(p, x, t)$$

and it is easily shown, along the lines given above for $u(p, t)$, that the continuous approximation to $\{E\theta(x), p, t\}$ obeys the backward equation.

Further, we can recover, from the backward equation, many familiar results. For example, as $t \to \infty$, $u(p, t)$ tends to our $u(p)$, the probability of ultimate fixation. Since the latter is not a function of t

$$\partial u(p, t)/\partial t \to 0 \qquad \text{as } t \to \infty$$

and letting $t \to \infty$ throughout the backward equation we find the familiar (Ch. 5)

$$M(p) \frac{du(p)}{dp} + \tfrac{1}{2}V(p) \frac{d^2 u(p)}{dp^2} = 0$$

ordinary rather than partial derivatives appearing since $u(p)$ is a function of p only.

Finally, consider mean sojourn times. From Chapter 3, the mean sojourn time at frequency x, given initial frequency p, is

$$\sum_{t=0}^{\infty} f(p, x, t)$$

which we may conveniently write as

$$\sum_{t=0}^{\infty} f(p, x, t)\delta t$$

with $\delta t = 1$ (generation). The continuous approximation for our mean sojourn time is, in the notation of Chapter 7, $T(p, x)\delta x$ (where $0 < x < 1$) and the continuous approximation for $f(p, x, t)$ is $\phi(p, x, t)\delta x$, where $\delta x = 1/(2N)$ in either case. Acting in the spirit of continuous variables, we replace our sum above by the corresponding integral, thus giving

$$T(p, x) = \int_0^\infty \phi(p, x, t) \, dt \qquad (0 < x < 1)$$

the δx on either side cancelling out. We now take the backward equation, written in terms of $\phi(p, x, t)$, and integrate both sides with respect to t from $t = 0$ to ∞, to obtain

$$\int_0^\infty \frac{\partial \phi(p, x, t)}{\partial t} \, dt = \int_0^\infty M(p) \frac{\partial \phi(p, x, t)}{\partial p} \, dt + \int_0^\infty \tfrac{1}{2} V(p) \frac{\partial^2 \phi(p, x, t)}{\partial p^2} \, dt$$

On the right-hand side, we now carry out integration with respect to t and differentiation with respect to p in the reverse order to that given, so that, for example, the first term on the right-hand side becomes

$$M(p) \frac{\partial}{\partial p} \int_0^\infty \phi(p, x, t) \, dt = M(p) \frac{\partial T(p, x)}{\partial p}$$

Since integration is the reverse of differentiation, the left-hand side is

$$[\phi(p, x, t)]_{t=0}^{t=\infty}$$

where x lies on the open interval, since we are dealing with mean sojourn times. With both boundaries exits, absorption is inevitable, so that $\phi(p, x, t) \to 0$ as $t \to \infty$. When $t = 0$, the allele frequency is p with probability one; thus, when $t = 0$,

$$\phi(p, x, t) = \delta(x - p) = \delta(p - x)$$

where δ is Dirac's function. Thus, we recover the result, familiar from Chapter 7, that

$$M(p) \frac{\partial T(p, x)}{\partial p} + \tfrac{1}{2} V(p) \frac{\partial^2 T(p, x)}{\partial p^2} = -\delta(p - x)$$

Clearly, there is much scope for a development of the subject using the backward equation as a unifying basis. The reader is referred to Maruyama

(1977) for a comprehensive treatment along these lines. In the present author's opinion, however, Maruyama's approach will appeal most to those sufficiently familiar with the subject to see the point of the inevitably very general and abstract arguments with which he begins.

Kimura-Ohta equation

We have shown that a wide range of relevant quantities obey the backward equation. An alternative equation, of comparable width of application, is due to Kimura & Ohta (1971a); we shall show a little later that their equation is closely related to the forward equation.

Suppose $\theta(x)$ is an arbitrary function of the allele frequency x; we wish to find the mean value of $\theta(x)$ at time t, given initial frequency p. Write x_t for the allele frequency at time t, $x_{t+\delta t}$ for the allele frequency at time $t + \delta t$ ($\delta t = 1$) and δx_t for $(x_{t+\delta t} - x_t)$. Then

$$E\theta(x_{t+\delta t}) = E\theta(x_t + \delta x_t)$$

We first consider the right-hand side, *conditional on the allele frequency being x_t in generation t*. We have

$$\theta(x_t + \delta x_t) = \theta(x_t) + \delta x_t \theta'(x_t) + \tfrac{1}{2}(\delta x_t)^2 \theta''(x_t) + \cdots$$

(primes denoting differentiation), the conditional mean of which is (with the usual notation and assumptions)

$$\theta(x_t) + M(x_t)\theta'(x_t) + \tfrac{1}{2}V(x_t)\theta''(x_t)$$

The unconditional mean is obtained, in the usual way, by averaging the conditional mean over all x_t.

Thus, with E still denoting unconditional mean,

$$E\theta(x_{t+\delta t}) = E\theta(x_t) + EM(x_t)\theta'(x_t) + E\tfrac{1}{2}V(x_t)\theta''(x_t)$$

Then writing, with $\delta t = 1$,

$$E\theta(x_{t+\delta t}) - E\theta(x_t) = \delta[E\theta(x_t)]/\delta t$$

and for simplicity replacing x_t by x, we have the continuous approximation

$$\frac{d}{dt} E\theta(x) = E[M(x)\theta'(x) + \tfrac{1}{2}V(x)\theta''(x)]$$

A special case of particular interest is

$$\theta(x) = x^n$$

so that

$$E\theta(x) = Ex^n = \mu'_n \qquad \text{say}$$

the nth moment about zero of x (for example, $\mu'_1 = Ex$, the mean of x, $\mu'_2 = Ex^2$). In the neutral case

$$M(x) = 0 \qquad V(x) = x(1-x)/(2N_e)$$

Also, by elementary calculus

$$\theta'(x) = nx^{n-1} \qquad \theta''(x) = n(n-1)x^{n-2}$$

whence

$$\begin{aligned}
\frac{d\mu'_n}{dt} &= E\left(\frac{x(1-x)}{4N_e} n(n-1)x^{n-2}\right) \\
&= \frac{n(n-1)}{4N_e} E(x^{n-1} - x^n) \\
&= \frac{n(n-1)}{4N_e} (\mu'_{n-1} - \mu'_n)
\end{aligned}$$

This equation was first obtained by Kimura 1952; it is given in Crow & Kimura (1970, p. 336), in a slightly different notation (and with terms μ'_{n-1}, μ'_n inadvertently interchanged).

With $n = 1$,

$$d\mu'_1/dt = 0$$

Thus we have the familiar result: the mean allele frequency never changes, being always equal to its initial value p.

With $n = 2$

$$\frac{d\mu'_2}{dt} = \frac{1}{2N_e}(p - \mu'_2)$$

since $\mu'_1 = p$. Let $X = \mu'_2 - p$. Then, since $dp/dt = 0$,

$$\frac{dX}{dt} = -\frac{X}{2N_e}$$

$$\frac{1}{X}\frac{dX}{dt} = \frac{d\log X}{dt} = -\frac{1}{2N_e}$$

$$X = Ce^{-t/(2N_e)} \qquad \mu'_2 = p + Ce^{-t/(2N_e)}$$

C being a constant. When $t = 0$, $x = p$, $\mu'_2 = p^2$, whence

$$C = p^2 - p = -pq \qquad (q = 1 - p)$$

The variance of the allele frequency, then, is

$$\mu'_2 - (\mu'_1)^2 = p - pqe^{-t/(2N_e)} - p^2 = pq(1 - e^{-t/(2N_e)})$$

the continuous approximation to the familiar

$$pq\left[1 - \left(1 - \frac{1}{2N_e}\right)^t\right]$$

Kimura used his equation to obtain a general approximate formula for μ'_n, which we consider a little later.

Initial, terminal and boundary conditions

We return now to the problem of finding $\phi(p, x, t)$ from the forward or backward equation. Now, neither of these equations has a unique solution; this is particularly obvious in the case of the backward equation, since many different quantities obey that equation. We shall, therefore, have to supplement these equations with additional conditions if we are to find $\phi(p, x, t)$ from either.

For example, $\phi(p, x, t)$ must depend on the set-up in generation 0; yet neither equation makes any statement on this. We therefore must state explicitly that, in generation 0, the allele frequency equals p with probability 1, anything else with probability 0. We have, then, the *initial condition* (i.e. condition when $t = 0$)

$$\phi(p, x, 0) = \delta(x - p)$$

where δ is Dirac's function.

Of course, we are free to choose p as we wish, but *any other conditions must emerge from the problem itself*. It is not possible to 'impose' several 'arbitrary' conditions, as would sometimes be possible in physics. For example, in the most famous of all physical problems described by a partial differential equation, the vibrating string (see e.g. Sagan 1961), the length of string allowed to vibrate and the initial position and velocity of the string are not 'part of the inherent physics' of the problem and are decided by the investigator to suit his particular needs. In our case, only the value of p can be decided in this 'arbitrary' way. This being so, the number of conditions we can find relevant to our problem depends to some extent on how much effort we devote to thinking about it. It is certainly possible to find more conditions than are really necessary to solve the problem, although, of course, the solution must obey all of them.

For example, when both boundaries are exits, ultimate absorption is inevitable, so that eventually all x on the open interval have probability zero. We have, then, the *terminal condition*

$$\lim_{t \to \infty} \phi(p, x, t) = 0$$

With two-way mutation, $\phi(p, x, t)$ tends to the stationary distribution $\phi(x)$, calculated as in Chapter 9, as $t \to \infty$. In no case, however, is a terminal condition (which would be very unusual in a physical problem) necessary for us to achieve a solution, although we can sometimes shorten the argument by invoking such a condition.

Other conditions relate to events at the boundaries 0 and 1, and are known as boundary conditions. For example, if both boundaries are exits, large probabilities cannot pile up in the neighbourhood of $x = 0$ or $x = 1$, in view of the flux of probability off the open interval. Thus, as pointed out in Chapter 8,

$$\lim_{x \to 0} \phi(p, x, t) = \text{finite} \qquad \lim_{x \to 1} \phi(p, x, t) = \text{finite}$$

which is the form in which the boundary conditions are given by Kimura (1964). Alternatively, in view of this finiteness,

$$\lim_{x \to 0} x(1 - x)\phi(p, x, t) = 0 \qquad \lim_{x \to 1} x(1 - x)\phi(p, x, t) = 0$$

which is the form given by Voronka & Keller (1975). Although these conditions are weaker than Kimura's (they will sometimes hold even in a case where $\phi(p, x, t)$ would not be finite as $x \to 0$ or 1), they do in fact suffice for a solution.

In the case of two-way mutation, Goldberg's boundary condition, i.e. zero flux across $x = 0$ or 1, will be appropriate.

We can also formulate boundary conditions in terms of p rather than x. Thus, with both boundaries exits, $p = 0$ or 1 means zero probability for all x on the open interval for all time. Symbolically

$$\phi(0, x, t) = 0 \qquad \phi(1, x, t) = 0$$

Notice that, while $\phi(p, x, t)$, $u(p, t)$, $v(p, t)$ all obey the backward equation, the boundary conditions for these three are not the same. We have

$$u(0, t) = 0 \qquad u(1, t) = 1 \qquad v(0, t) = 1 \qquad v(1, t) = 0$$

Hence $\phi(p, x, t)$, $u(p, t)$, $v(p, t)$ have to be found by a separate procedure for each, so that the obedience of all three to the backward equation is not helpful as might first appear.

For two-way mutation we argue as follows. If we start with $p = 0$ or 1, we know (from Ch. 6) that the allele frequency leaves these boundaries instantaneously. Hence it cannot matter all that much, for the probability distribution at any time after $t = 0$, whether we started with $p = 0$ exactly rather than p very close to 0. Thus

$$\lim_{p \to 0} \frac{\partial \phi(p, x, t)}{\partial p} = \text{finite} \qquad \text{when } t > 0$$

Similarly,

$$\lim_{p \to 1} \frac{\partial \phi(p, x, t)}{\partial p} = \text{finite} \qquad \text{when } t > 0$$

An alternative approach

A different approach (Kimura 1955a, Crow & Kimura 1956) would appear, at first sight, to dispense with boundary conditions altogether. The idea is to calculate the moments of the distribution from the Kimura-Ohta equation given in the previous section, taken in conjunction with the initial condition

$$\mu'_n = p^n \qquad \text{when } t = 0$$

Then $\phi(p, x, t)$ is that solution of the forward equation whose moments agree with the moments already calculated and which also agrees with our initial condition $\phi(p, x, 0) = \delta(x - p)$.

This approach amounts to a concealed application of boundary conditions, as we now show, very much in outline.

We attempt to derive the Kimura-Ohta equation from the forward equation. Suppose, for definiteness, that both boundaries are exits. Let $\theta(x)$ be an arbitrary function of x. In obtaining an expression for $E\theta(x)$, we note that the probability density function $\phi(p, x, t)$ covers only x between 0 and 1, so that we must bring in $u(p, t)$, $v(p, t)$ to allow for the contributions of $x = 0$, 1 to $E\theta(x)$. However, at this stage of the argument, we do not make use of our earlier formulae for $u(p, t)$, $v(p, t)$. Then, for convenience dropping p throughout, we have, as a matter of definition,

$$E\theta(x) = \lim_{\varepsilon \to 0} \int_{\varepsilon}^{1-\varepsilon} \theta(x)\phi(x, t) \, dx + \theta(1)u(t) + \theta(0)v(t)$$

Thus

$$\frac{d}{dt} E\theta(x) = \lim_{\varepsilon \to 0} \int_{\varepsilon}^{1-\varepsilon} \theta(x) \frac{\partial \phi(x, t)}{\partial t} \, dx + \theta(1)u'(t) + \theta(0)v'(t)$$

(primes denoting differentiation with respect to t). We now replace $\partial\phi(x, t)/\partial t$ by the right-hand side of the forward equation. After two integrations by parts, we recover the Kimura-Ohta equation, provided we assume that, in the limit as $\varepsilon \to 0$,

$$\left[\theta(x) \left(-M(x)\phi(x, t) + \tfrac{1}{2} \frac{\partial}{\partial x} V(x)\phi(x, t) \right) \right]_{x=\varepsilon}^{x=1-\varepsilon} + \theta(1)u'(t)$$
$$+ \theta(0)v'(t) - [\tfrac{1}{2}\theta'(x)V(x)\phi(x, t)]_{x=\varepsilon}^{x=1-\varepsilon} = 0$$

Since the Kimura-Ohta equation was shown earlier, by a different proof, to be correct, this equation necessarily implies the correctness of the boundary condition just given. Notice that this condition follows automatically once we take it that $\phi(x, t)$ and $\partial\phi(x, t)/\partial x$ remain finite as $x \to 0$ or 1. For then, we have the familiar

$$u'(t) = \lim_{x \to 1} \left(M(x)\phi(x, t) - \tfrac{1}{2} \frac{\partial}{\partial x} V(x)\phi(x, t) \right)$$

$$v'(t) = \lim_{x \to 0} \left(-M(x)\phi(x, t) + \tfrac{1}{2} \frac{\partial}{\partial x} V(x)\phi(x, t) \right)$$

$\phi(p, x, t)$ IN THE NEUTRAL, NO-MUTATION CASE

and also

$$V(x)\phi(x, t) \to 0 \qquad \text{as } x \to 0 \text{ or } 1$$

One can carry out an analogous argument for the case of two-way mutation. This has an interesting result. It turns out that, after supposing zero flux across the boundaries, we are left with

$$\lim_{\varepsilon \to 0} [\tfrac{1}{2}\theta'(x)V(x)\phi(x, t)]_{x=\varepsilon}^{x=1-\varepsilon} = 0$$

as a necessary condition for the Kimura–Ohta equation to hold. Since $\theta(x)$, and hence $\theta'(x)$, is arbitrary, it follows that

$$\lim_{x \to 0} V(x)\phi(x, t) = 0 \qquad \lim_{x \to 1} V(x)\phi(x, t) = 0$$

even in the two-way mutation case, despite the fact that, sometimes, $\phi(x, t) \to \infty$ as $x \to 0$ or 1. Thus for our stationary distribution of Chapter 9,

$$\phi(x) = \frac{\Gamma(U + V)}{\Gamma(U)\Gamma(V)} x^{V-1}(1 - x)^{U-1}$$

with $U < 1$, $V < 1$, $\phi(x) \to \infty$ as $x \to 0$ or 1; but

$$V(x)\phi(x) = \frac{x(1 - x)}{2N_e} \phi(x) = \frac{1}{2N_e} \frac{\Gamma(U + V)}{\Gamma(U)\Gamma(V)} x^V(1 - x)^U$$

which tends to 0 as $x \to 0$ or 1. In practice, the boundary conditions

$$\lim_{x \to 0} V(x)\phi(x, t) = 0 \qquad \lim_{x \to 1} V(x)\phi(x, t) = 0$$

are more easily applied than the zero flux condition when obtaining $\phi(p, x, t)$ from the forward equation in the two-way mutation case.

Finding $\phi(p, x, t)$ in the neutral, no-mutation case

(1) *Initial and boundary conditions.* We shall now solve the backward equation in the neutral, no-mutation case. We have

$$M(p) = 0 \qquad V(p) = p(1 - p)/(2N_e)$$

and the backward equation becomes

$$\frac{\partial \phi}{\partial t} = \frac{p(1-p)}{4N_e} \frac{\partial^2 \phi}{\partial p^2}$$

For the moment, we consider the case $0 < x < 1$. Then we have the boundary conditions

$$\phi(0, x, t) = 0 \quad \text{and} \quad \phi(1, x, t) = 0$$

(valid for all x on the open interval and for all t). We shall take it that the continuity of $\phi(p, x, t)$, considered as a function of p, extends to $p = 0$ and 1 (this, in fact, helps to make our earlier Taylor expansion valid); in other words, the boundary conditions can also be written

$$\lim_{p \to 0} \phi(p, x, t) = 0 \quad \text{and} \quad \lim_{p \to 1} \phi(p, x, t) = 0$$

We shall also require the initial condition

$$\phi(p, x, 0) = \delta(x - p) = \delta(p - x)$$

We shall not discuss, in any rigorous way, whether these conditions suffice to yield a unique solution to the backward equation but will give informal arguments in justification of the procedure that we shall use.

(2) *Form of the solution.* We start by considering the *mathematical form* of the solution. Now from the theory developed in Chapter 3, viewed in the light of Feller's formula for the latent roots given in that chapter, $f(p, x, t)$, the exact probability, given initial frequency p, that the allele frequency is x at time t, x *lying between* 0 *and* 1, has the form

$$f(p, x, t) = \sum_i R_i$$

where every R_i has the form

(a term depending on t only) × (a term depending on p only)

× (a term depending on x only)

It is, at least, highly plausible that we should seek a solution of the backward equation that has this form.

$\phi(p, x, t)$ IN THE NEUTRAL, NO-MUTATION CASE

Now, for the exact probability, the term depending on t only is λ_i^t, where the λ_i are all different from one another, $\lambda_1 > \lambda_2 > \lambda_3 > \cdots$. Suppose the R_i are listed in order of diminishing λ_i; that is R_1 is the R_i depending on λ_1, R_2 is the R_i depending on λ_2, and so on.

At a very late stage of the process (t very large)

$$\lambda_1^t \gg \lambda_2^t \gg \lambda_3^t \gg \cdots$$

and we have, effectively,

$$f(p, x, t) = R_1$$

all other R_i being negligible in comparison. At a somewhat earlier stage, we shall have to include R_2 but can still ignore $R_3, R_4, \ldots$; effectively

$$f(p, x, t) = R_1 + R_2$$

while at a somewhat earlier stage yet

$$f(p, x, t) = R_1 + R_2 + R_3$$

and so on.

Very plausibly, our solution of the backward equation will contain terms analogous to the λ_i and these analogues can be used to order our analogues to the R_i, which, for simplicity, we will also write as R_i. Then, at a very late stage,

$$\phi(p, x, t) = R_1$$

effectively. This being so, R_1 must obey the backward equation and boundary conditions. At a somewhat earlier stage

$$\phi(p, x, t) = R_1 + R_2$$

effectively. Thus $(R_1 + R_2)$ obeys the backward equation and boundary conditions. If, then, we write R_2 in the form

$$(R_1 + R_2) - (R_1)$$

it follows that R_2 must obey the backward equation and boundary conditions. That the same must be true for *every* R_i follows by a simple extension of the argument.

DIFFUSION METHODS

We now sum up. Consideration of the exact solution indicates that $\phi(p, x, t)$ can be put in the form $\Sigma_i R_i$, where every R_i obeys the backward equation and boundary conditions and every R_i takes the form

(a term depending on t only) × (a term depending on p only)

× (a term depending on x only)

(3) *Some properties of the terms making up R_i.* Consider then any one of the R_i's. We write $T(t)$ for the function of t only and $S(x)$ for the function of x only. The function of p only will conveniently be written

$$p(1-p)X(p)$$

This assures that R_i will obey the boundary conditions ($R_i = 0$ when $p = 0$ or 1), save perhaps in cases where $X(p)$ is not finite when $p \to 0$ or 1.
Thus

$$R_i = p(1-p)X(p)T(t)S(x)$$

and since R_i obeys the backward equation

$$\frac{\partial R_i}{\partial t} = \frac{p(1-p)}{4N_e} \frac{\partial^2 R_i}{\partial p^2}$$

we have (primes denoting differentiation with respect to the immediately following variable in parentheses)

$$p(1-p)X(p)T'(t)S(x) = \frac{p(1-p)}{4N_e}[-2X(p) + 2(1-2p)X'(p)$$
$$+ p(1-p)X''(p)]T(t)S(x)$$

Thus

$$\frac{T'(t)}{T(t)} = \frac{p(1-p)X''(p) + 2(1-2p)X'(p) - 2X(p)}{4N_e X(p)}$$

The left-hand side is a function of t only, the right-hand side a function of p only; this is possible only if each side equals the same constant, $-k_i$, say (the minus sign has, as yet, no real significance, since we have not yet established the sign of k_i, but we shall show, very shortly, that k_i must be positive).

$\phi(p, x, t)$ IN THE NEUTRAL, NO-MUTATION CASE

(4) *Finding $T(t)$.* We have then

$$T'(t)/T(t) = -k_i$$

that is

$$\frac{1}{T(t)}\frac{dT(t)}{dt} = \frac{d \log T(t)}{dt} = -k_i$$

so

$$T(t) = \text{constant} \times e^{-k_i t}$$

From the terminal condition, every R_i and hence every $T(t)$ must tend to 0 as $t \to \infty$, whence every k_i is positive, as we confirm below (without using the terminal condition). Note that e^{-k_i} is our approximation to λ_i.

(5) *Finding $X(p)$.* To determine $X(p)$ is more difficult. We have

$$\frac{p(1-p)X''(p) + 2(1-2p)X'(p) - 2X(p)}{4N_e X(p)} = -k_i$$

so

$$p(1-p)X''(p) + 2(1-2p)X'(p) - (2 - 4N_e k_i)X(p) = 0$$

which is the *hypergeometric equation*

$$p(1-p)X''(p) + [\gamma - (\alpha + \beta + 1)p]X'(p) - \alpha\beta X(p) = 0$$

for the special case

$$\gamma = 2 \qquad \alpha + \beta = 3 \qquad \alpha\beta = 2 - 4N_e k_i$$

We wish to solve our hypergeometric equation, i.e. find $X(p)$. Now the equation is

(a) ordinary (only one independent variable, namely p);
(b) homogeneous (all terms involve the dependent variable or derivatives);
(c) linear (linear in the dependent variable and derivatives);
(d) second-order (highest derivative is a second derivative).

It follows from the theory of differential equations that all solutions are special cases of

$$X(p) = Ag(p) + Bh(p)$$

where A, B are constants and $g(p)$, $h(p)$ are distinct functions of p, which are distinct solutions of the equation (in special cases *one* of these functions may be a constant).

We attempt to express one of these functions, $g(p)$ say, as a series in ascending powers of p,

$$g(p) = C_0 + C_1 p + C_2 p^2 + \cdots + C_\lambda p^\lambda + \cdots$$
$$= C_0\left(1 + \frac{C_1}{C_0} p + \frac{C_2}{C_0} p^2 + \cdots + \frac{C_\lambda}{C_0} p^\lambda + \cdots\right)$$

with no decision yet on whether the series terminates or is infinite. The constants $C_0, C_1, C_2, \ldots$ will be found by substituting our series for $g(p)$ in the hypergeometric equation. To do this, we shall have to differentiate $g(p)$ term by term. For a terminated series, this is justified by the usual rule for the derivative of a sum. For an infinite series, intended to be valid over the range $0 < p < 1$, our term-by-term differentiation will, from standard theory of power series, be justified *post hoc* if the series we obtain converges (i.e. has a finite sum) when $|p| < 1$.

On substituting in the hypergeometric equation, we obtain, after some rather tedious algebra,

$$(\alpha\beta C_0 - \gamma C_1) + [(\alpha + 1)(\beta + 1)C_1 - (1 + 1)(\gamma + 1)C_2]p$$
$$+ [(\alpha + 2)(\beta + 2)C_2 - (1 + 2)(\gamma + 2)C_3]p^2 + \cdots$$
$$+ [(\alpha + \lambda)(\beta + \lambda)C_\lambda - (1 + \lambda)(\gamma + \lambda)C_{\lambda+1}]p^\lambda + \cdots = 0$$

(in this context, λ is a dummy variable).

For this to hold over all p, the coefficients of $p^0, p^1, p^2, \ldots, p^\lambda, \ldots$ must all be zero. Hence

$$\alpha\beta C_0 - \gamma C_1 = 0 \qquad \frac{C_1}{C_0} = \frac{\alpha\beta}{\gamma}$$

$$(\alpha + 1)(\beta + 1)C_1 - (1 + 1)(\gamma + 1)C_2 = 0 \qquad \frac{C_2}{C_1} = \frac{(\alpha + 1)(\beta + 1)}{2(\gamma + 1)}$$

$$(\alpha + 2)(\beta + 2)C_2 - (1 + 2)(\gamma + 2)C_3 = 0 \qquad \frac{C_3}{C_2} = \frac{(\alpha + 2)(\beta + 2)}{3(\gamma + 2)}$$

and generally, for $\lambda = 0, 1, 2, 3, \ldots,$

$$(\alpha + \lambda)(\beta + \lambda)C_\lambda - (1 + \lambda)(\gamma + \lambda)C_{\lambda+1} = 0 \qquad \frac{C_{\lambda+1}}{C_\lambda} = \frac{(\alpha + \lambda)(\beta + \lambda)}{(1 + \lambda)(\gamma + \lambda)}$$

Since $g(p)$ will contribute to $X(p)$ in the form $Ag(p)$, the constant C_0 can be regarded as absorbed into the constant A, so that we have

$$g(p) = 1 + \frac{\alpha\beta}{\gamma}p + \frac{\alpha(\alpha+1)\beta(\beta+1)}{1 \times 2 \times \gamma(\gamma+1)}p^2 + \frac{\alpha(\alpha+1)(\alpha+2)\beta(\beta+1)(\beta+2)}{1 \times 2 \times 3 \times \gamma(\gamma+1)(\gamma+2)}p^3 + \cdots$$

This is known as *Gauss's hypergeometric function*, denoted by $F(\alpha, \beta, \gamma, p)$. We shall state a few properties of this function, the first two of which are obvious from the series as written out above (we shall be concerned only with the case γ positive).

(i) The series is unaffected if α and β are interchanged, i.e.

$$F(\alpha, \beta, \gamma, p) = F(\beta, \alpha, \gamma, p)$$

(ii) If α or β is zero or a negative integer, the series terminates.
(iii) Otherwise, the series is infinite and converges if $|p| < 1$; when $p = 1$, the series converges if $\gamma > \alpha + \beta$.
(iv) Provided $\gamma > \alpha + \beta$

$$F(\alpha, \beta, \gamma, 1) = \frac{\Gamma(\gamma)\Gamma(\gamma - \alpha - \beta)}{\Gamma(\gamma - \alpha)\Gamma(\gamma - \beta)}$$

(v) $F(\alpha, \beta, \gamma, p) = (1 - p)^{\gamma - \alpha - \beta} F(\gamma - \alpha, \gamma - \beta, \gamma, p)$
(vi) $\dfrac{d}{dp} F(\alpha, \beta, \gamma, p) = \dfrac{\alpha\beta}{\gamma} F(\alpha + 1, \beta + 1, \gamma + 1, p)$

Note that, directly from the definition

$$p(1 - p)F(\alpha, \beta, \gamma, p) = 0 \qquad \text{when } p = 0$$

We recall that, in R_i, our function of p is $p(1 - p)X(p)$, and that this must be zero when $p = 0$ to fulfil the boundary condition at $p = 0$. Thus, is we put $X(p) = Ag(p) = AF(\alpha, \beta, \gamma, p)$ this boundary condition is satisfied. Before considering the other boundary condition, we just note that, in our general solution for the hypergeometric equation,

$$X(p) = Ag(p) + Bh(p)$$

$h(p)$ turns out to be

$$\frac{(\alpha - 1)(\beta - 1)}{(\gamma - 1)} F(\alpha, \beta, \gamma, p) \log p + p^{-1}(1 + D_1 p + D_2 p^2 + D_3 p^3 + \cdots)$$

where the D's are constants. For the method of derivation, see almost any book on differential equations; we shall not give details here, since $h(p)$ is clearly irrelevant to our problem. For, recalling that

$$\lim_{p \to 0} p \log p = 0$$

we see that

$$\lim_{p \to 0} p(1 - p)h(p) = 1$$

in disagreement with our boundary condition. In order then that $X(p)$ fulfils this condition, we must put the constant B equal to zero and we are left with

$$X(p) = AF(\alpha, \beta, \gamma, p)$$

We consider now

$$p(1 - p)F(\alpha, \beta, \gamma, p)$$

when $p = 1$. We use (property (v) above)

$$F(\alpha, \beta, \gamma, p) = (1 - p)^{\gamma - \alpha - \beta} F(\gamma - \alpha, \gamma - \beta, \gamma, p)$$

In our case $\gamma = 2$, $\alpha + \beta = 3$, and so

$$p(1 - p)F(\alpha, \beta, \gamma, p) = pF(2 - \alpha, 2 - \beta, 2, p)$$

Note that $2 > (2 - \alpha) + (2 - \beta) = 1$. Thus, when $p = 1$ (see properties (iii) and (iv) above)

$$pF(2 - \alpha, 2 - \beta, 2, p) = \frac{\Gamma(2)\Gamma(1)}{\Gamma(\alpha)\Gamma(\beta)}$$

which, from standard properties of the gamma function, is zero only if α or β is zero or a negative integer; it will be recalled that our series terminates in that case.

Thus, to fulfil the boundary condition at $p = 1$, we must have α or β zero or a negative integer; since $F(\alpha, \beta, \gamma, p)$ is symmetrical in α, β is does not matter whether we take α or β here; we choose β.

To sum up this section:

$$X(p) = \text{constant} \times F(\alpha, \beta, \gamma, p)$$

where $\beta = 0$ or a negative integer, $\alpha + \beta = 3$, $\alpha\beta = 2 - 4N_e k_i$, $\gamma = 2$.

(6) *A fairly complete solution.* So far, we have considered properties that every R_i must have. There will be an infinitude of R_i's having these properties, each corresponding to a different value of $\beta = 0, -1, -2, -3, \ldots$. We write α_1, β_1 for the α, β appearing in R_1; α_2, β_2 for the α, β appearing in R_2; and so on. We let i take values $1, 2, 3, 4, \ldots$, whence $\beta_i = 1 - i$, $\alpha_i = 2 + i$ (since $\alpha_i + \beta_i = 3$ for all i). Then

$$F(\alpha_i, \beta_i, \gamma, p) = F(2 + i, 1 - i, 2, p)$$

Also, from

$$\alpha_i \beta_i = 2 - 4N_e k_i$$

$(2 + i)(1 - i) = 2 - 4N_e k_i \qquad 2 - i - i^2 = 2 - 4N_e k_i \qquad k_i = i(i + 1)/(4N_e)$

We recall that a constant multiplier appeared in both our $T(t)$ and $X(p)$; we merge these constants into a single constant, say C_i in R_i. Further, R_i contains a term, which we have not really considered so far, since it cancelled out earlier, which is a function of x only; call the version of this function, which appears in R_i, $S_i(x)$. Finally, merge C_i and $S_i(x)$ into a single quantity $D_i(x)$ say.

Putting all this together, we have

$$\phi(p, x, t) = \sum_{i=1}^{\infty} D_i(x) p(1 - p) F(2 + i, 1 - i, 2, p) e^{-i(i+1)t/(4N_e)}$$

Note that, while the sum over any subset of the i would accord with the backward equation and boundary conditions, we have summed over *all* our i.

DIFFUSION METHODS

To justify this, and also to find an expression for the $D_i(x)$, we appeal to the initial condition.

(7) *Fitting the initial condition.* At this point we introduce the Gegenbauer polynomials $T^1_{i-1}(Z)$ defined by

$$T^1_{i-1}(Z) = \frac{i(i+1)}{2} F\left(2+i, 1-i, 2, \frac{1-Z}{2}\right)$$

Let $Z = 1 - 2p$, so that $p = (1-Z)/2$ and $p(1-p) = (1-Z^2)/4$. Therefore

$$\phi(p, x, t) = \sum_{i=1}^{\infty} D_i(x)\tfrac{1}{4}(1-Z^2)\frac{2}{i(i+1)} T^1_{i-1}(Z)e^{-i(i+1)t/(4N_e)}$$

It will be convenient to replace the (dummy) variable i by j, giving

$$\phi(p, x, t) = \sum_{j=1}^{\infty} \tfrac{1}{2}D_j(x) \frac{1}{j(j+1)} (1-Z^2)T^1_{j-1}(Z)e^{-j(j+1)t/(4N_e)}$$

Naturally, we would like to put $t = 0$ and equate the resulting expression to Dirac's delta function, i.e.

$$\tfrac{1}{2}\sum_{j=1}^{\infty} D_j(x) \frac{1}{j(j+1)} (1-Z^2)T^1_{j-1}(Z) = \delta(p-x)$$

It is by no means obvious, however, that the delta function can be expressed in this way, so that it is not immediately clear that our last equation makes sense. In fact, all is well, but only because *every* $T^1_{j-1}(Z)$, from $j = 1$ to ∞, appears on the left-hand side. It is out of the question to drop even a single one of them. Thus our equation, which, as we show below, fixes the $D_j(x)$ uniquely, must be taken as it stands and no further conditions can be considered. We might hope that any conditions not considered so far have been taken care of automatically, as is indeed the case.

(8) *Final steps.* Multiply both sides of our last equation by $T^1_{i-1}(Z)$ and integrate with respect to Z from -1 to 1. By a standard property of the Gegenbauer polynomials

$$\int_{-1}^{1} (1-Z^2)T^1_{i-1}(Z)T^1_m(Z)\,dZ = 0 \qquad \text{if } (i-1) \neq m$$

$$= 2(i+1)i/(2i+1) \qquad \text{if } (i-1) = m$$

Using these 'orthogonality relations' we obtain, on the left-hand side,

$$\tfrac{1}{2}\sum_{j=1}^{\infty} D_j(x) \frac{1}{j(j+1)} \int_{-1}^{1} (1-Z^2) T^1_{i-1}(Z) T^1_{j-1}(Z) \, dZ$$

$$= \tfrac{1}{2} D_i(x) \frac{1}{i(i+1)} \frac{2(i+1)i}{2i+1} = D_i(x) \frac{1}{2i+1}$$

since the only non-zero term in the sum has $j = i$.

On the right-hand side, since $Z = 1 - 2p$,

$$\int_{-1}^{1} T^1_{i-1}(Z) \delta(p-x) \, dZ = \int_{1}^{0} \frac{i(i+1)}{2} F(2+i, 1-i, 2, p) \delta(p-x)(-2) \, dp$$

$$= \int_{0}^{1} i(i+1) F(2+i, 1-i, 2, p) \delta(p-x) \, dp$$

$$= i(i+1) F(2+i, 1-i, 2, x)$$

from the usual properties of the Dirac function. Therefore

$$D_i(x) = i(i+1)(2i+1) F(2+i, 1-i, 2, x)$$

and we have, at last,

$$\phi(p, x, t) = \sum_{i=1}^{\infty} p(1-p) i(i+1)(2i+1) F(2+i, 1-i, 2, p)$$

$$\times F(2+i, 1-i, 2, x) e^{-i(i+1)t/(4N_e)}$$

$$= 6p(1-p) e^{-t/(2N_e)} + 30p(1-p)(1-2p)(1-2x) e^{-3t/(2N_e)} + \cdots$$

(Kimura 1955a). For very large t, we have the close approximation

$$6p(1-p) e^{-t/(2N_e)}$$

which is independent of x. Thus, at a *very late* stage, all allele frequencies in the open interval are equally probable (Wright 1931–see our Ch. 3).

Our derivation of $\phi(p, x, t)$ was very lengthy. It will probably help the reader to grasp the argument if he or she attempts to derive our result from the forward equation; details closely resemble those appearing in our proof. Convenient boundary conditions are

$$\lim_{x \to 0} x(1-x) \phi(p, x, t) = 0 \qquad \lim_{x \to 1} x(1-x) \phi(p, x, t) = 0$$

DIFFUSION METHODS

Seek a solution of the form $\Sigma_i R_i$, where every R_i takes form

$$T(t)S(x)X(p)$$

It will emerge that

$$T(t) = \text{constant} \times e^{-k_i t}$$

$$S(x) = \text{constant} \times F(\alpha, \beta, \gamma, x)$$

with α, β, γ and k_i as above. To complete the analysis, use the initial condition

$$\phi(p, x, 0) = \delta(x - p)$$

Finding $v(p, t)$ and $u(p, t)$ in the neutral, no-mutation case

It follows directly from the definition of $F(\alpha, \beta, \gamma, x)$ that

$$\lim_{x \to 0} F(\alpha, \beta, \gamma, x) = 1$$

Thus

$$\frac{dv(p, t)}{dt} = \frac{1}{4N_e} \lim_{x \to 0} \phi(p, x, t)$$

$$= \frac{1}{4N_e} \sum_{i=1}^{\infty} p(1-p)i(i+1)(2i+1)F(2+i, 1-i, 2, p)e^{-i(i+1)t/(4N_e)}$$

whence, by elementary calculus,

$$v(p, t) = -\sum_{i=1}^{\infty} p(1-p)(2i+1)F(2+i, 1-i, 2, p)e^{-i(i+1)t/(4N_e)} + \text{constant}$$

The constant equals $\lim_{t \to \infty} v(p, t)$, the probability of ultimate loss, which is $(1-p)$, since the sum tends to 0 as $t \to \infty$. Thus

$$v(p, t) = (1-p) - \sum_{i=1}^{\infty} p(1-p)(2i+1)F(2+i, 1-i, 2, p)e^{-i(i+1)t/(4N_e)}$$

The constant can also be found, with more difficulty, using the initial condition $v(p, 0) = 0$ rather than our terminal condition.

Before considering $u(p, t)$ it will be convenient to introduce the *Legendre polynomials*. By *Murphy's formula*, $P_i(Z)$, the Legendre polynomial of order i, equals

$$F\left(-i, i+1, 1, \frac{1-Z}{2}\right)$$

It will be convenient (although non-standard!) to regard this as the definition of $P_i(Z)$. We shall, when necessary, quote standard properties of the Legendre polynomials without proof. For example

$$P_i(-Z) = (-1)^i P_i(Z)$$

and thus

$$F\left(-i, i+1, 1, \frac{1+Z}{2}\right) = (-1)^i F\left(-i, i+1, 1, \frac{1-Z}{2}\right)$$

Let $p = (1-Z)/2$, $1-p = (1+Z)/2$. So

$$F(-i, i+1, 1, 1-p) = (-1)^i F(-i, i+1, 1, p)$$

Differentiating both sides with respect to p and noting

$$\frac{d}{dp} F(\alpha, \beta, \gamma, p) = \frac{\alpha\beta}{\gamma} F(\alpha+1, \beta+1, \gamma+1, p)$$

we find

$$-i(i+1)F(1-i, i+2, 2, 1-p)(-1) = (-1)^i(-i)(i+1)F(1-i, i+2, 2, p)$$

$$F(1-i, i+2, 2, 1-p) = (-1)^{i-1} F(1-i, i+2, 2, p)$$

or

$$F(1-i, i+2, 2, p) = (-1)^{i-1} F(1-i, i+2, 2, 1-p)$$

Hence

$$F(1-i, i+2, 2, 1) = (-1)^{i-1} F(1-i, i+2, 2, 0) = (-1)^{i-1}$$

Then

$$\frac{du(p,t)}{dt} = \frac{1}{4N_e} \lim_{x \to 1} \phi(p, x, t)$$

$$= \frac{1}{4N_e} \sum_{i=1}^{\infty} p(1-p)i(i+1)(2i+1)(-1)^{i-1}$$

$$\times F(2+i, 1-i, 2, p)e^{-i(i+1)t/(4N_e)}$$

Therefore

$$u(p,t) = p + \sum_{i=1}^{\infty} p(1-p)(2i+1)(-1)^i F(2+i, 1-i, 2, p)e^{-i(i+1)t/(4N_e)}$$

The probability that the population is polymorphic is

$$1 - u(p,t) - v(p,t) = \sum_{i=1}^{\infty} p(1-p)(2i+1)[1-(-1)^i]$$

$$\times F(2+i, 1-i, 2, p)e^{-i(i+1)t/(4N_e)}$$

Kimura rewrites these expressions in terms of the Legendre polynomials; this is convenient for computation. We have, from the definition of $P_i(Z)$,

$$P_0(Z) = 1 \qquad P_1(Z) = Z$$

and all other $P_i(Z)$ may be calculated successfully, for any given value of Z, from the standard formula

$$(i+1)P_{i+1}(Z) - (2i+1)ZP_i(Z) + iP_{i-1}(Z) = 0$$

It turns out that

$$u(p,t) = p + \sum_{i=1}^{\infty} \tfrac{1}{2}(-1)^i [P_{i-1}(r) - P_{i+1}(r)]e^{-i(i+1)t/(4N_e)}$$

$$v(p,t) = 1 - p - \sum_{i=1}^{\infty} \tfrac{1}{2}[P_{i-1}(r) - P_{i+1}(r)]e^{-i(i+1)t/(4N_e)}$$

and so

$$1 - u(p, t) - v(p, t) = \tfrac{1}{2}\sum_{i=1}^{\infty}[1 - (-1)^i][P_{i-1}(r) - P_{i+1}(r)]e^{-i(i+1)t/(4N_e)}$$

$$= \sum_{j=0}^{\infty}[P_{2j}(r) - P_{2j+2}(r)]e^{-(2j+1)(2j+2)t/(4N_e)}$$

where $r = 1 - 2p$.

We may note finally that $u(p, t)$, $v(p, t)$ could have been found directly from the backward equation, rather than indirectly through $\phi(p, x, t)$. Given, however, that $\phi(p, x, t)$, which of course is of great interest in its own right, has been found, the method we have used to derive $u(p, t)$, $v(p, t)$ is much the quickest. We shall, therefore, just indicate the argument for the direct approach.

Taking a clue from the terminal condition, we write

$$u(p, t) = p + \psi(p, t)$$

Then $\psi(p, t)$ obeys the backward equation; the boundary conditions $\psi(0, t) = 0$, $\psi(1, t) = 0$ give the correct boundary conditions for $u(p, t)$. From Chapter 3 $\psi(p, t)$ takes the form $\Sigma_i R_i$ where every R_i has form $AT(t)X(p)$, A being a constant. The procedure we used for $\phi(p, t)$ will supply $T(t)$ and $X(p)$ also in the present case. We find

$$u(p, t) = \frac{1-Z}{2} + \sum_{j=1}^{\infty}\tfrac{1}{2}A_j(1 - Z^2)\frac{1}{j(j+1)}T^1_{j-1}(Z)e^{-j(j+1)t/(4N_e)}$$

where $Z = 1 - 2p$. Putting $u(p, 0) = 0$, multiplying through by $T^1_{i-1}(Z)$, integrating with respect to Z and using the orthogonality relations, we obtain

$$2Ai/(2i + 1) = \int_{-1}^{1} T^1_{i-1}(Z)(Z - 1)\, dZ$$

$$= -2\int_{0}^{1} i(i + 1)pF(2 + i, 1 - i, 2, p)\, dp$$

$$= -2(-1)^{i-1} = 2(-1)^i$$

the integral here being a special case of an integral to be discussed below.

Voronka–Keller formulae

Unfortunately, none of the preceding formulae are computationally convenient when t is small, requiring, in that case, the calculation of a very large number of terms to secure numerical accuracy. However, Voronka & Keller (1975) have derived alternative formulae for $\phi(p, x, t)$, which are very accurate solutions of the forward equation when t is small and are very convenient for computation. From these formulae, they also obtained convenient expressions for $u(p, t)$ and $v(p, t)$. While their formulae look very different from Kimura's, numerically they are very close to the corresponding Kimura formulae when $t < 2N_e$. The two approaches, then, are complementary.

Voronka & Keller's very interesting derivation involves advanced mathematical techniques – Zauderer (1983) is helpful for understanding these – so we shall quote from their results without giving proofs. They obtain two expressions for $\phi(p, x, t)$, one, ϕ_L, appropriate except for x very close to unity, the other, ϕ_R, appropriate except for x very close to zero; ϕ_L and ϕ_R are numerically equivalent for most of the range of x.

In the neutral, no-mutation case, their results (after correction of a misprint in their formula (5.16)) come out as follows:

$$\phi_L = \left\{ \left(\frac{4N_e}{t}\right)^2 p^{1/4}(1-p)^{1/4} x_l^{3/4}(1-x)^{-3/4} \sum_{i=0}^{\infty} \left(\frac{16N_e^2 x x_l}{t^2}\right)^i \Big/ [i!(i+1)!] \right\}$$

$$\times e^{-(4N_e/t)\{\frac{1}{2}x^{1/2}R(p,0) + [\frac{1}{4}R(x,0) - \frac{1}{4}R(p,0)]^2\}}$$

where

$$R(Z, 0) = 2\cos^{-1}(1 - 2Z) \qquad x_l = \tfrac{1}{16}[R(p, 0)]^2$$

and

$$\phi_R = \left\{ \left(\frac{4N_e}{t}\right)^2 p^{1/4}(1-p)^{1/4}(1-x_r)^{3/4} x^{-3/4} \sum_{i=0}^{\infty} \right.$$

$$\left. \left(\frac{16N_e^2 (1-x)(1-x_r)}{t^2}\right)^i \Big/ [i!(i+1)!] \right\}$$

$$\times e^{-(4N_e/t)\{\frac{1}{2}(1-x)^{1/2}R(1,p) + [\frac{1}{4}R(x,0) - \frac{1}{4}R(p,0)]^2\}}$$

where

$$R(1, Z) = 2\cos^{-1}(2Z - 1) \qquad 1 - x_r = \tfrac{1}{16}[R(1, p)]^2$$

and $R(Z, 0)$ is as above.

As $x \to 0$, $(1-x) \to 1$, $R(x, 0) \to 0$

$$\phi_L \to \left(\frac{4N_e}{t}\right)^2 p^{1/4}(1-p)^{1/4} x_l^{3/4} e^{-4N_e x_l/t}$$

since, in the sum over i, only the term in $i = 0$, which is 1 even when $x \to 0$, remains. Therefore

$$\frac{dv(p, t)}{dt} = \frac{1}{4N_e} \lim_{x \to 0} \phi(p, x, t)$$

$$= \frac{4N_e}{t^2} p^{1/4}(1-p)^{1/4} x_l^{3/4} e^{-4N_e x_l/t}$$

$$v(p, t) = p^{1/4}(1-p)^{1/4} x_l^{-1/4} e^{-4N_e x_l/t} + \text{constant}$$

where the constant must be taken as zero in order to fit the initial condition $v(p, 0) = 0$. We have then

$$v(p, t) = \left(\frac{p(1-p)}{C'}\right)^{1/4} e^{-4N_e C'/t}$$

where $C' = x_l = \frac{1}{4}[\cos^{-1}(1-2p)]^2$, as given, without proof, in Chapter 4. Similarly, as $x \to 1$, $R(x, 0) \to 2\pi$; and since $\pi - \cos^{-1} Z = \cos^{-1}(-Z)$

$$\phi_R \to \left(\frac{4N_e}{t}\right)^2 p^{1/4}(1-p)^{1/4}(1-x_r)^{3/4} e^{-4N_e(1-x_r)/t}$$

giving

$$u(p, t) = \left(\frac{p(1-p)}{C}\right)^{1/4} e^{-4N_e C/t}$$

where

$$C = 1 - x_r = \tfrac{1}{4}[\cos^{-1}(2p-1)]^2$$

Moments of the distribution

As usual, let μ'_n be the nth moment about zero of x,

$$\mu'_n = Ex^n$$

DIFFUSION METHODS

We write $\mu'_n(t)$ for the value of μ'_n in generation t. In generation 0, $x = p$ with probability 1, so that $\mu'_n(0) = p^n$.

We continue with the neutral, no-mutation case and suppose first that $N_e = N$. Then, given n small, exact moments can be found fairly easily. One way of doing this is to seek linear functions of the moments that either stay constant or go down by a constant factor every generation. If v_i is such a function, with constant factor λ_i,

$$v_i(t+1) = \lambda_i v_i(t)$$

$$v_i(t) = \lambda_i^t v_i(0)$$

We may write

$$v_1 = \mu'_1 = \text{E}x$$

which we know never alters, so

$$v_1(t) = v_1(0) = \mu'_1(0) = p$$

We write

$$v_2 = \mu'_1 - \mu'_2 = \text{E}x - \text{E}x^2 = \text{E}[x(1-x)]$$

which we know goes down by

$$1 - \frac{1}{2N}$$

every generation, so

$$v_2(t) = \left(1 - \frac{1}{2N}\right)^t v_2(0) = \left(1 - \frac{1}{2N}\right)^t (p - p^2)$$

With $v_2(t)$ and $v_1(t)$ thus known, $\mu'_2(t)$ is easily calculated as

$$\mu'_1(t) - v_2(t) = v_1(t) - v_2(t) = p - p(1-p)\left(1 - \frac{1}{2N}\right)^t$$

It may be shown that

$$v_3 = \mu'_1 - 3\mu'_2 + 2\mu'_3$$

goes down by

$$\left(1 - \frac{1}{2N}\right)\left(1 - \frac{2}{2N}\right)$$

every generation, whence $\mu'_3(t)$; and that

$$v_4 = \mu'_1 - \frac{12N - 7}{2N - 1}\mu'_2 + \frac{20N - 12}{2N - 1}\mu'_3 - \frac{10N - 6}{2N - 1}\mu'_4$$

or, if N is large, approximately

$$\mu'_1 - 6\mu'_2 + 10\mu'_3 - 5\mu'_4$$

goes down by

$$\left(1 - \frac{1}{2N}\right)\left(1 - \frac{2}{2N}\right)\left(1 - \frac{3}{2N}\right)$$

every generation, whence $\mu'_4(t)$. These properties of v_3 and v_4 follow from standard properties of the binomial distribution. The resulting formulae for $\mu'_3(t)$, $\mu'_4(t)$ are given in Crow & Kimura (1970, p. 335), the notation used there being $\mu_3^{(t)}$, $\mu_4^{(t)}$.

Unfortunately, this method rapidly becomes unmanageable once n rises above 4. There is, however, a continuous version of this method, by which an approximate formula can be found for any μ'_n; see note 5 for one version of this method (compare Maruyama 1977, Ch. 7).

It will, however, be easiest for us to find $\mu'_n(t)$ from

$$\mu'_n(t) = \int_0^1 x^n \phi(p, x, t) \, dx + 1^n u(p, t) + 0^n v(p, t)$$

$$= \sum_{i=1}^{\infty} p(1 - p)(2i + 1)F(2 + i, 1 - i, 2, p)$$

$$\times \left(\int_0^1 i(i + 1)x^n F(2 + i, 1 - i, 2, x) \, dx\right) e^{-i(i+1)t/(4N_e)}$$

$$+ p + \sum_{i=1}^{\infty} p(1 - p)(2i + 1)(-1)^i F(2 + i, 1 - i, 2, p) e^{-i(i+1)t/(4N_e)}$$

$$= p + \sum_{i=1}^{\infty} p(1 - p)(2i + 1)F(2 + i, 1 - i, 2, p)$$

$$\times \left((-1)^i + \int_0^1 i(i + 1)x^n F(2 + i, 1 - i, 2, x) \, dx\right) e^{-i(i+1)t/(4N_e)}$$

Using the result quoted earlier, i.e.

$$\frac{d}{dx} F(\alpha, \beta, \gamma, x) = \frac{\alpha\beta}{\gamma} F(\alpha + 1, \beta + 1, \gamma + 1, x)$$

we see that

$$\int F(2 + i, 1 - i, 2, x) \, dx = -\frac{1}{i(i+1)} F(1 + i, -i, 1, x)$$

Hence, using integration by parts,

$$\int_0^1 i(i+1) x^n F(2+i, 1-i, 2, x) \, dx$$

$$= [-x^n F(1+i, -i, 1, x)]_0^1 + n \int_0^1 x^{n-1} F(1+i, -i, 1, x) \, dx$$

The first term on the right equals

$$-F(1+i, -i, 1, 1) = -F(-i, i+1, 1, 1)$$
$$= -(-1)^i F(-i, i+1, 1, 0) = -(-1)^i$$

(using the relationships noted earlier). The second term on the right, on putting $x = (1 - Z)/2$, becomes

$$\frac{n}{2} \int_{-1}^1 \left(\frac{1-Z}{2}\right)^{n-1} F\left(1+i, -i, 1, \frac{1-Z}{2}\right) dZ = \frac{n}{2^n} \int_{-1}^1 (1-Z)^{n-1} P_i(Z) \, dZ$$

where $P_i(Z)$ is the Legendre polynominal of order i. Thus

$$\mu'_n(t) = p + \sum_{i=1}^\infty p(1-p)(2i+1) F(2+i, 1-i, 2, p) \frac{n}{2^n}$$

$$\times \left(\int_{-1}^1 (1-Z)^{n-1} P_i(Z) \, dz \right) e^{-i(i+1)t/(4N_e)}$$

since the terms in $(-1)^i$ cancel. From the standard properties of the Legendre polynomials (see note 6)

$$\frac{n}{2^n} \int_{-1}^1 (1-Z)^{n-1} P_i(Z) \, dZ = 0 \qquad \text{when } i \geq n$$

$$= (-1)^i \frac{(n-1)(n-2)(n-3) \cdots (n-i)}{(n+1)(n+2)(n+3) \cdots (n+i)} \qquad \text{when } i < n$$

Thus

$$\mu'_n(t) = p + \sum_{i=1}^{n-1} p(1-p)(2i+1)(-1)^i F(2+i, 1-i, 2, p)$$

$$\times \frac{(n-1)(n-2)(n-3)\cdots(n-i)}{(n+1)(n+2)(n+3)\cdots(n+i)} e^{-i(i+1)t/(4N_e)}$$

For example,

$$\mu'_1(t) = p$$
$$\mu'_2(t) = p - p(1-p)e^{-t/(2N_e)}$$
$$\mu'_3(t) = p - \tfrac{3}{2}p(1-p)e^{-t/(2N_e)} + \tfrac{1}{2}p(1-p)(1-2p)e^{-3t/(2N_e)}$$
$$\mu'_4(t) = p - \tfrac{9}{5}p(1-p)e^{-t/(2N_e)} + p(1-p)(1-2p)e^{-3t/(2N_e)}$$
$$- \tfrac{1}{5}p(1-p)(1-5p+5p^2)e^{-6t/(2N_e)}$$

In practice, μ_n, the moments about the mean,

$$\mu_n = E(x - Ex)^n$$

are more useful, when $n \geqslant 2$, than are our μ'_n, the moments about zero. When n is small, however, it is easy to convert from one to the other. Thus, by a standard result,

$$\mu_2 = \mu'_2 - \mu'^2_1$$
$$\mu_3 = \mu'_3 - 3\mu'_1\mu'_2 + 2\mu'^3_1$$
$$\mu_4 = \mu'_4 - 4\mu'_1\mu'_3 + 6\mu'^2_1\mu'_2 - 3\mu'^4_1$$

Sometimes one requires the moments of the distribution of allele numbers rather than of allele frequencies. These are given by

$$\mu_n(\text{allele number}) = (2N)^n \mu_n(\text{allele frequency})$$

Fisher's view on the reliability of diffusion methods

Fisher (1930b) considered that diffusion methods are appropriate for intermediate allele frequencies only and that these methods can give very misleading answers when alelle frequencies are extreme. Fisher's view has been contested and there can, indeed, be no doubt that he overstated the situation. Nevertheless, Fisher was not entirely mistaken on this point, as we now show.

Suppose $N_e = N$. Consider the approximate (i.e. diffusion) distribution of the frequency of A, conditional on that frequency taking given value p one generation earlier. This conditional distribution, which we introduced very early in our discussions, lies at the heart of many of our calculations for the neutral, no-mutation case; its moments about zero, μ'_n, are obtained by putting $t = 1$ in our formulae above. The corresponding moments about the mean, μ_n, for $n = 2$ to 4 inclusive, are then easily found. Finally, to facilitate comparison with the exact conditional distribution (which, given $N_e = N$, is of course binomial), we convert μ'_1, μ_2, μ_3, μ_4 into the corresponding moments, say M'_1, M_2, M_3, M_4, for number, rather than frequency, of A. For example, the mean number M'_1 of A alleles comes out as

$$M'_1 = 2Np$$

and the variance M_2 of the number of A alleles is (with $q = 1 - p$)

$$\begin{aligned} M_2 &= (2N)^2 \mu_2 = (2N)^2(\mu'_2 - \mu'^2_1) \\ &= (2N)^2(p - pqe^{-1/(2N)} - p^2) \\ &= (2N)^2 pq(1 - e^{-1/(2N)}) \\ &= (2N)^2 pq\left[\frac{1}{2N} - \frac{1}{2}\left(\frac{1}{2N}\right)^2 + \frac{1}{6}\left(\frac{1}{2N}\right)^3 - \cdots\right] \end{aligned}$$

which, when N is at all large, will be very close to $2Npq$. Thus our approximate distribution agrees with the binomial distribution in mean and variance; of course, the approximate distribution was designed to agree in this way.

On the other hand, we find that, to a close approximation when N is at all large,

$$M_3 = \tfrac{3}{2} 2Npq(q - p)$$

$$M_4 = 3(2N)^2 p^2 q^2 + 2Npq(3 - 19pq)$$

whereas for the binomial the corresponding moments are $2Npq(q - p)$ and $3(2N)^2 p^2 q^2 + 2Npq(1 - 6pq)$ respectively.

However, *when p is intermediate*, these discrepancies do not matter. For, as N becomes large, the coefficient of skewness $M_3/(M_2)^{3/2}$ and coefficient of kurtosis $[M_4/(M_2)^2] - 3$ both tend to zero. Thus our approximate distribution should, when p is intermediate and N large, be close to the normal distribution, as indeed follows at once from the work of Voronka & Keller (1975, Eqn 3.12). Thus approximate and exact distributions tend to the same limit as N becomes large.

When allele frequencies are extreme, the discrepancies are more serious. Thus, when $p = 1/(2N)$ and N is large, we may take q as equal to 1 and obtain the close approximations

$$M'_1 = 1 \qquad M_2 = 1 \qquad M_3 = 1.5 \qquad M_4 = 6$$

whereas the corresponding moments for the Poisson distribution are 1, 1, 1, 4 respectively. Thus our continuous approximation rather distorts the situation in this case.

To see the nature of this distortion, we consider the probability distribution corresponding to our moments 1, 1, 1.5, 6. This can be found from Voronka & Keller's ϕ_L given above. Put $t = 1$, $N_e = N$ large, $p = 1/(2N)$. Then p is small and we have close approximations

$$1 - p = 1 \qquad 2\cos^{-1}(1 - 2p) = 4\sqrt{p} \qquad x_I = p$$

Also, with p so small, the allele frequency x will still be small after one generation, giving approximations

$$1 - x = 1 \qquad 2\cos^{-1}(1 - 2x) = 4\sqrt{x}$$

In simplifying expressions, it is helpful to remember that $2Np = 1$. Let $s = 2Nx$, the number of A alleles. The probability that x, and therefore s, equals zero, i.e.

$$p^{1/4}(1-p)^{1/4}x_I^{-1/4}e^{-4Nx_I}$$

reduces to e^{-2}. For $s > 0$, s appears as a continuous variable; since ϕ_L is the probability density function for x, the probability density function for s is, by standard theory

$$\phi_L \frac{dx}{ds} = \frac{\phi_L}{2N}$$

which reduces to

$$4e^{-2(s+1)} \sum_{i=0}^{\infty} \frac{(4s)^i}{i!(i+1)!}$$

DIFFUSION METHODS

No difficulties arise through the fact that the distribution of s is part discrete, part continuous; we handle this distribution in accordance with informed common sense. For example, we naturally expect that

$$e^{-2} + \int_0^\infty 4e^{-2(s+1)} \sum_{i=0}^\infty \frac{(4s)^i}{i!(i+1)!} ds = 1$$

and may be verified by term-by-term integration, using integration by parts. The mean of the distribution is

$$0 \times e^{-2} + \int_0^\infty s 4e^{-2(s+1)} \sum_{i=0}^\infty \frac{(4s)^i}{i!(i+1)!} ds$$

which comes out as 1, and similarly we find that the second, third and fourth moments about the mean are 1, 1.5, 6 respectively, thus confirming the correctness of our procedure.

How well do the probabilities that $s = 0, 1, 2, 3, \ldots$ obtained from this distribution approximate the corresponding probabilities obtained from the Poisson distribution? As shown above, the probability that $s = 0$ comes out as e^{-2}, whereas, N being large, it should be e^{-1}. For $s > 0$, we ignore the term e^{-2} and find the probability that s alleles are present by integrating between limits $s - \frac{1}{2}$, $s + \frac{1}{2}$, since s increases in jumps of 1. Results are given in Table 10.1, where the integral between limits 0 and $\frac{1}{2}$ is also included.

Although our continuous approximation distorts all probabilities to some extent, the principal distortion is the probability of about 0.26 for the

Table 10.1 Probability distribution of number of A alleles given one A present one generation earlier.

Value or range	Corresponding number of A	Probability from diffusion methods	Probability from Poisson distribution
0	0	0.1353	0.3679
0.0–0.5	–	0.2590	–
0.5–1.5	1	0.3587	0.3679
1.5–2.5	2	0.1609	0.1839
2.5–3.5	3	0.0590	0.0613
3.5–4.5	4	0.0192	0.0153
4.5–5.5	5	0.0057	0.0031
5.5–6.5	6	0.0016	0.0005
6.5–7.5	7	0.0004	0.0001
7.5–8.5	8	0.0001	0.0000
Total		0.9999	1.0000

biologically meaningless interval 0 to 0.5 (which, in terms of allele frequency, is 0 to $1/(4N)$); this interval has been a continual problem to us. Clearly, much of the probability that should attach to frequency zero has become associated with this irrelevant interval, thus leading to a marked underestimate of the flux of probability into frequency zero over a single generation.

This underestimation will interact with other distortions, noted in Chapter 6, produced by diffusion methods. Consider the mean sojourn time at the *initial* frequency. We noted that diffusion methods ignore the fact that, in the Wright–Fisher model, a generation must be spent at the initial frequency but not necessarily at any other specific frequency between 0 and 1. It is natural, therefore, to add 1 to the diffusion estimate ($= 2$ generations) of the mean sojourn time at the initial frequency, as suggested by Robertson (quoted in Pollak & Arnold 1975); this turns out to be correct provided the initial frequency is intermediate, as shown by Pollak & Arnold. For extreme frequencies, however, the situation is more complicated; the diffusion approach, in ignoring the extra generation noted above, gives a mean sojourn time that is too small, but in underestimating the flux into zero gives a mean sojourn time that is too large. There seems to be no way of predicting in advance how those opposing factors will balance out. We noted earlier Fisher's finding that, for initial frequency $1/(2N)$, the mean sojourn time is 2.24 generations; thus we have to add 0.24 to the diffusion value and generally the quantity to be added is less than 1 when initial frequencies are extreme. Thus, by a supreme irony, the distortion introduced by diffusion methods is *less* at extreme than at intermediate frequencies in this particular case.

Two-way mutation: no selection

We now consider a two-allele model, without selection. A mutates to a at rate u, a mutates at A at rate v. We have the backward equation

$$\frac{\partial \phi}{\partial t} = [v - (u+v)p] \frac{\partial \phi}{\partial p} + \frac{pq}{4N_e} \frac{\partial^2 \phi}{\partial p^2}$$

and boundary conditions

$$\lim_{p \to 0} \frac{\partial \phi}{\partial p} = \text{finite} \qquad \lim_{p \to 1} \frac{\partial \phi}{\partial p} = \text{finite}$$

The initial condition, as usual, is

$$\phi(p, x, 0) = \delta(p - x)$$

DIFFUSION METHODS

We seek a solution of form $\sum_i R_i$, where every R_i has form

$$T(t)X(p)S(x)$$

Substitution in the backward equation leads to

$$T(t) = \text{constant} \times e^{-k_i t}$$

$$p(1-p)X''(p) + (B - Ap)X'(p) + \Lambda X(p) = 0$$

where

$$A = 4N_e(u+v) \qquad B = 4N_e v \qquad \Lambda = 4N_e k_i$$

which is the hypergeometric equation (see earlier), with $\gamma = B$, $\alpha + \beta + 1 = A$, $\alpha\beta = -\Lambda$ (however, in the present case, γ will very rarely be an integer). This equation has solutions

$$F(\alpha, \beta, \gamma, p) \qquad \text{and} \qquad p^{1-\gamma} F(\alpha - \gamma + 1, \beta - \gamma + 1, 2 - \gamma, p)$$

unless γ is an integer, in which case the first solution stands but the second solution takes a different form.

Since

$$\frac{d}{dp} F(\alpha, \beta, \gamma, p) = \frac{\alpha\beta}{\gamma} F(\alpha+1, \beta+1, \gamma+1, p)$$

which remains finite as $p \to 0$, $F(\alpha, \beta, \gamma, p)$ obeys the boundary condition for $p \to 0$. On the other hand, the second solution above, when differentiated with respect to p, gives an expression that, when p is very small, is dominated numerically by a term $1/p^\gamma$ and therefore tends to $\pm\infty$ as $p \to 0$. Thus the second solution is irrelevant (as also holds when γ is an integer) and we need consider only

$$\text{constant} \times F(\alpha, \beta, \gamma, p)$$

If α or β is zero, $F(\alpha, \beta, \gamma, p)$ is unity for all p, so that its derivative with respect to p is always zero and thus remains zero (i.e. finite) as $p \to 1$. We consider, then,

$$\lim_{p \to 1} \frac{\alpha\beta}{\gamma} F(\alpha+1, \beta+1, \gamma+1, p)$$

TWO-WAY MUTATION: NO SELECTION

Now

$$(\gamma + 1) - (\alpha + 1) - (\beta + 1) = B - A = -4N_e u$$

is negative, so that $F(\alpha + 1, \beta + 1, \gamma + 1, p)$, when an infinite series, does not converge when $p = 1$. By *Abel's theorem*, it follows that $\lim_{p \to 1} F(\alpha + 1, \beta + 1, \gamma + 1, p)$ is infinite in that case. Thus, for the boundary condition for $p \to 1$ to be satisfied, $F(\alpha + 1, \beta + 1, \gamma + 1, p)$ must be a terminated series, which will be the case if $(\alpha + 1)$ or $(\beta + 1)$ is zero or a negative integer. Combining this result with that given above for the case α or $\beta = 0$, we see that the solution of our hypergeometric equation obeying the boundary conditions is

$$\text{constant} \times F(\alpha, \beta, \gamma, p)$$

where $\gamma = B$, $\alpha + \beta + 1 = A$, $\alpha\beta = -4N_e k_i$ and α or β is zero or a negative integer. As noted earlier, we can choose α or β here; we choose β.

Thus possible values for β are $0, -1, -2, -3, \ldots$. On putting $i = 0, 1, 2, 3, \ldots$, we have

$$\beta_i = -i \qquad \alpha_i = A - \beta_i - 1 = A - 1 + i \qquad 4N_e k_i = -\alpha_i \beta_i = i(A + i - 1)$$

and so

$$k_i = i\left(u + v + \frac{i-1}{4N_e}\right)$$

Therefore

$$T(t) = \text{constant} \times e^{-i[u+v+(i-1)/(4N_e)]t}$$

and

$$X(p) = \text{constant} \times F(A - 1 + i, -i, B, p)$$
$$= \text{constant} \times F(-i, A - 1 + i, B, p)$$

a *Jacobi polynomial*, $J_i(a, b, p)$ say, with $a = A - 1$, $b = B$, where, as a matter of definition,

$$J_i(a, b, p) = F(-i, a + i, b, p)$$

We have, then

$$\phi(p, x, t) = \sum_{i=0}^{\infty} S_i(x) F(A - 1 + i, -i, B, p) e^{-i[u+v+(i-1)/(4N_e)]t}$$

To evaluate $S_i(x)$ we put the initial condition in the form

$$\phi(p, x, 0) = \sum_{j=0}^{\infty} S_j(x) J_j(a, b, p) = \delta(p - x)$$

multiply both sides by

$$p^{b-1}(1-p)^{a-b} J_i(a, b, p)$$

and integrate with respect to p from 0 to 1. Now, it can be shown that

$$\int_0^1 p^{b-1}(1-p)^{a-b} J_i(a, b, p) J_j(a, b, p) = 0 \qquad \text{when } i \neq j$$

Thus

$$S_i(x) \int_0^1 p^{b-1}(1-p)^{a-b} [J_i(a, b, p)]^2 \, dp$$

$$= \int_0^1 p^{b-1}(1-p)^{a-b} J_i(a, b, p) \delta(p - x) \, dp$$

$$= x^{b-1}(1-x)^{a-b} J_i(a, b, x)$$

$$= x^{B-1}(1-x)^{A-B-1} F(A - 1 + i, -i, B, x)$$

It may also be shown that

$$\int_0^1 p^{b-1}(1-p)^{a-b} [J_i(a, b, p)]^2 \, dp = \frac{[\Gamma(B)]^2 \Gamma(A - B + i) i! \Gamma(A + 2i - 1)}{\Gamma(A - 1 + i) \Gamma(B + i) \Gamma(A + 2i)}$$

So

$$S_i(x) = \frac{x^{B-1}(1-x)^{A-B-1} F(A - 1 + i, -i, B, x) \Gamma(A - 1 + i) \Gamma(B + i) \Gamma(A + 2i)}{i! [\Gamma(B)]^2 \Gamma(A - B + i) \Gamma(A + 2i - 1)}$$

NOTES AND EXERCISES

This result is due to Crow & Kimura (1956), who obtained it in a slightly different form, which we can recover by using

$$F(A - 1 + i, -i, B, Z) = \frac{\Gamma(A - B + i)\Gamma(B)}{\Gamma(A - B)\Gamma(B + i)}$$
$$\times (-1)^i F(A - 1 + i, -i, A - B, 1 - Z)$$

to give

$$\phi(p, x, t) = \sum_{i=0}^{\infty} x^{B-1}(1 - x)^{A-B-1} F(A - 1 + i, -i, A - B, 1 - x)$$
$$\times F(A - 1 + i, -i, A - B, 1 - p)$$
$$\times \frac{\Gamma(A - B + i)\Gamma(A + 2i)\Gamma(A + i - 1)}{i![\Gamma(A - B)]^2 \Gamma(B + i)\Gamma(A + 2i - 1)} e^{-i[u+v+(i-1)/(4N_e)]t}$$

Returning to our earlier form, we note that, as $t \to \infty$, all terms vanish except those for which $i = 0$, leaving

$$\phi = x^{B-1}(1 - x)^{A-B-1} F(A - 1, 0, B, x) F(A - 1, 0, B, p) e^0$$
$$\times \frac{\Gamma(A - 1)\Gamma(B)\Gamma(A)}{0!\Gamma(B)\Gamma(B)\Gamma(A - B)\Gamma(A - 1)}$$

which, on recalling that $A = 4N_e(u + v)$, $B = 4N_e v = V$, say, $A - B = 4N_e u = U$, say, and that $F(\alpha, \beta, \gamma, Z) = 1$ when $\beta = 0$, reduces to

$$\frac{\Gamma(U + V)}{\Gamma(U)\Gamma(V)} (1 - x)^{U-1} x^{V-1}$$

our stationary distribution of Chapter 9.

Selection

As already stated, to find $\phi(p, x, t)$ when selection is acting is a very formidable problem, well beyond the scope of this book. An excellent account is given in Crow & Kimura (1970, pp. 396–414).

Notes and exercises

1 *Direct derivation of the forward equation.* Let $h(s|x - s)$ be the probability that the allele frequency increases by amount s in the time interval t to

$t + \delta t$, conditional on allele frequency $(x - s)$ at time t (s takes sign). Then, in our usual notation

$$f(p, x, t + \delta t) = \sum_s f(p, x - s, t)h(s|x - s)$$

We write, as usual, $\phi(p, x, t)$ for the continuous analogue of $f(p, x, t)$, with the usual approximation

$$f(p, x, t) = \phi(p, x, t)\delta x$$

where $\delta x = 1/(2N)$. Similarly, we write $g(s|x - s)$ for the continuous analogue of $h(s|x - s)$, with approximation

$$h(s|x - s) = g(s|x - s)\delta s$$

where $\delta s = 1/(2N)$. We naturally expect that these approximations will become increasingly accurate as N becomes large. Suppressing, for the moment, any anxieties we might feel over possible distortions introduced by the use of continuous approximations, we have

$$\phi(p, x, t + \delta t) = \int_s \phi(p, x - s, t)g(s|x - s)\,ds$$

appropriate to the case N large. Writing ϕ as short for $\phi(p, x, t)$ and g as short for $g(s|x)$, we have

$$\phi(p, x - s, t)g(s|x - s) = \phi g - s\frac{\partial}{\partial x}(\phi g) + \frac{s^2}{2!}\frac{\partial^2}{\partial x^2}(\phi g) - \cdots$$

Now, since g is a probability density function,

$$\int_s g\,ds = 1 \qquad \int_s sg\,ds = Es \qquad \int_s s^2 g\,ds = Es^2$$

Es and Es^2 being functions of x. Given slow evolution, $\int_s s^r g\,ds$ is negligible when $r > 2$. Thus (assuming that the order in which we carry out differentiation and integration may be inverted)

$$\phi(p, x, t + \delta t) = \phi(p, x, t) - \frac{\partial}{\partial x}[(Es)\phi] + \tfrac{1}{2}\frac{\partial^2}{\partial x^2}[(Es^2)\phi]$$

i.e.

$$\frac{\phi(p, x, t + \delta t) - \phi(p, x, t)}{\delta t} = -\frac{\partial}{\partial x}\left(\frac{Es}{\delta t}\phi\right) + \tfrac{1}{2}\frac{\partial^2}{\partial x^2}\left(\frac{Es^2}{\delta t}\phi\right)$$

On defining, as in Chapter 8,

$$M = \lim_{\delta t \to 0} \frac{Es}{\delta t} \qquad V = \lim_{\delta t \to 0} \frac{Es^2}{\delta t}$$

and letting $\delta t \to 0$, we have the forward equation

$$\frac{\partial \phi}{\partial t} = -\frac{\partial}{\partial x}(M\phi) + \tfrac{1}{2}\frac{\partial^2}{\partial x^2}(V\phi)$$

where M and V are written in terms of x.

2 The relationship

$$T(p, x) = \int_0^\infty \phi(p, x, t)\, dt$$

where, as usual, $T(p, x)/(2N)$ approximates the mean sojourn time at frequency x, given initial frequency p, leads to an interesting relationship between mean sojourn times and probability of ultimate fixation. Writing, as usual, $u(p, t)$ for the probability that fixation has occurred by the time we reach generation t, we have (Ch. 8)

$$\frac{du(p, t)}{dt} = \frac{1}{4N_e} \lim_{x \to 1} \phi(p, x, t)$$

Now $u(p, 0) = 0$. Thus the probability of ultimate fixation $u(p, \infty)$ is

$$u(p, \infty) - u(p, 0) = \int_0^\infty \frac{1}{4N_e} \lim_{x \to 1} \phi(p, x, t)\, dt = \frac{1}{4N_e} \lim_{x \to 1} T(p, x)$$

Although Fisher gave no formal proof of this result, he used it to calculate the probability of ultimate fixation of an advantageous allele under genic selection for the special case $p = 1/(2N)$, $N_e = N$ (Fisher 1930a, b).

Similarly, the probability of ultimate loss is

$$\frac{1}{4N_e} \lim_{x \to 0} T(p, x)$$

Clearly, $T(p, x)$ remains finite as $x \to 0$ or 1; this will be helpful below.

3 If we take the forward equation and integrate both sides with respect to t from $t = 0$ to $t = \infty$ we have, in cases where $x = 0$ and $x = 1$ are exit boundaries,

$$-\frac{\partial}{\partial x}[MT(p, x)] + \tfrac{1}{2}\frac{\partial^2}{\partial x^2}[VT(p, x)] = [\phi(p, x, t)]_{t=0}^{t=\infty} = -\delta(x - p)$$

where $\delta(x - p)$ is the Dirac function. To obtain $T(p, x)$ we shall need two additional conditions. We have, in fact, the boundary conditions $T(p, x)$ finite when $x \to 0$ or 1; it follows that

$$\lim_{x \to 0}[x(1 - x)T(p, x)] = 0 \qquad \lim_{x \to 1}[x(1 - x)T(p, x)] = 0$$

which boundary conditions are more convenient in practice. Thus $T(p, x)$ can be found from the forward just as well as from the backward equation.

Fisher (1930a, b) considered the case $N_e = N$, $p = 1/(2N)$. For $x > p$, we have

$$-\frac{\partial}{\partial x}[MT(p, x)] + \tfrac{1}{2}\frac{\partial^2}{\partial x^2}[VT(p, x)] = 0$$

as given, essentially, by Fisher, whose treatment was exceptionally obscure here; see Ewens (1963a) for an elucidation, using a different approach from ours. Fisher appreciated that his solution of this equation would tend to ∞ as $x \to 0$, but mistakenly attributed this to a supposed comprehensive failure of the diffusion approach when x is very small, rather than to a change in mathematical form of $T(p, x)$ when $x < p$. Thus, while he could still use the boundary condition $T(p, x)$ finite as $x \to 1$, the corresponding condition for $x \to 0$ was not available to him. For the neutral case, he abandoned diffusion methods in favour of the branching process approach we discussed in Chapter 7. For genic selection, he noted that, when x is close to his p, $T(p, x)$ will be much the same as the known value in the neutral case. This supplies the missing condition. Thus for genic selection, $p = 1/(2N)$, $N_e = N$, he found $T(p, x)$ for $x \geq p$.

4 In the neutral case, the non-unit latent roots λ_i are approximated, in the diffusion approach, by e^{-k_i}, where the k_i take values $i(i + 1)/(4N_e)$ ($i = 1, 2, \ldots$). These values of the k_i, determined by the boundary conditions, are known as *eigenvalues*. It is rather confusing that, in modern terminology, the older term 'latent root' has been replaced by 'eigenvalue'.

5 *Direct method for finding moments μ'_i (neutral, no-mutation case).* Let

$$e_n = \sum_{i=1}^{n} a_i \mu'_i \qquad (n = 1, 2, 3, \ldots)$$

where $a_1 = 1$ and the other a_i's are such that

$$de_n/dt = -k_n e_n$$

where k_n is a constant. Writing $e_n(t)$ for the value of e_n in generation t, we have

$$e_n(t) = e_n(0) e^{-k_n t}$$

From the Kimura–Ohta equation, we know that

$$\frac{d\mu'_n}{dt} = \frac{n(n-1)}{4N_e} (\mu'_{n-1} - \mu'_n)$$

Then

$$\frac{de_n}{dt} = \sum_{i=1}^{n} a_i \frac{d\mu'_i}{dt} = \sum_{i=1}^{n} a_i \frac{i(i-1)}{4N_e} (\mu'_{i-1} - \mu'_i)$$

so that

$$\sum_{i=1}^{n} a_i \frac{i(i-1)}{4N_e} (\mu'_{i-1} - \mu'_i) = -k_n \sum_{i=1}^{n} a_i \mu'_i$$

Equating coefficients of μ'_n gives

$$k_n = (n-1)n/(4N_e)$$

and on equating coefficients for the remaining μ'_i individually, we find

$$\frac{a_{i+1}}{a_i} = \frac{(i-n)(i+n-1)}{i(i+1)} \qquad \text{(with } a_1 = 1\text{)}$$

Thus $e_1 = \mu'_1$, $e_2 = \mu'_1 - \mu'_2$, $e_3 = \mu'_1 - 3\mu'_2 + 2\mu'_3$ and so on. Using the initial condition

$$\mu'_n(0) = p^n$$

we find

$$e_1(0) = p \qquad e_2(0) = p - p^2 \qquad e_3(0) = p - 3p^2 + 2p^3$$

and generally

$$e_n(0) = pF(1 - n, n, 2, p)$$

where F is Gauss's hypergeometric function. The $e_n(t)$ then follow at once from

$$e_n(t) = e_n(0)e^{-(n-1)nt/(4N_e)}$$

Let $f_n = e_{n+1}$, where n now takes values 0, 1, 2, 3, ... Then

$$f_0(t) = f_0(0) = p \qquad f_n(t) = f_n(0)e^{-n(n+1)t/(4N_e)}$$

where

$$f_n(0) = e_{n+1}(0) = pF(-n, n+1, 2, p)$$

and on using

$$F(\alpha, \beta, \gamma, p) = (1-p)^{\gamma-\alpha-\beta}F(\gamma-\alpha, \gamma-\beta, \gamma, p)$$

we have

$$f_n(0) = p(1-p)F(2+n, 1-n, 2, p)$$

$$f_n(t) = p(1-p)F(2+n, 1-n, 2, p)e^{-n(n+1)t/(4N_e)}$$

We now write

$$\mu'_n(t) = p + \sum_{i=1}^{n-1} b_i f_i(t) = p + \sum_{i=1}^{n-1} b_i p(1-p)F(2+i, 1-i, 2, p)e^{-i(i+1)t/(4N_e)}$$

where the b_i's are constants; the first term, p, follows from the terminal condition: $\lim_{t \to \infty} \mu'_n(t) = p$ (for all n).

We still have to find the b_i. Putting $t = 0$ and recalling the initial condition $\mu'_n(0) = p^n$, we have

$$\sum_{i=1}^{n-1} b_i p(1-p)F(2+i, 1-i, 2, p) = p^n - p$$

On putting $Z = 1 - 2p$, rewriting in terms of the Gegenbauer polynomials, replacing i by j, multiplying through by $T^1_{i-1}(Z)$, integrating with respect to Z from -1 to 1 and using the orthogonality properties we find

$$b_i = (2i + 1) \int_{-1}^{1} \left[\left(\frac{1-Z}{2}\right)^n - \left(\frac{1-Z}{2}\right) \right] T^1_{i-1}(Z) \, dZ$$

$$= (2i + 1) \int_0^1 i(i + 1)(p^n - p) F(2 + i, 1 - i, 2, p) \, dp$$

(see the main text for the method of integration).

6 Let $P_i(Z)$ be the Legendre polynomial of order i. To evaluate

$$J = \frac{n}{2^n} \int_{-1}^{1} (1 - Z)^{n-1} P_i(Z) \, dZ$$

we use the standard result for any function $f(Z)$

$$\int_{-1}^{1} f(Z) P_i(Z) \, dZ = \frac{(-1)^i}{2^i i!} \int_{-1}^{1} (Z^2 - 1)^i f^{(i)}(Z) \, dZ$$

where $f^{(i)}(Z)$ is the ith derivative of $f(Z)$. By experimenting with a few values of i and n, the reader will easily discover that, when $f(Z) = (1 - Z)^{n-1}$,

$$f^{(i)}(Z) = 0 \qquad \qquad \text{if } i \geq n$$

$$= \frac{(n-1)!}{(n-i-1)!} (-1)^i (1 - Z)^{n-i-1} \qquad \text{if } i < n$$

Hence, if $i \geq n$, $J = 0$. If $i < n$

$$J = \frac{n!}{2^{n+i} i! (n - i - 1)!} \int_{-1}^{1} (Z^2 - 1)^i (1 - Z)^{n-i-1} \, dZ$$

Since $(Z^2 - 1) = (Z - 1)(Z + 1)$ the integrand may be written as

$$(Z - 1)^i (Z + 1)^i (1 - Z)^{n-i-1} = (-1)^i (1 - Z)^{n-1} (1 + Z)^i$$

On making the substitution $x = (1 - Z)/2$, we find

$$J = (-1)^i \frac{n!}{i!(n-i-1)!} \int_0^1 x^{n-1} (1 - x)^{(i+1)-1} \, dx$$

DIFFUSION METHODS

From the standard properties of the beta function, the integral is $(n-1)!i!/(n+i)!$, whence

$$J = (-1)^i \frac{n!(n-1)!}{(n-i-1)!(n+i)!}$$

$$= (-1)^i \frac{n(n-1)(n-2)\cdots(n-i)}{n(n+1)(n+2)\cdots(n+i)}$$

as may be seen by writing out the factorial terms in full.

7 Consider the case of irreversible mutation. Suppose a mutates to A at rate v, with no reverse mutation. Write x for the current, and p for the initial, frequency of A. The backward equation is

$$\frac{\partial \phi}{\partial t} = v(1-p)\frac{\partial \phi}{\partial p} + \frac{p(1-p)}{4N_e}\frac{\partial^2 \phi}{\partial p^2}$$

Seek a solution of the form

$$\phi = \Sigma(1-p)X(p)T(t)S(x)$$

the boundary conditions being

$$\phi(1, x, t) = 0 \qquad \partial\phi/\partial p \text{ finite at } p = 0$$

The initial condition is, as usual,

$$\phi(p, x, 0) = \delta(x-p) = \delta(p-x)$$

Verify the solution, given in Crow & Kimura (1970), who used the forward equation as their starting point,

$$\phi(p, x, t) = \sum_{i=0}^{\infty} \frac{(c+1+2i)\Gamma(c+1+i)\Gamma(c+i)}{i!(i+1)![(\Gamma(c)]^2}$$

$$\times (1-p)F(-i, i+c+1, c, p)x^{c-1}$$

$$\times F(-i, i+c+1, c, x)e^{-(i+1)(c+i)t/(4N_e)}$$

where $c = 4N_e v$ and $0 < x < 1$.

8 Consider Voronka & Keller's (1975) ϕ_L (in the neutral, no-mutation case) when $p = 1/(2N)$, N being large. Noting that

$$\cos^{-1}(1-2x) \geqslant 2\sqrt{x}$$

with near-equality when x is small (<0.01, say), prove that, when t is very large but still very much less than $2N_e$, ϕ_L is closely approximated by

$$2N\left(\frac{2}{t}\right)^2\left(\frac{N_e}{N}\right)^2 e^{-(2N_e/tN)s}$$

where $s = 2Nx$, the number of A alleles (for s extremely large, the approximation will be poor in the *relative* sense, but this is immaterial since both exact and approximate expressions will then be exceedingly small). Thus the distribution of the number of A alleles will be closely approximated by a continuous distribution with probability density function

$$\phi_L \frac{dx}{ds} = \left(\frac{2}{t}\right)^2\left(\frac{N_e}{N}\right)^2 e^{-(2N_e/tN)s}$$

except for $s = 0$, which has probability very close to

$$1 - (2N_e/tN)$$

Now consider the distribution of the number of A alleles, considering only cases where A is still present ($s > 0$). This will have probability density function

$$\left[\left(\frac{2}{t}\right)^2\left(\frac{N_e}{N}\right)^2 e^{-(2N_e/tN)s}\right] \div \left[1 - \left(1 - \frac{2N_e}{tN}\right)\right] = \frac{2N_e}{tN} e^{-(2N_e/tN)s}$$

Thus, considering only those neutral mutants with descendants still extant t generations after the mutant first arose (t very large, but $t \ll 2N_e$), the probability that the number of descendants of a single mutant present at that time exceeds a given number R equals

$$\int_R^\infty \frac{2N_e}{tN} e^{-(2N_e/tN)s} \, ds = \left[e^{-(2N_e/tN)s}\right]_{s=\infty}^{s=R} = e^{-(2N_e/tN)R}$$

as obtained for the case $N_e = N$ by Fisher (1930b), using branching process methods. Now Fisher believed that, in nature, N_e would be close to N. Hence his famous statement: 'in the absence of favourable selection, the number of individuals having a gene derived from a single mutation cannot greatly exceed the number of generations since its occurrence'. As he pointed out, the probability that s would exceed $3\frac{1}{2}t$ is e^{-7}, i.e. less than

1 in 1000. It is interesting that Fisher's conclusion was quoted in Ford's (1931) widely read *Mendelism and Evolution* and presumably reached many biologists through that source – see also Ford (1940).

Note that Fisher's conclusion is not appropriate when N_e is small compared to N. Thus if $N_e = 0.1N$, the probability that s exceeds $3\frac{1}{2}t$ is $e^{-0.7} = 0.4966$.

11
General comments and conclusions

> What Song the Syrens sang, or what name Achilles assumed when he hid himself among women, though puzzling Questions, are not beyond all conjecture.
>
> Sir Thomas Browne, *Urn Burial*

A note on methods

We comment first on analytical methods. For most problems, exact solutions have not been obtained. Further, it seems likely that, in most cases, exact solutions would be monstrously complicated and hence very difficult to interpret. Inevitably, then, approximate methods will usually be required. We have shown that all the methods that have been used can mislead the unwary.

For example, it is natural to suppose that, when the effective population size N_e is very large, the deterministic approach will be reliable in all cases; to suppose, in other words, that as $N_e \to \infty$, the true solution to any problem tends to the solution obtained deterministically. To test this assumption, consider the probability of fixation of an allele with initial frequency p_0 and selective advantage α under genic selection. Since we are supposing N_e very large, we may take $N_e\alpha$ large. Hence, to a close approximation when α is small, the probability of fixation (see Ch. 5) equals

$$1 - e^{-4N_e\alpha p_0}$$

For this to be close to the deterministic value, unity, we require $4N_e\alpha p_0$ to be large. Clearly, this condition cannot hold for *all* p_0, since p_0 could be as small as $1/(2N)$.

Generally (Maruyama 1977), for any given problem, the deterministic approach may well yield very inaccurate solutions for some range or ranges of allele frequency, even though N_e is very large; see, for example, our results for mean sojourn times (Ch. 7).

At first sight, this result is surprising. For, if we take the forward equation and put V, the variance of the change in allele frequency in a single generation, equal to zero from the start, we obtain an equation that represents a deterministic process (see Ch. 6, note 9). It is natural to guess, wrongly, that the case N_e very large, giving V very small, is almost equivalent to the case $V = 0$. However, to put the matter very informally, the corresponding equations look very different! Specifically, one equation in second-order (highest derivative appearing is a second derivative) whereas the other is first-order (highest derivative appearing is a first derivative). It is, therefore, unreasonable to assume, without further discussion, that these equations will yield essentially equivalent solutions. Precisely the same argument applies to the backward equation.

Unfortunately, a detailed discussion would be very technical, so we shall just give an outline of Maruyama's argument. We are treating the case V very small as a small disturbance, or *perturbation*, of the case $V = 0$. In that the order of the forward (or backward) equation drops as we pass from one case to the other, our perturbation is said to be *singular* in type. It follows from the theory of singular perturbations (see e.g. O'Malley 1974, Zauderer 1983) that, given an arbitrarily small V, we may find a range or ranges of allele frequency for which the deterministic solution is altogether unsatisfactory. This result does not, of course, imply an overall condemnation of the deterministic approach, which has often proved very productive. Rather, it indicates that results obtained deterministically cannot be regarded as secure unless some attempt is made to assess the relevance of drift for the problem under consideration, as we noted in Chapter 9 for a model with two stable deterministic equilibria.

We may note in passing that we can also use perturbation theory when comparing different stochastic models. Thus we can treat the case of very small selective advantage as a perturbation of the neutral case. Here, however, the order of the forward (or backward) equation is the same in either case. Our perturbation here is said to be *regular* in type; the complications noted above for singular perturbations do not arise. As we have found on several occasions, solutions in the case of an advantageous allele always tend to the corresponding solutions for a neutral allele when the selective advantage tends to zero.

Consider now the branching process approach. We have had to justify this case by case; sometimes this has meant a fairly lengthy preliminary discussion. In compensation, branching process methods, when appropriate, yield results of high accuracy. We have noted a crucial problem in genetic conservation where these accurate results are essential but not obtainable by diffusion methods. When calculating the probability of fixation of a newly arisen advantageous mutant, we noted (with Ewens 1969) that the branching process approach is best when the selective advantage is large, whereas the diffusion approach is best when the selective advantage is small. Clearly, no single method is optimal in all circumstances.

The attraction of diffusion methods lies in their remarkable flexibility. An obvious advantage is that, when using these methods, it is sufficient to specify the mean and variance of the change in allele frequency over a single generation, it being unnecessary to find the corresponding probability distribution in full, which would sometimes be very difficult. The happy circumstance that this variance often takes the form $x(1 - x)/$(a constant), conveniently written $x(1 - x)/(2N_e)$, where x is the allele frequency in the parental generation, means that a very wide range of biological situations can be covered, provided that we can calculate corresponding values of the effective population size N_e. It would probably be agreed that the introduction of the concept of N_e, followed by formulae for N_e in several highly relevant situations, is the greatest of Wright's many contributions to our subject.

As we have shown, it is fairly easy to spot situations where the diffusion approach might give inaccurate results. However, sometimes the results are very much better than one has any right to expect. This is particularly true for the probability of ultimate fixation under selection. Stationary distributions, taken in conjunction with the famous Wright-Kimura 'subterminal' formulae, come out surprisingly well and seem to be accurate enough for practical purposes. On the other hand, results for some problems (e.g. the probability that a new mutant will ever reach frequency x, where $0 < x < 1$) are less satisfactory. Thus the effort devoted to checking results obtained by diffusion methods is fully justified.

The forward and backward equations often appear difficult and mysterious to the beginner. It is hoped that our gradual approach to these equations will have helped. Of course, for *some* situations, we can obtain the probability distribution of allele frequency (or at least relative probabilities) without using these equations. It should not, however, be assumed that, in general, such alternative methods are easy; much ingenuity may be required to achieve a solution - see Wright (1931, 1952) and especially Wright & Kerr (1954).

It was fortunate that Fisher and Wright used different methods, since occasional discrepancies between their results led to mutual correction of errors arising through inappropriate approximations; see the excellent account in Provine (1986).

General summary and conclusions

Finally, we gather together our main conclusions. Clearly, the neutral theory cannot explain the evolution of a natural population unless N_e, the effective size of that population, is fairly small. The time required for a major change in allele frequency, arising from chance only, is of order N_e generations. Thus, if N_e were really large, very little change in genetic composition would occur, even on an evolutionary timescale. Note that this difficulty does not arise for evolution under natural selection.

We have not discussed the theory of population subdivision in full. However, our limited discussion has shown that, in some circumstances at least, the outcome is affected very little by such subdivision, even when migration between subpopulations is very restricted; given random mating in every subpopulation, the whole species behaves like one random mating population. In fact, this result holds in a wide range of circumstances. Maruyama has carried out an extensive investigation for the neutral case. He concludes (Maruyama 1972b):

> All likely two-dimensional populations behave like random-mating populations as far as the fate of neutral genes are concerned... many theories of population genetics which are based on the assumption of random mating can be applied to most geographically structured populations without involving serious errors.

Thus, for neutral alleles, N_e will usually be the effective size of the whole species concerned (probably the same applies for evolution under natural selection, except of course in cases where selective values vary from one subpopulation to another).

The neutralist must suppose, then, that N_e is usually very much smaller than the census size N. Obviously enough, the world N will often be exceedingly large (it is a sobering experience to make a rough count of the number of *Drosophila* alive in one's department!). Large differences in fecundity, frequent subpopulation extinction and replacement, and periodic drastic reductions in numbers would all help to make N_e small relative to N, but it is difficult to decide the extent to which such 'reducing' factors actually operate. Thus it is often argued that speciation will (usually) be associated with a marked reduction in number, but this argument could be mistaken; see e.g. Ridley (1985) for a discussion of this contentious issue.

Even with fairly small N_e (say 10^5 to 10^7), the probability of ultimate fixation of a newly arisen advantageous mutant will greatly exceed the corresponding probability for a neutral mutant, unless the selective advantage is very small. Hence advantageous alleles will contribute nearly all of the total of alleles ultimately fixed, unless such advantageous alleles appear only very infrequently. The neutralist, then, must assume that the overwhelming majority of non-harmful mutants are neutral, or almost neutral. Further, given such infrequent appearance of advantageous mutants, hitch-hiking effects (Maynard Smith & Haigh 1974, Thomson 1977) will only rarely affect,

GENERAL SUMMARY AND CONCLUSIONS

in any significant way, the rate of spread of neutral alleles in outbreeding species, and the standard neutralist scheme, which we discussed in Chapter 7, stands. We stress again that, on this scheme, a neutral mutant that is eventually fixed spends, on average, about the same time (approximately $2N_e/N$ generations) at every frequency.

Equilibrium distributions for neutral alleles are most likely to be attained when N_e is small. Given, then, our neutralist assumptions, we may tentatively suppose that these distributions are relevant to natural populations. As we indicated briefly in Chapter 9, results from natural populations accord reasonably well with this supposition, provided we allow for the fact that many mutations will be disadvantageous; see Kimura (1983) for further discussion. Of course, significant deviations from expectation, which are sometimes reported, could merely indicate absence of equilibrium and do not constitute evidence against the neutral theory as such!

We have stressed that, if the low values of N_e assumed by neutralists are correct, some traditional views on the operation of natural selection must be modified. Consider, for example, the spread, in a large population, of an advantageous allele, arising by mutation at rate u. The time required for such an allele to spread through the population will depend on (a) the time T_1 until at least one mutant appears that happens to survive chance extinction when rare and (b) the subsequent time T_2 for the allele to reach fixation. Traditionally, it was supposed that $T_1 \ll T_2$. This supposition is certainly justified if (following Fisher) we assume that $N_e u \geqslant 1$, but it is not, in general, justified if $N_e u$ is small. Thus with $N_e = 3.3 \times 10^6$, $u = 10^{-8}$ and genic selection with selective advantage $\alpha = 0.01$, on average T_1 will be about 758 generations and T_2 about 2474 generations. With the same N_e and u, but $\alpha = 0.001$, the corresponding times are 7580 and 20 130 generations.

As a further example, consider the long-term frequency of a recessive lethal, arising by mutation at rate v. Here the ever-popular deterministic 'balance between mutation and selection' argument gives very misleading results when $4N_e v < 1$, in which case the most probable long-term frequency is actually zero.

Now consider mean times for changes in allele frequency under (directional) natural selection. We have confined ourselves, for simplicity, to genic selection. Further, we considered a locus in isolation from others, ignoring possible effects of linkage and epistasis. Nevertheless, our results give a qualitative indication of events in more complicated cases of mild directional selection (with some exceptions in some cases of extreme epistasis). First, an advantageous allele will behave much like a neutral at extreme allele frequencies. Secondly, when N_e is large, deterministic theory will give mean times correctly for most of the range of intermediate allele frequency, unless selective advantages are very small indeed. Lying between the 'neutral' ranges and the 'deterministic' range will be ranges where drift, although not all-important, will still affect mean times. Thirdly, and most importantly, when N_e is at all large, mean times for most of the range of frequencies will be much smaller for an advantageous allele than for a neutral, again unless selective advantages are very small. It follows that mean fixation times under natural selection will, in general, be much smaller than mean fixation times under pure drift, the difference being particularly marked when N_e is very large. Combining this result with our results for probability of ultimate fixation, we see that (for most species at least), unless world N_e is less than world N by several orders of magnitude, natural selection will determine both the direction and speed of evolutionary change.

For most processes discussed, we have given alternative scenarios, with no final decision on which of these is typical. The reader may feel that he has merely acquired

> The two-edged tongue of mighty Zeno, who,
> Say what one would, could argue it untrue.

(see Plutarch, *Life of Pericles*). In the present author's opinion, such pessimism would be mistaken. One task for the theoretician is to argue ahead of his or her data, thus helping the experimentalist to decide which data are worth collecting. The stimulus that theory has repeatedly given to observational and experimental studies in population genetics is probably the strongest justification of that theory.

References

Abramowitz, M. & I. A. Stegun 1965. *Handbook of mathematical functions.* New York: Dover.

Aquadro, C. F., K. M. Lado & W. A. Noon 1988. The *rosy* region of *Drosophila melanogaster* and *Drosophila simulans.* I. Contrasting levels of naturally occurring DNA restriction map variation and divergence. *Genetics* **119**, 875-88.

Avery, P. J. 1977. The effect of random selection coefficients on populations of finite size – some particular models. *Genet. Res.* **29**, 97-112.

Bartlett, M. S. 1955. *An introduction to stochastic processes.* Cambridge: Cambridge University Press.

Bartlett, M. S. & J. B. S. Haldane 1934. The theory of inbreeding in autotetraploids. *J. Genet.* **29**, 175-80.

Birley, A. J. 1984. Restriction endonuclease map variation and gene activity in the ADH region in a population of *Drosophila melanogaster. Heredity* **52**, 103-12.

Brookfield, J. F. Y. 1986. The population biology of transposable elements. *Phil. Trans. R. Soc. B* **312**, 217-26.

Burke, T. & M. W. Bruford 1987. DNA fingerprinting in birds. *Nature* **327**, 149-52.

Burrows, P. M. & C. C. Cockerham 1974. Distributions of time to fixation of neutral genes. *Theor. Pop. Biol.* **5**, 192-207.

Carr, R. N. & R. F. Nassar 1970. Effects of selection and drift on the dynamics of finite populations. I. Ultimate probability of fixation of a favorable allele. *Biometrics* **26**, 41-9.

Charlesworth, B. 1985. The population genetics of transposable elements. In *Population genetics and molecular evolution,* T. Ohta & K. Aoki (eds), 213-32. Berlin: Springer.

Charlesworth, B. & D. Charlesworth 1983. The population dynamics of transposable elements. *Genet. Res.* **42**, 1-27.

Clarke, B. 1971. Natural selection and the evolution of proteins. *Nature* **232**, 487.

Cook, R. D. & D. L. Hartl 1975. Stochastic selection in large and small populations. *Theor. Pop. Biol.* **7**, 55-63.

Cox, D. R. & H. D. Miller 1965. *The theory of stochastic processes.* London: Methuen.

Cross, S. R. H. & A. J. Birley 1986. Restriction endonuclease map variation in the *Adh* region in populations of *Drosophila melanogaster. Biochem. Genet.* **24**, 415-33.

Crow, J. F. 1954. Breeding structure of populations. II. Effective population number. In *Statistics and mathematics in biology,* O. Kempthorne, T. A. Bancroft, J. W. Gowen & J. L. Lush (eds), 543-56. Ames, IA: Iowa State College Press.

Crow, J. F. & M. Kimura 1956. Some genetic problems in natural populations. In *Proc. third Berkeley symp. on mathematical statistics and probability,* Vol. 4, 1-22.

Crow, J. F. & M. Kimura 1970. *An introduction to population genetics theory.* New York: Harper & Row.

Crow, J. F. & N. E. Morton 1955. Measurement of gene frequency drift in small populations. *Evolution* **9**, 202-14.

REFERENCES

Ellingboe, A. H. 1982. Genetical aspects of active defence. In *Active defence mechanisms in plants*, R. K. S. Wood (ed.), 179-92. New York: Plenum.

Eriksson A. W., H. Forsius, H. R. Nevanlinna, P. L. Workman & R. K. Norio (eds) 1980. *Population structure and genetic disorders*. London: Academic Press.

Ewens, W. J. 1963a. The diffusion equation and a pseudo-distribution in genetics. *J. R. Stat. Soc. B* **25**, 405-12.

Ewens, W. J. 1963b. Numerical results and diffusion approximations in a genetic process. *Biometrika* **50**, 241-50.

Ewens, W. J. 1963c. The mean time for absorption in a process of genetic type. *J. Austr. Math. Soc.* **3**, 375-83.

Ewens, W. J. 1964a. The pseudo-transient distribution and its uses in genetics. *J. Appl. Prob.* **1**, 141-56.

Ewens, W. J. 1964b. The maintenance of alleles by mutation. *Genetics* **50**, 891-8.

Ewens, W. J. 1965. The adequacy of the diffusion approximation to certain distributions in genetics. *Biometrics* **21**, 386-94.

Ewens, W. J. 1967a. The probability of survival of a new mutant in a fluctuating environment. *Heredity* **22**, 438-43.

Ewens, W. J. 1967b. Random sampling and the rate of gene replacement. *Evolution* **21**, 657-63.

Ewens, W. J. 1967c. The probability of fixation of a mutant: the two locus case. *Evolution* **21**, 532-40.

Ewens, W. J. 1969. *Population genetics*. London: Methuen.

Ewens, W. J. 1972. The sampling theory of selectively neutral alleles. *Theor. Pop. Biol.* **3**, 87-112.

Ewens, W. J. 1973. Conditional diffusion processes in population genetics. *Theor. Pop. Biol.* **4**, 21-30.

Ewens, W. J. 1979. *Mathematical population genetics*. Berlin: Springer.

Ewens, W. J. 1982. On the concept of the effective population size. *Theor. Prop. Biol.* **21**, 373-8.

Ewens, W. J. 1983. The role of models in the analysis of molecular genetic data, with particular reference to restriction fragment data. In *Statistical analysis of DNA sequence data*, B. S. Weir (ed.), 45-73. New York: Marcel Dekker.

Ewens, W. J. & P. M. Ewens 1966. The maintenance of alleles by mutation - Monte Carlo results for normal and self-sterility populations. *Heredity* **21**, 371-8.

Ewens, W. J., R. S. Spielman & H. Harris 1981. Estimation of genetic variation at the DNA level from restriction endonuclease data. *Proc. Nat. Acad. Sci.* **78**, 3748-50.

Feller, W. 1951. Diffusion processes in genetics. In *Proc. second Berkeley symp. on mathematical statistics and probability*, J. Neyman (ed.) 227-46. Berkeley: University of California Press.

Feller, W. 1952. The parabolic differential equation and the associated semi-group of transformations. *Ann. Math.* **55**, 468-519.

Feller, W. 1954. Diffusion processes in one dimension. *Trans. Am. Math. Soc.* **77**, 1-31.

Feller, W. 1957. *An introduction to probability theory and its applications*, Vol. I, 2nd edn. New York: Wiley.

Fincham, J. R. S., P. R. Day & A. Radford 1979. *Fungal genetics*, 4th edn. Oxford: Blackwell.

Fisher, R. A. 1921. Review of A. L. Hagedoorn and A. C. Hagedoorn 'The Relative Value of the Processes Causing Evolution'. *Eug. Rev.* **13**, 467-70.

Fisher, R. A. 1922. On the dominance ratio. *Proc. R. Soc. Edin.* **42**, 321-41.

REFERENCES

Fisher, R. A. 1929. Letter to S. Wright. Reprinted in *Natural selection, heredity, and eugenics*, J. H. Bennett (ed.), 273. Oxford: Clarendon.

Fisher, R. A. 1930a. The distribution of gene ratios for rare mutations. *Proc. R. Soc. Edin.* **50**, 205-20.

Fisher, R. A. 1930b. *The genetical theory of natural selection*. Oxford: Clarendon (1958, 2nd edn. New York: Dover).

Fisher, R. A. 1949. *The theory of inbreeding*. Edinburgh: Oliver & Boyd.

Fisher, R. A. 1965. *The theory of inbreeding*, 2nd edn. Edinburgh: Oliver & Boyd.

Ford, E. B. 1931. *Mendelism and evolution*. London: Methuen.

Ford, E. B. 1940. Polymorphism and taxonomy. In *The new systematics*, J. S. Huxley (ed.), 493-513. Oxford: Oxford University Press.

Gale, J. S. 1980. *Population genetics*. Glasgow: Blackie.

Gale, J. S. 1987. Factors delaying the spread of a virulent mutant of a fungal pathogen: some suggestions from population genetics. In *Populations of plant pathogens: their dynamics and genetics*, M. S. Wolfe & C. E. Caten (eds), 55-62. Oxford: Blackwell.

Gale, J. S. & M. J. Lawrence 1984. The decay of variability. In *Crop genetic resources: conservation and evaluation*, J. H. W. Holden & J. T. Williams (eds), 77-101. London: George Allen & Unwin.

Gamow, G. 1966. *Thirty years that shook physics*. New York: Doubleday.

Gillespie, J. H. 1973. Natural selection with varying selection coefficients - a haploid model. *Genet. Res.* **21**, 115-20.

Haigh, J. & J. Maynard Smith 1972. Population size and protein variation in man. *Genet. Res.* **19**, 73-89.

Haldane, J. B. S. 1924a. A mathematical theory of natural and artificial selection. Part I. *Trans. Camb. Phil. Soc.* **23**, 19-41.

Haldane, J.B.S. 1924b. A mathematical theory of natural and artificial selection. II. The influence of partial self-fertilisation, inbreeding, assortative mating and selective fertilisation on the composition of Mendelian populations, and on natural selection. *Proc. Camb. Phil. Soc. (Biol. Sci.)* (later *Biol. Rev.*) **1**, 158-63.

Haldane, J. B. S. 1927. A mathematical theory of natural and artificial selection. V. Selection and mutation. *Proc. Camb. Phil. Soc.* **23**, 838-44.

Haldane, J. B. S. 1930. Theoretical genetics of autopolyploids. *J. Genet.* **22**, 359-72.

Haldane, J. B. S. 1932. *The causes of evolution*. London: Longmans Green.

Haldane, J. B. S. 1939. The equilibrium between mutation and random extinction. *Ann. Eug.* **9**, 400-5.

Hardy, G.H. 1952. *A course of pure mathematics*, 10th edn. Cambridge: Cambridge University Press.

Harris, T. E. 1963. *The theory of branching processes*. Berlin: Springer.

Hartl, D. L. & R. D. Cook 1973. Balanced polymorphisms of quasi-neutral alleles. *Theor. Pop. Biol.* **4**, 163-72.

Hartl, D. L. & R. D. Cook 1976. Stochastic selection and the maintenance of genetic variation. In *Population genetics and ecology*, S. Karlin & E. Nevo (eds), 593-615. New York: Academic Press.

Hartl, D. L. & D. E. Dykhuizen 1985. The neutral theory and the molecular basis of preadaptation. In *Population genetics and molecular evolution*, T. Ohta & K. Aoki (eds), 107-24. Berlin: Springer.

Hill, W. G. 1970. Theory of limits to selection with line crossing. In *Mathematical topics in population genetics*, K. I. Kojima (ed.), 210-45. Berlin: Springer.

Hill, W. G. & A. Robertson 1966. The effect of linkage on limits to artificial selection. *Genet. Res.* **8**, 269-94.

REFERENCES

Humphreys, M. O. & J. S. Gale 1974. Variation in wild populations of *Papaver dubium*. VIII. The mating system. *Heredity* **33**, 33-41.

Jeffreys, A. J. 1979. DNA sequence variants in the $^G\gamma$-, $^A\gamma$-, δ- and β-globin genes of man. *Cell* **18**, 1-10.

Jeffreys, A. J., V. Wilson & S. L. Thein 1985a. Hypervariable 'minisatellite' regions in human DNA. *Nature* **314**, 67-73.

Jeffreys, A. J., V. Wilson & S. L. Thein 1985b. Individual-specific 'fingerprints' of human DNA. *Nature* **316**,76-9.

Jeffreys, A. J., V. Wilson, S. L. Thein, D. J. Weatherall & B. A. J. Ponder 1986. DNA 'fingerprints' and segregation analysis of multiple markers in human pedigrees. *Am. J. Hum. Genet.* **39**, 11-24.

Jensen, L. 1973. Random selection advantages of genes and their probability of fixation. *Genet. Res.* **21**, 215-19.

Karlin, S. & B. Levikson 1974. Temporal fluctuations in selection intensities: case of small population size. *Theor. Pop. Biol.* **6**, 383-412.

Karlin, S. & U. Lieberman 1974. Random temporal variation in selection intensities: case of large population size. *Theor. Pop. Biol.* **6**, 355-82.

Karlin, S. & H. M. Taylor 1975. *A first course in stochastic processes*, 2nd edn. New York: Academic Press.

Karlin, S. & H. M. Taylor 1981. *A second course in stochastic processes*. New York: Academic Press.

Kemeny, J. G. & J. L. Snell 1960. *Finite Markov chains*. New York: Van Nostrand.

Kemeny, J. G., J. L. Snell & G. L. Thompson 1965. *Introduction to finite mathematics*. Englewood Cliffs, NJ: Prentice-Hall.

Khazanie, R. G. & H. E. McKean 1966. A Mendelian Markov process with binomial transition probabilities. *Biometrika* **53**, 37-48.

Kimura, M. 1955a. Solution of a process of random genetic drift with a continuous model. *Proc. Nat. Acad. Sci.* **41**, 144-50.

Kimura, M. 1955b. Stochastic processes and distribution of gene frequencies under natural selection. *Cold Spring Harbor Symp. Quant. Biol.* **20**, 33-53.

Kimura, M. 1957. Some problems of stochastic processes in genetics. *Ann. Math. Stat.* **28**, 882-901.

Kimura, M. 1962. On the probability of fixation of mutant genes in a population. *Genetics* **47**, 713-19.

Kimura, M. 1964. Diffusion models in population genetics. *J. Appl. Prob.* **1**, 177-232.

Kimura, M. 1968a. Evolutionary rate at the molecular level. *Nature* **217**, 624-6.

Kimura, M. 1968b. Genetic variability maintained in a finite population due to mutational production of neutral and nearly neutral isoalleles. *Genet. Res.* **11**, 247-69.

Kimura, M. 1969. The number of heterozygous nucleotide sites maintained in a finite population due to steady flux of mutations. *Genetics* **61**, 893-903.

Kimura, M. 1970a. The length of time required for a selectively neutral mutant to reach fixation through random frequency drift in a finite population. *Genet.Res.* **15**, 131-3.

Kimura, M. 1970b. Stochastic processes in population genetics, with special reference to distribution of gene frequencies and probability of gene fixation. In *Mathematical topics in population genetics*, K. I. Kojima (ed.), 178-209. Berlin: Springer.

Kimura, M. 1971. Theoretical foundations of population genetics at the molecular level. *Theor. Pop. Biol.* **2**, 174-208.

REFERENCES

Kimura, M. 1979. Model of effectively neutral mutations in which selective constraint is incorporated. *Proc. Nat. Acad. Sci.* **76**, 3440-4.

Kimura, M. 1983. *The neutral theory of molecular evolution.* Cambridge: Cambridge University Press.

Kimura, M. 1986. DNA and the neutral theory. *Phil. Trans. R. Soc. B* **312**, 343-54.

Kimura, M. 1987. Molecular evolutionary clock and the neutral theory. *J. Mol. Evol.* **26**, 24-33.

Kimura, M. & J. F. Crow 1963. The measurement of effective population number. *Evolution* **17**, 279-88.

Kimura M. & J. F. Crow 1964. The number of alleles that can be maintained in a finite population. *Genetics* **49**, 725-38.

Kimura, M. & T. Maruyama 1971. Pattern of neutral polymorphism in a geographically structured population. *Genet. Res.* **18**, 125-31.

Kimura, M. & T. Ohta 1969a. The average number of generations until fixation of a mutant gene in a finite population. *Genetics* **61**, 763-71.

Kimura, M. & T. Ohta 1969b. The average number of generations until extinction of an individual mutant gene in a finite population. *Genetics* **63**, 701-9.

Kimura, M. & T. Ohta 1971a. *Theoretical aspects of population genetics.* Princeton, NJ: Princeton University Press.

Kimura, M. & T. Ohta 1971b. Protein polymorphism as a phase of molecular evolution. *Nature* **229**, 467-9.

Kojima, K. I. & T. M. Kelleher 1962. Survival of mutant genes. *Am. Nat.* **96**, 329-46.

Kreitman, M. 1983. Nucleotide polymorphism at the alcohol dehydrogenase locus of *Drosophila melanogaster*. *Nature* **304**, 412-17.

Langley, C. H., J. F. Y. Brookfield & N. Kaplan 1983. Transposable elements in Mendelian populations. I. A theory. *Genetics* **104**, 457-71.

Langley, C. H., E. Montgomery & W. F. Quattlebaum 1982. Restriction map variation in the *Adh* region of *Drosophila*. *Proc. Nat. Acad. Sci.* **79**, 5631-5.

Levene, H. 1949. On a matching problem in genetics. *Ann. Math. Stat.* **20**, 91-4.

Levikson, B. & S. Karlin 1975. Random temporal variation in selection intensities acting on infinite diploid populations: diffusion method analysis. *Theor. Pop. Biol.* **8**, 292-300.

Lewin, B. 1987. *Genes*, 3rd edn. New York: Wiley.

Li, C. C. 1955. *Population genetics*. Chicago: University of Chicago Press.

Li, W. H., C. C. Luo & C. I. Wu 1985. Evolution of DNA sequences. In *Molecular evolutionary genetics*, R. J. MacIntyre (ed.), 1-94. New York: Plenum.

Li, W. H., M. Tanimura & P. M. Sharp 1987. An evaluation of the molecular clock hypothesis using mammalian DNA sequences. *J. Mol. Evol.* **25**, 330-42.

Littler, R. A. 1975. Loss of variability at one locus in a finite population. *Math. Biosci.* **25**, 151-63.

Mackie, A. G. 1965. *Boundary value problems*. Edinburgh: Oliver & Boyd.

Malécot, G. 1948. *Les mathématiques de l'hérédité* (1969, *The mathematics of heredity*, D. M. Yermanos (transl.) San Francisco: Freeman).

Maruyama, T. 1970a. On the rate of decrease of heterozygosity in circular stepping stone models of populations. *Theor. Pop. Biol.* **1**, 101-19.

Maruyama, T. 1970b. Effective number of alleles in a subdivided population. *Theor. Pop. Biol.* **1**, 273-306.

Maruyama, T. 1970c. On the fixation probability of mutant genes in a subdivided population. *Genet. Res.* **15**, 221-5.

REFERENCES

Maruyama, T. 1972a. The average number and variance of generations at particular gene frequency in the course of fixation of a mutant gene in a finite population. *Genet. Res.* **19**, 109–13.

Maruyama, T. 1972b. Rate of decrease of genetic variability in a two-dimensional continuous population of finite size. *Genetics* **70**, 639–51.

Maruyama, T. 1977. *Stochastic problems in population genetics*. Berlin: Springer.

Maruyama, T. & M. Kimura 1980. Genetic variability and effective population size when local extinction and recolonization of sub-populations are frequent. *Proc. Nat. Acad. Sci.* **77**, 6710–14.

Maynard Smith, J. & J. Haigh 1974. The hitch-hiking effect of a favourable gene. *Genet. Res.* **23**, 23–35.

Montgomery, E. A. & C. H. Langley 1983. Transposable elements in Mendelian populations. II. Distribution of three copia-like elements in a natural population of *Drosophila melanogaster*. *Genetics* **104**, 473–83.

Moran, P. A. P. 1958a. Random processes in genetics. *Proc. Camb. Phil. Soc.* **54**, 60–71.

Moran, P. A. P. 1958b. A general theory of the distribution of gene frequencies. I. Overlapping generations. II. Non-overlapping generations. *Proc. R. Soc. B* **149**, 102–16.

Moran, P. A. P. 1960a. The survival of a mutant gene under selection. *J. Austr. Math. Soc.* **1**, 121–6.

Moran, P. A. P. 1960b. The survival of a mutant gene under selection. II. *J. Austr. Math. Soc.* **1**, 485–91.

Moran, P. A. P. 1961. The survival of a mutant under general conditions. *Proc. Camb. Phil. Soc.* **57**, 304–14.

Moran, P. A. P. 1962. *The statistical processes of evolutionary theory*. Oxford: Clarendon.

Moran, P. A. P. 1968. *An introduction to probability theory*. Oxford: Clarendon.

Nagylaki, T. 1977. *Selection in one- and two-locus systems*. Berlin: Springer.

Narain, P. & E. Pollak 1977. On the fixation probability of a gene under random fluctuations in selection intensities in small populations. *Genet. Res.* **29**, 113–21.

Neel, J. V. 1979. History and the Tay-Sachs allele. In *Genetic diseases among Ashkenazi Jews*, R. M. Goodman & A. G. Motulsky (eds), 285–99. New York: Raven.

Nei, M. 1968. The frequency distribution of lethal chromosomes in finite populations. *Proc. Nat. Acad. Sci.* **60**, 517–24.

Nei, M. 1987. *Molecular evolutionary genetics*. New York: Columbia University Press.

Nei, M. & M. Murata 1966. Effective population size when fertility is inherited. *Genet. Res.* **8**, 257–60.

Norman, M. F. 1975a. Diffusion approximation of non-Markovian processes. *Ann. Prob.* **3**, 358–64.

Norman, M. F. 1975b. Approximation of stochastic processes by Gaussian diffusions, and applications to Wright-Fisher genetic models. *SIAM J. Appl. Math.* **29**, 225–42.

Ohta, T. 1974. Mutational pressure as the main cause of molecular evolution and polymorphism. *Nature* **252**, 351–4.

Ohta, T. 1976. Role of very slightly deleterious mutations in molecular evolution and polymorphism. *Theor. Pop. Biol.* **10**, 254–75.

REFERENCES

Ohta, T. 1977. Extension to the neutral mutation random drift hypothesis. In *Molecular evolution and polymorphism*, M. Kimura (ed.), 148-67. Mishima: National Institute of Genetics.

Ohta, T. 1980. *Evolution and variation of multigene families*. Berlin: Springer.

Ohta, T. 1983. On the evolution of multigene families. *Theor. Pop. Biol.* **23**, 216-40.

O'Malley, R. E. 1974. *Introduction to singular perturbations*. New York: Academic Press.

Pais, A. 1982. '*Subtle is the Lord*'. *The science and the life of Albert Einstein*. Oxford: Oxford University Press.

Parlevliet, J. E. 1981. Stabilizing selection in crop pathosystems: an empty concept or a reality? *Euphytica* **30**, 259-69.

Piva, M. & P. Holgate 1977. The eigenvectors of a finite population model. *Ann. Hum. Genet.* **41**, 103-6.

Pollak, E. 1966. On the survival of a gene in a subdivided population. *J. Appl. Prob.* **3**, 142-55.

Pollak, E. & B. C. Arnold 1975. On sojourn times at particular gene frequencies. *Genet. Res.* **25**, 89-94.

Provinc, W. B. 1986. *Sewall Wright and evolutionary biology*. Chicago: University of Chicago Press.

Renshaw, E. 1982. Change of genotype frequencies under selection in a self-pollinated population. *Heredity* **48**, 139-44.

Ridley, M. 1985. *The problems of evolution*. Oxford: Oxford University Press.

Roach, G. F. 1970. *Green's functions*. New York: Van Nostrand.

Robbins, R. B. 1917. Some applications of mathematics to breeding problems. I. *Genetics* **2**, 489-504.

Robbins, R. B. 1918a. Some applications of mathematics to breeding problems. II. *Genetics* **3**, 73-92.

Robbins, R. B. 1918b. Some applications of mathematics to breeding problems. III. *Genetics* **3**, 375-89.

Robertson, A. 1960. A theory of limits in artificial selection. *Proc. R. Soc. B* **153**, 234-49.

Robertson, A. 1962. Selection for heterozygotes in small populations. *Genetics* **47**, 1291-300.

Robertson, A. 1970. A theory of limits in artificial selection with many linked loci. In *Mathematical topics in population genetics*, K. I. Kojima (ed.), 246-88. Berlin: Springer.

Robertson, A. 1977. Artificial selection with a large number of linked loci. In *Proc. Int. Conf. on quantitative genetics*, E. Pollak, O. Kempthorne & T. B. Bailey (eds.), 307-22. Ames, IA: Iowa State University Press.

Sagan, H. 1961. *Boundary and eigenvalue problems in mathematical physics*. New York: Wiley.

Schaffer, H. E. 1970. Survival of mutant genes as a branching process. In *Mathematical topics in population genetics*, K. I. Kojima (ed.), 317-36. Berlin: Springer.

Slatkin, M. 1981. Fixation probabilities and fixation times in a subdivided population. *Evolution* **35**, 477-88.

Sneddon, I. N. 1961. *Special functions of mathematical physics and chemistry*. Edinburgh: Oliver & Boyd.

Spiegel, M. R. 1971. *Calculus of finite differences and difference equations* (Schaum's Outline Series). New York: McGraw-Hill.

REFERENCES

Stephenson, G. 1968. *An introduction to partial differential equations for science students*. London: Longman.

Stewart, F. M. 1976. Variability in the amount of heterozygosity maintained by neutral mutations. *Theor. Pop. Biol.* **9**, 188–201.

Takahata, N. 1981. Genetic variability and rate of gene substitution in a finite population under mutation and fluctuating selection. *Genetics* **98**, 427–40.

Tavaré, S. 1979. Sojourn times for conditioned Markov chains in genetics. *Theor. Pop. Biol.* **15**, 108–13.

Thomson, G. 1977. The effect of a selected locus on linked neutral loci. *Genetics* **85**, 753–88.

Voronka, R. & J. B. Keller 1975. Asymptotic analysis of stochastic models in population genetics. *Math.Biosci.* **25**, 331–62.

Watson, H. W. & F. Galton 1874. On the probability of the extinction of families. *J. Anthropol. Inst. GB Irel.* **4**, 138–44.

Watterson, G. A. 1962. Some theoretical aspects of diffusion theory in population genetics. *Ann. Math. Stat.* **33**, 939–57.

Watterson, G. A. 1975. On the number of segregating sites in genetical models without recombination. *Theor. Pop. Biol.* **7**, 256–76.

Wetton, J. H., R. E. Carter, D. T. Parkin & D. Walters 1987. Demographic study of a wild house sparrow population by DNA fingerprinting. *Nature* **327**, 147–9.

Wilson, A. C., S. S. Carlson & T. J. White 1977. Biochemical evolution. *Annu. Rev. Biochem.* **46**, 573–639.

Wright, S. 1931. Evolution in Mendelian populations. *Genetics* **16**, 97–159.

Wright, S. 1937. The distribution of gene frequencies in populations. *Proc. Nat. Acad. Sci.* **23**, 307–20.

Wright, S. 1938a. Size of population and breeding structure in relation to evolution. *Science* **87**, 430–1.

Wright, S. 1938b. The distribution of gene frequencies under irreversible mutation. *Proc. Nat. Acad. Sci.* **24**, 253–9.

Wright, S. 1939a. The distribution of self-sterility alleles in populations. *Genetics* **24**, 538–52.

Wright, S. 1939b. *Statistical genetics in relation to evolution*. Paris: Hermann.

Wright, S. 1945. The differential equation of the distribution of gene frequencies. *Proc. Nat. Acad. Sci.* **31**, 382–9.

Wright, S. 1952. The theoretical variance within and among subdivisions of a population that is in a steady state. *Genetics* **37**, 312–21.

Wright, S. 1969. *Evolution and the genetics of populations*, Vol. 2. Chicago: University of Chicago Press.

Wright S. & W. E. Kerr 1954. Experimental studies of the distribution of gene frequencies in very small populations of *Drosophila melanogaster*. II. Bar. *Evolution* **8**, 225–40.

Zauderer, E. 1983. *Partial differential equations of applied mathematics*. New York: Wiley.

Index

Abel's theorem 385
Abramowitz, M. 262, 320, 343, 403
Absorption
  definition 56
  certain if mutation absent 56–7
Absorption time
  definition 78–80, 213; *see also* probability distribution of absorption time; mean absorption time: variance of absorption time
Alcohol dehydrogenase 8
Alu element 336
Animal and plant breeding 7
Aquadro, C. F. 312, 403
Arnold, B. C. 75, 199, 237, 248, 267, 383, 409
Autotetraploids 103–5
Avery, P. J. 182, 403
Avogadro, A. 12

Backward equation
  derivation 348–50
  derivation of mean sojourn time 351
  extreme frequencies covered 350–1
  for mean of function of frequency 351
  initial, terminal and boundary conditions 355–7
  probability distribution of frequency 359–70, 373, 383–7, 393
  probability of ultimate fixation 351
Bartlett, M. S. 1, 58, 103, 403
Beta-thalassaemia 5
Birley, A. J. 312, 403
Boundaries, events at 200–3
  definition of entrance, exit and regular boundaries 201–2
  in absence of mutation 201
  in infinite allele model 202–3
  with two-way mutation 201–2
Boundary conditions
  mean absorption time 216, 238, 242
  mean sojourn time 215, 237, 242, 276, 390
  probability distribution of frequency 288, 356–7, 359–60, 362, 365–7, 383–5
  probability of fixation 153, 157, 173, 177
  to give consistency 358
Branching process approach
  accuracy 114–15, 117–19, 398–9
  probability of fixation 130–43, 165
  probability of survival 107–30, 144–51
Brookfield, J. F. Y. 336, 403, 407
Brownian motion 286
Bruford, M. W. 36, 403
Burke, T. 36, 403
Burrows, P. M. 248, 403

Carlson, S. S. 410
Carr, R. N. 178, 403
Carter, R. E. 410
Cauchy distribution 186
Charlesworth, B. 336, 403
Charlesworth, D. 336, 403
Chebyshev's inequality 32
Clarke, B. 6, 189, 403
Clarkia 286
Cockerham, C. C. 248, 403
Computer checking of results 2, 106, 117–19, 164, 169, 178, 194, 198–200, 233–4, 236, 248–9, 320–2, 329
Continuous approximations
  errors arising from 194, 197–200, 201–3, 280–4
  of allele frequency 44, 194, 196–200, 279–84, 288, 290–1
  of time 44–5, 194, 196–200, 201–2, 279, 284–6, 287–9, 295–6, 298–300
  representation of initial frequency 203–6; *see also* diffusion methods; probability flux
Cook, R. D. 182, 403, 405
Cox, D. R. 58, 304–5, 403
Cross, S. R. H. 312–403
Crow, J. F. 22, 25, 27, 31, 36, 41–2, 45, 55, 202, 285, 304, 325, 340, 354, 357, 377, 387, 394, 403, 407

Day, P. R. 404
Deterministic models
  change of frequency under selection 50–1, 259–65
  forward and backmutation 346
  misleading results 49–51, 259–60, 305, 339, 342–3, 344–5, 397–8
  mutation and selection 49–50, 337–9, 342–3
  relevance 13–14, 17–20

411

INDEX

Difference equations 156–7
Diffusion methods
  accuracy 45–6, 103, 108–10, 194–203, 209–10, 236, 379–83
  diffusion process, definition 45
  for Wright-Fisher model 197–203
  for Moran's model 47, 196–7, 199–200
  general features 44–5
  general use of 46, 347–96, 399
  representation of initial frequency 203–6
  representation of extreme frequencies 281–4, 288; *see also* continuous approximations; infinite allele model; probability flux; stationary distributions
Dirac, P. A. M. 206
Dirac's delta function
  derivative of Heaviside's function 299, 301–2
  description of 205–6, 210
  introduction to 203–5
  manipulation of 206–8, 210–12
  'spotlight' property 206, 208, 244–5
  use in finding mean sojourn time 242–7, 253–7, 273–5, 351–2, 390
  use in representing initial frequency 203–6, 355, 368–9, 383, 386
  use in representing deterministic process 207, 211–12, 298–9
Discrepancy between effective and census size, effect of
  on fixation probability 130–43, 167–72
  on survival probability 110, 119–30, 147–51
DNA fingerprinting 36
Dobzhansky, T. 1
Drosophila 140, 312, 336, 400
Dykhuizen, D. E. 6, 405

Ecological geneticists 1
Edwards, A. W. F. 45
Effective population size
  eigenvalue, definition of 74
  inbreeding, definition of 42
  in experiments 30
  Moran's model 49
  necessary conditions for equality with census size 22–7, 30
  smallness, general effect of 277–8, 401–2
  unreliability of estimates 36, 172
  variance, definition of 33; *see also* discrepancy between effective and census size
Effective population size, factors affecting
  fecundity differences 7, 24–31, 33–4, 36, 52, 119–23, 140, 144, 400
  inbreeding 22–4, 34–6

in poppies 53–4
periodic reduction in numbers 7, 36–9, 138–40, 144, 400
self-incompatibility 25
sex ratio 25, 34
speciation 400
subpopulation extinction and replacement 7, 138, 140, 144, 400
Einstein, A. 8, 286
Ellingboe, A. H. 143, 404
Epistasis 141–2, 401
Eriksson, A. W. 278
Escherichia coli 6
Euler-Maclaurin formula 99, 223–4
Ewens, P. M. 329, 404
Ewens, W. J. 12, 22, 25, 27, 34, 42, 44, 45–6, 49, 70, 75, 78, 82, 87, 91, 96, 103, 110, 131, 138–41, 143, 164, 197, 202, 219–20, 232, 237, 243, 247–8, 251, 256–7, 260, 275, 283, 285, 303, 312, 316, 318, 325, 326–8, 329–31, 333–4, 345, 390, 398, 404

Feller, W. 2, 44–5, 62, 152, 200, 237, 285, 360, 404
Fincham, J. R. S. 143, 404
Fisher, R. A. 1, 6, 15, 20–1, 45, 56, 58, 67, 70, 72–3, 75, 82, 85, 99–100, 102–3, 113–114, 117, 125–7, 134, 137, 140, 143, 164, 172, 185–6, 188–9, 192, 224, 227, 230, 232–3, 236–7, 247–8, 257, 269, 273, 285, 290, 335, 379–80, 383, 389–90, 395–6, 399, 401, 404–5.
Fixation time, definition of 78–80; *see also* probability distribution of fixation time; mean fixation time; variance of fixation time
Fluctuating viabilities, probability of fixation under 182–3
Fokker, A. D. 348
Ford, E. B. 1, 396, 405
Forsius, H. 404
Forward equation
  derivation 347–8, 387–9
  deterministic process 211–12, 398
  initial, terminal and boundary conditions 355–7, 369–70
  probability distribution of frequency 369–70, 374–5
  special treatment of extreme frequencies 348, 350, 370–3, 375
Frequency spectrum
  applications 335–6
  definition 331
  properties 331–4
Functions continuous but not always differentiable 196–7, 208

412

INDEX

Fungal pathogens 8, 143–4, 278

Gale, J. S. 5, 53, 124, 143, 405–6
Galton, F. 113, 410
Gamow, G. 206, 405
Gauss's hypergeometric function
  definition 365
  properties 365
Gegenbauer polynomials
  definition 368
  orthogonality relations 368, 393
Gene conversion 268–9
Gene, modern definition of 8
Generating function
  for probabilities 111–50
  for mean sojourn times 225–34
Genetic conservation 7, 58, 106–7, 123–4, 147–9, 213, 250, 284
Genic selection (=haploid model)
  definition 129
  mean fixation time 263–5
  mean sojourn time 256–7
  mean time in frequency range 258–63
  probability of fixation under 130–43, 150–1
  probability of survival under 127–30, 143–4, 150–1
Gillespie, J. H. 182, 405
Girling, A. J. 306
Goldberg, S. 288, 357
Green's function 243; *see also* mean sojourn time

Haigh, J. 143, 400, 405, 408
Haldane, J. B. S. 1, 15, 35, 70, 103, 113–14, 134, 233–4, 236, 265, 403, 405
Haploid model *see* genic selection
Hardy, G. H. 290, 405
Hardy-Weinberg law 12
Harmonic mean 39, 138–40, 151, 306
Harris, H. 404
Harris, T. E. 126, 405
Hartl, D. L. 6, 182, 403, 405
Heaviside's unit function 299, 301–2
Heterozygotes, mean frequency of 39–42, 54, 311–13, 323–5
Hill, W. G. 7, 405
Histone 268
Hitch-hiking 400–1
Holgate, P. 69, 409
Humphreys, M. O. 53, 406
Hypergeometric equation
  definition 363
  solutions 363–6, 384

Inbreeding, theory of 58, 74, 103–5, 156
Infinite allele model
  definition 304
  effective number alleles present 325, 329
  frequency spectrum 331–4
  mean frequency of heterozygotes 325
  mean number alleles in frequency range 329–31
  mean number alleles present 325–9, 332–3, 345
  stationary properties 325
Infinite sites model
  mean number heterozygous sites 335–6
  mean number polymorphic sites 334–5
Initial conditions
  probability distribution of allele frequency 355, 357, 360, 383, 386
  moments 357, 391
Initial frequency, representation of 203–6
Introns
  as components of genes 7–8
  evolution 3, 5, 185
Inversion of limiting operations 290

Jacobi polynomial
  definition 385
  orthogonality properties 386
Jeffreys, A. J. 36, 269, 312, 406
Jensen, L. 182, 406
Jones, R. C. 244

Kaplan, N. 407
Karlin, S. 45, 95, 175, 182, 209, 238, 285, 350, 406–7.
Kelleher, T. M. 122, 407
Keller, J. B. 69, 108, 356, 374, 381, 410
Kemeny, J. G. 9, 58, 82, 89–90, 155, 406
Kerr, W. E. 399, 410
Khazanie, R. G. 103, 406
Kimura, M. 2, 3, 4, 6, 7, 15, 22, 25, 27, 29, 31, 36, 42, 45, 54, 69–70, 73–5, 108, 138, 164, 166–9, 172–3, 177, 179, 183–6, 188–93, 202, 206–7, 209, 211, 236–7, 252, 267, 277, 282, 285, 290–1, 297, 300–1, 304, 307 312–13, 318–23, 325, 328–9, 334, 340, 345, 347, 353–7, 369, 372, 374, 377, 387, 394, 399, 401, 403, 406–8
Kimura's diploid formula for fixation probability
  accuracy 178
  conclusions 178–80
  statement of 177
Kimura's haploid formula for fixation probability
  accuracy 163–4, 168–9, 178, 192–3
  conclusions 170–2
  statement of 168
Kimura-Ohta equation

INDEX

direct derivation 353
derivation from forward equation 358–9
when calculating moments 354–5, 357, 391
Kojima, K. I. 122, 407
Kolmogotov, A. N. 45, 126, 348
Kreitman, M. 140, 312, 407

Lado, K. M, 403
Langley, C. H. 312, 336, 407–8
Latent roots (eigenvalues)
clash of terminology 62, 390
definition 62
dominant latent root, definition 71
dominant latent root, derivation 95–6
Feller's formula 62–3, 66, 73, 91–5, 360
introduction to 60–7
utility of 71–8, 360–2
when selfing autotetraploids 62, 390
Lawrence, M. J. 124, 405
Leader sequences 8
Legendre polynomials
applications 372–3, 378, 393–4
definition 371
properties 371, 372, 393
Levene, H. 39, 407
Levikson, B. 182, 406–7.
Lewin, B. 5, 8, 407
Lewontin, R. C. 339
Li, C. C. 22, 24, 407
Li, W. H. 3, 184, 268, 407
Lieberman, U. 182, 406
Linear independence 66
linear pressure, definition of 19
Linkage 141–2, 401
Littler, R. A. 251, 407
Long-headed poppy (*Papaver dubium*) 27, 32, 36, 53–4
Luo, C. C. 407

Mackay, I. J. 27, 32, 53
McKean, H. E. 103, 406
Mackie, A. G. 2, 208, 407
Malécot G. 22–3, 325, 407
Man 279, 312
Markov chain approach
definition 58
guide to solution of diffusion equations 360–2
theory 58–105
use in checking approximations 58; *see also* computer checking of results
Markovian property 44, 54–5, 58, 304–5
Maruyama, T. 7, 45–6, 70, 75, 137–8, 165–7, 207, 237, 243–4, 251, 257, 295, 307, 312, 352–3, 377, 397–8, 400, 407–8
Maynard Smith, J. 143, 400, 405, 408
Mean absorption time

accuracy of formulae 232–3, 248–9, 275
boundary conditions 216
calculation of 82–4, 88, 90–1, 214, 218–19, 221–2, 230–3, 237–8, 246–7, 248–9, 269–73, 274–5
definition 57, 80, 213
directly proportional to effective population size 247, 250
implications for genetic conservation 250
in finding mean number polymorphic sites 334–5
Moran's model 221–2
multiple alleles 251
representation as area under curve 239–40
rise with frequency of rarer allele 249–50
unnoticed appearance in early work 82
utility as descriptive quality 84
when selfing autotetraploids 105
Wright-Fisher model, 231–3, 269–73, 275
Mean fixation time
accuracy of formulae 236
calculation of 82–4, 88, 91, 214, 222, 235–7, 252–3, 263–5
definition 80, 82, 213
for neutral mutants 236, 252, 277
for genic selection 263–5, 401–2
greatly reduced by selection 265
Moran's model 222
utility as descriptive quantity 84
Wright-Fisher model 235–7
Mean interval between successive fixations 184
Mean number of alleles present 325–9, 345
Mean number of heterozygous sites 335–6
Mean number of polymorphic sites 334–5
Mean recurrence time 305–6
Mean sojourn time
accuracy of formulae 193, 197–9, 223–5, 234, 248, 383
boundary conditions 215
calculation of 86–8, 215–30, 233–5, 237–48, 251–7, 269–76, 279
definition 57, 213–14
derivation from diffusion equations 275–6, 351–2, 390
matrix representation 90–1
Moran's model 219–21
neutral mutant in modified process 251, 401
relation between times in modified and unmodified process 87, 198, 214
relation to fixation probability 218, 389
under natural selection 253–7, 401–2
uniqueness of solutions 217
when selfing autotetraploids 104–5
Wright-Fisher model 222–30, 233–4, 269–74

INDEX

Mean time spent over range of frequencies
   genic selection 257–65, 276
   neutral alleles 15, 251–2; *see also* mean fixation time
Mendel, G. 42
Migration *see* population subdivision
Miller, H. D. 58, 304–5, 403
Mitochondria 5
Modified process
   biological significance 75
   definition 75
   latent roots 76–8
   probability distribution of frequency 75–9
   probability distribution of fixation time 78–82; *see also* mean fixation time; variance of fixation time; mean sojourn time; variance of sojourn time; mean time spent over range of frequencies
Moments of distributions
   definition 32
   from Kimura-Ohta equation 354–5, 391–3
   initial condition for 357
   neutral no-mutation case 375–82
   relevance 32–3, 52
   use in checking approximations 108, 375–81
Montgomery, E. 336, 407–8.
Moran, P. A. P. 44–6, 49, 70, 75, 82, 122, 128, 156, 158, 160–2, 164–5, 167–8, 192, 237, 282, 285, 288, 344, 408
Moran's method 70, 158–67, 190, 192
Moran's model
   as check on diffusion methods 47
   definition 46–7
   effective population size 49
   expectation of life 48, 55
   generation equivalent 48
   mean absorption time 221–2, 232
   mean fixation time 222
   mean sojourn time 199, 219–22, 247
   probability of fixation 155–9, 169, 190–2
   probability of reaching given frequency 195–7, 214
   variance of allele frequency 48–9
Morphological characters, evolution of 3
Morton, N. E. 27, 53, 403
Multigene families, concerted evolution of 267–9
Muons 8
Murata, M. 33, 408
Murphy's formula 371
Mutation and selection, limitations of deterministic approach 20, 49–50, 278, 339, 342–3, 401; *see also* stationary distributions
Mutation rate 193

Nagylaki, T. 67, 408
Narain, P. 182, 408
Nassar, R. F. 178, 403
Natural populations
   polymorphism 4, 312–13
   selection in 1, 3
   selection for morphological characters in 1, 3
   variance of allele frequency in 32
Neel, J. V. 279, 408
Nei, M. 3, 33, 184, 340, 408
Neutral theory
   conditions for theory to work 6–7, 171–2, 277, 312–13, 322–4, 399–402
   described 4
   evidence for 4–6
   rate of evolution under 183–5
Nevanlinna, H. R. 404
Newtonian mechanics 8
Non-uniform convergence 51
Noon, W. A. 403
Norio, R. K. 404
Norman, M. F. 25, 44, 55, 305, 346, 408

Ohta, T. 75, 184, 236–7, 252, 267, 269, 277, 313, 334, 353, 407–9.
O'Malley, R. E. 398, 409
Owen, A. R. G. 158

Pais, A. 348, 409
Parkin, D. T. 410
Parlevliet, J. E. 144, 409
Perturbation theory 398
Piva, M. 69, 409
Planck, M. 348
Plant and animal breeding 7, 58
Pollak, E. 75, 137, 182, 199, 237, 248, 267, 383, 408–9
Ponder, B. A. J. 406
Population subdivision
   early controversy on 6–7
   effect on decay of variability 74–5
   effect on probability of fixation 137–8, 165–7, 169, 180–2
   general effect 400; *see also* effective population size
Principal components of frequency 60–7, 104
Probability distribution of absorption time 57, 78–82; *see also* mean absorption time; variance of absorption time
Probability distribution of allele frequency
   definition 14–15
   irreversible mutation 394
   mean (basic ideas) 15–20, 32, 51–2
   need to calculate 278–9
   neutral no-mutation case 359–75

problems in presentation 69
two-way mutation 383–7
variance (basic ideas) 20–39, 46–9, 52; *see also* branching process approach; continuous approximations; diffusion methods; Markov chain approach; Moran's model; probability flux; stationary distributions; Wright-Fisher model
Probability distribution of fixation time 78–82; *see also* mean fixation time; variance of fixation time
Probability flux
  approximation to 287
  components 290–2
  definition 286–7, 296–7
  errors arising in 293–4, 296–7
  expressions for M and V 297–300
  flux into extreme frequencies 288–9, 300–1
  formula 292–6
  history of concept 290
  in deterministic process 298–9
  use when finding stationary distributions 304, 306–7
Probability generating functions 111–50
Probability of absorption 56–7
Probability of ever reaching given frequency 89–90, 195–9, 214, 399
Probability of monomorphism 322–4
Probability of survival
  advantageous mutants 127–30, 143–4
  neutral mutants 107–26, 147–50
Probability of ultimate fixation
  derivation by branching process method 130–43, 165, 398
  effect of intensity of selection on 170–2
  Kimura's method for finding 172–8, 190–3, 351, 398–9
  Moran's method for finding 70, 158–65
  Moran's model 155–9, 190–2
  neutral alleles 10–11, 70
  recessive advantageous allele 178–9, 181–2
  under natural selection 10–11, 70–1, 130–43, 152–94, 255
  unique solution for 152–5
  utility as descriptive quantity 69–71
  with fluctuating viabilities 182–3
Proteins, evolution of 3–5, 184–5
Provine, W. B. 399, 409
Pseudogenes, evolution of 3–4, 185

Quasi-monomorphism, definition of 323
Quattlebaum, W. F. 407

Radford, A. 404
RAF 55
Rate of evolution

introns 3, 5, 185
morphological characters 3, 185
proteins 3, 5, 184–5
pseudogenes 3–4, 185
synonymous DNA changes 3–4, 185
theory 183–9
under natural selection 6
Renshaw, E. 67, 409
Restriction endonuclease 303
Ridley, M. 400, 409
Roach, G. F. 274, 409
Robbins, R. B. 156, 409
Robertson, A. 7, 20, 383, 405, 409
rRNA 268

Sagan, H. 2, 356, 409
Schaffer, H. E. 143, 409
Schwarz, L. 206
Sea urchin 268
Selfing, partial 22–4, 34–6
Self-sterility alleles 2
Sharp, P. M. 407
Simple models, appropriateness of 12–13
Slatkin, M. 180, 409
Sneddon, I. N. 2, 409
Snell, J. L. 58, 82, 89–90, 155, 406
Spiegel, M. R. 2, 156, 223, 409
Spielman, R. S. 404
Stationary distributions
  accuracy of results 320–2, 399
  basic ideas 303–6, 401
  boundary conditions 288, 357, 359
  calculation of 306–7, 313–15, 337, 340–2, 344, 387
  effective number of bases 310–12
  effective number of alleles 324
  extreme allele frequencies 316–20, 341–2, 345
  harmful recessives 340–3, 401
  mean 307–8, 316
  mean heterozygosity 311–13, 323–4
  probability of monomorphism at nucleotide site 322–4
  under selection 336–43
  variance 308–10, 316; *see also* infinite alleles model frequency spectrum
Stegun, I. A. 262, 320, 343, 403
Stephenson, G. 2, 410
Stewart, F. M. 324, 410
Stochastic clock scale 167
Stochastic models
  definition 13
  meaning of probability under 13
Strong selectionist view 6, 185, 189, 277
Student's t-distribution 186
'Subterminal' formulae 301, 318–19, 341, 399

Synonymous changes in DNA, evolution of 3–5, 185

Takahata, N. 183, 410
Talmud 202
Tanimura, M. 407
Tavaré, S. 75, 267, 410
Taylor, H. M. 45, 95, 175, 209, 238, 285, 350, 406
Tay-Sachs disease 279
Terminal condition
  for probability distribution of allele frequency 356, 363
  for moments 392
Thein, S. L. 406
Thompson, G. L. 406
Thomson, G. 400, 410
Trailer sequences 8
Transfer RNA 4–5
Transposable elements
  as source of mutation 7, 193
  frequency spectrum 336

Unequal crossing over 268–9

Variance of absorption time 84–6, 89, 265–7
Variance of fixation time 84–6, 89–90, 265–7
Variance of sojourn time 89–90, 265
Vibrating string 356
Voronka, R. 69, 108, 356, 374, 381, 410
Voronka-Keller formulae 108–10, 374–5, 394–6

Walters, D. 410
Watson, H. W. 113, 410

Watterson, G. A. 44–5, 55, 82, 233, 237, 247, 283, 288, 290, 305, 334, 345, 410
Weatherall, D. J. 406
Wetton, J. H. 36, 410
White, T. J. 410
Wilson, A. C. 184, 410
Wilson, V. 406
Workman, P. L. 404
Wright, S. 1, 2, 6, 7, 15, 21–3, 25, 27, 30, 33–5, 39, 42, 45, 70, 72–4, 82, 95–7, 99, 164, 202, 237, 278, 282, 285, 290, 301, 307, 310, 318–19, 321, 337, 340, 342–3, 346, 369, 399, 410
Wright-Fisher model
  definition 21–2
  diploid population size two 58–75, 78–91
  events at boundaries 200–3
  fixation probabilities 127–34, 138–43, 159–67, 169, 192–3
  for checking diffusion methods 380–3
  latent roots 62–3, 66, 71–5, 91–6
  mean absorption time 231–3, 247–8, 269–73
  mean fixation time 235–7
  mean sojourn time 197–9, 222–30, 233–4, 247, 269–74
  one-way mutation 209–10
  probability of reaching given frequency 197–9
  relevance 22–36, 43–4
  survival probabilities 108–19, 124–7
  very late stage 72–4, 96–103
Wu, C. I. 407

Zauderer, E. 374, 398, 410